Essential Oi

GW00645452

Essential oils are also known as volatile oils, ethereal oils, aetherolea, or simply as the 'oil of' the plant from which they are extracted, such as the oil of clove. An oil is 'essential' in the sense that it contains the characteristic fragrance of the plant that it is taken from. Essential oils do not form a distinctive category for any medicinal, pharmacological, or culinary purpose - and they are not essential for health, although they have been used medicinally in history. Although some are suspicious or dismissive towards the use of essential oils in healthcare or pharmacology, essential oils retain considerable popular use, partly in fringe medicine and partly in popular remedies. Therefore it is difficult to obtain reliable references concerning their pharmacological merits.

Medicinal applications proposed by those who sell or use medical oils range from skin treatments to remedies from cancer - and are generally based on historical efficacy. Having said this, some essential oils such as those of juniper and agathosma are valued for their diuretic effects. Other oils, such as clove oil or eugenol were popular for many hundreds of years in dentistry and as antiseptics and local anaesthetics. However as the use of essential oils has declined in evidence based medicine, older text-books are frequently our only sources for information! Modern works are less inclined to generalise; rather than referring to 'essential oils' as a class at all, they prefer to discuss specific compounds, such as methyl salicylate, rather than 'oil of wintergreen.'

Nevertheless, interest in essential oils has considerably revived in recent decades, with the popularity of aromatherapy, alternative health stores and massage. Generally, the oils are volatized or diluted with a carrier oil to be used in massage, or diffused in the air by a nebulizer, heated over a candle flame, or burned as incense. Their usage goes way back, and the earliest recorded mention of such methods used to produce essential oils was made by Ibn al-Baitar (1188-1248), an Andalusian physician, pharmacist and chemist. Different oils were claimed to have differing properties; some to have an uplifting and energizing effect on the mind such as grapefruit and jasmine, whilst others such as rose lavender have a reputation as de-stressing and relaxing - and also, usefully, as an insect repellent.

The oils themselves are usually extracted by 'distillation', often by using steam -but some other processes include 'expression' or 'solvent extraction'. Distillation involves raw plant material (be that flowers, leaves, wood, bark,

Odorographia

A Natural History of Raw
Materials and Drugs used in the
Perfume Industry

by

John Charles Sawer

British Library Cataloguing-in-Publication Data
A catalogue record for this book is available from the
British Library

roots, seeds or peel) put into an alembic (distillation apparatus) over water. As the water is heated, the steam passes through the plant material, vaporizing the volatile compounds. The vapours flow through a coil, where they condense back to liquid, which is then collected in the receiving vessel. 'Expression' differs in that it usually merely uses a mechanical or cold press to extract the oil. Most citrus peel oils are made in this way, and due to the relatively large quantities of oil in citrus peel and low cost to grow and harvest the raw materials, citrus-fruit oils are cheaper than most other essential oils. 'Solvent extraction' is perhaps the most difficult of the three methods, and is generally used for flowers, which contain too little volatile oil to undergo expression. Instead, a solvent such as hexane or supercritical carbon dioxide is used to extract the oils.

These techniques have allowed essential oils to be used in all manner of products; from perfumes to cosmetics, soaps - and as flavourings for food and drinks as well as adding scent to incense and household cleaning products. The science, history and folkloric tradition of essential oils is incredibly fascinating - and a still much debated area. We hope the reader is inspired by this book to find out more.

PREFACE.

An endeavour has here been made to collect together into one Manual the information which has hitherto been only obtainable by reference to an immense number of Works and Journals, English and Foreign, in many cases inaccessible to readers interested in the subject.

Information has also been obtained first-hand from some of the largest growers and manufacturers of Grasse, Nice, and localities in the Straits Settlements and West Indies.

In dealing with such a multitude of subjects, concision and abruptness of style are unavoidable—otherwise the Work would be inordinately bulky and expensive; but copious references are given for the convenience of such readers who may desire to examine more fully into the botanical and chemical details of the matter abstracted.

The physical and chemical properties of pure products are described in order to enable purchasers and consumers to detect adulterations, few substances being subjected to such extensive adulterations as drugs.

Methods of extraction and abstracts of Patent Specifications are given for the benefit of those who may wish to learn the rudiments

a 2

of a profitable business,—and at the same time they may be suggestive to inventors, in the way of making improvements thereon. This information is given, as far as possible, up to date. To another class of readers residing in our Colonies some hints are given as to the profitable nature of the cultivation of certain drug-producing plants. The Government of Victoria has recently called attention to this subject, money awards amounting to £25,000 having been voted to pioneers of certain agricultural industries, of which Flower-farming of this description is one *.

The author is still engaged upon studies in this department, and hopes to publish another volume in due course.

* Report No. 2, Royal Commission on Vegetable Products of Victoria. Melbourne, 1892.

INTRODUCTION.

No EXTERNAL sense is so intimately connected with the internal sense of perception as that of smell, and none are more capable of receiving such delicate impressions. No sensation can be remembered in so lively a manner as those which are recalled by peculiar odours, which are frequently known to excite our emotion to a degree which influences our physical and moral propensities. Truly, the memories of the past, fond or sad, are recalled to sensitive minds by music, but not so keenly as by some particular perfume—possibly because the bulk of humanity is deficient (morally or physically) in the power of appreciating musical harmony. The idea of the harmony of musical tones can be conveyed from one mind to another by printed characters, although the poetry or soul of it cannot be conveyed to a mind unconstituted for its reception, but it is impossible to convey in any way whatever the idea of an odour, perfume, or flavour, except by *comparison.*

The acuteness of the sensation of smell in animals is marvellous. The distance at which a dog tracks his master is scarcely credible. Birds of prey scent the battle-field at prodigious distances. Pliny even affirms that crows have so acute a sense of approaching corruption, that they can scent death three days before actual dissolution, and sometimes pay the moribund a visit before his time to avoid disappointment. This may have originated a superstition existing in most countries, that such a visit forebodes death.

The sense of smell is probably the leading sensorial endowment of most insects, and it is acute in some fishes, as, for example, the shark, which is the most active, if not the most intelligent, of fishes. The olfactory membrane of the shark, if spread out, would cover some twelve square feet.

As insects breathe in a very different manner from the larger animals, namely, by a number of spiracles along each side of the body, it becomes a question of some difficulty where their organs of smell are situated. We cannot easily conceive of smell being produced except by a current of air, in which odoriferous particles are diffused, passing through a moistened channel; and yet an opinion has been adopted by Cuvier, Duméril, Lehmann, and others that the spiracles, or breathing-holes, of insects are their organs of smell, and chiefly by reason that the inspiration of air seems to be an indispensable condition of smelling. If it should be argued that this organ must be near the mouth to serve as a guide as to the quality of food, Lehmann answers ('De usu Antennarum,' p. 31) that this is not so requisite in insects, because they are usually so much smaller than their food, and frequently even reside in what they eat, and many therefore smell as advantageously with the tail as with the head. De Blainville decides more positively than the facts seem to authorize that the antennæ are the organs of smell ('De l'Organisation des Animaux'). The modification, he remarks, of the skin which invests them is, in general, olfactory only in a small degree, this power appearing to be more vivid in the thickest portion of the organ, where it is more soft and tender, as in the carrion beetles (*Necrophaga*), which possess so delicate a sense of smelling. From spiders being destitute of antennæ he thinks it very difficult to conceive where the seat of their organ of smell is placed, if, indeed, they possess one, which he is disposed to doubt. Crabs and lobsters, on the other hand, whose scent is very delicate, are furnished with two pairs of antennæ. The varied effects of different odours on bees were experimentally ascertained by Huber in numerous instances (Huber on Bees), amongst which he says :—" We sprinkled some powdered musk on a drop of honey, into which some bees thrust their

suckers as if by stealth; for they kept as far back from it as possible; but although they often appeared to suck it, we did not
perceive it to become less in a quarter of an hour, long before
which it would have disappeared had it not been mixed with musk."
Pounded asafœtida, whose odour is so disagreeable to us, upon
being mixed with honey and put at the entrance of a hive, did not
seem to annoy the bees, for they greedily sucked all the honey,
neither attempting to withdraw, nor vibrating their wings, till
they only left the particles of the gum. Huber found (ibid. p. 269)
that the odour of their own poison had a very remarkable effect
on bees. The sting of one was extracted and presented to some
workers before the entrance of a hive. Although they had
previously been quiet and tranquil, they became all at once much
agitated. None flew away, but two or three darted against the
sting, and one furiously assailed the experimenters. That it was
the odour of the sting-poison alone which produced these violent
emotions was obvious from their appearing insensible to its
presence when it lost its scent by drying. In another instance bees
were confined in a glass tube and irritated with an awn of barley,
till they protruded their stings and left some poison on the sides
of the glass. The mouth of the tube was then presented to a group
of bees at the entrance of a hive and it soon produced the agitation
of rage evidently unaccompanied with fear.

Bomare relates an experiment (Dict. raisonné d'Hist. Nat., art.
Punaise) to prove that the bed-bug (*Cimex lectularius*) is not attracted, as popularly supposed, by heat, but by smell. He put a
bug into an empty bed-chamber, and throwing himself upon the
bed, perceived that the insect was not long in smelling him out
and making a direct course towards his face. These insects form
a very extensive family (Cimicidæ), and it is by no means im-
probable that they and many other insects employ the offensive
odours which nature has enabled them to discharge to produce
effects of terror upon their enemies. The fœtor of the various
species of bugs is always similar, though their food is so various.
The pretty little beetles called "ladybirds" (Coccinellidæ), of

which children are so fond, emit a similar, though not quite so offensive an odour. The rove-beetles (Staphylinidæ), in addition to their threatening and formidable attitudes, emit a very disagreeable odour, though it is not quite so bad as that of others (Silphidæ) which feed on carrion. The "church-yard beetle" (*Blaps mortisaga*) has been noted for the same circumstance since the time of Pliny (Hist. Nat. xxix. 6). Some bees (Andrenidæ) have a strong smell of garlic, which may probably be disagreeable to their various enemies (Kirby, ' Monographia,' i. p. 136). The very beautiful caterpillar of the swallow-tailed butterfly (*Papilio Machaon*, Linn.) possesses a curious organ supposed to be intended for the similar purpose of defence ; it is of a dark orange colour, and is always concealed within one of the black rings on the shoulders, unless the creature be irritated, when it darts it out to the extent of about an inch and at the same time emits a strong odour resembling fennel. This may be intended to intimidate the ichneumons from depositing their parasite eggs in its body, or to warn off the thrushes or other creatures from devouring it.

Although the sense of smell in man is very inferior to that developed in animals yet it is marvellously delicate, as some recent experiments made by Professor Vallentine prove :—He found that a current of air containing $\frac{1}{2,000,000}$ of a milligramme of oil of rose could be perceived by the sense of smell. He also ascertained that the amount of odoriferous air which must pass over the olfactory membrane in order to excite the sense of smell was from $\frac{1}{10}$ to $\frac{1}{5}$ of a pint; he calculated, therefore, that the actual amount of oil of rose necessary to excite the sense of smell was about $\frac{1}{120,000}$ of a grain. The power of perceiving the presence of mercaptan is even more wonderful; Messrs. Fischer and Pentzoldt of Erlangen have found that air containing $\frac{1}{23,000,000,000}$ of a milligramme of this compound to the cubic centimètre of air could be appreciated, and it was estimated that only $\frac{1}{460,000,000}$ of a milligramme of mercaptan is necessary to excite a sensation of smell.

The pleasant odour of the soil has been traced by Berthelot, the distinguished French chemist, to a minute trace of a camphorated body the exact constitution of which has not yet been determined. It is so odorous that even a trillionth of a milligramme gives a perceptible smell.

It may seem surprising that the delicate organs of the sense of smell should remain unimpaired over a number of years and uninjured by thoughtless and careless usage, but, as a matter of fact, they do suffer in acuteness by such causes, although not to such an extent as the organs of sight.

The sense of smell is not equally developed in mankind; on the average it is much more delicate in males than in females, but the degree of keenness ranges widely as between individuals; thus, in an experiment on record, three male observers were able to detect 1 part of prussic acid in 2 million parts of water, though in this proportion its presence was not revealable by a chemical test; others, of both sexes, could not detect prussic acid in solutions of almost overpowering strength.

Few people have perfect sight, the focal range of one eye generally differing from that of the other; many people are conscious of this defect, and to balance the inconvenience wear a glass in one eye; many are unconsciously partially colour-blind (some totally so, as proved by recorded examinations of men, applying for work as engine-drivers, not being able to distinguish between red and green); also very many are unconsciously deficient as regards their olfactory sense and the sense of taste which is dependent upon it.

On placing a sapid substance in the mouth and at the same time closing the nostrils, the sensation of taste is suspended, consequently persons with an imperfect sense of smell have also an imperfect sense of taste. Some substances have a strong taste without any or without much scent. Odour which accompanies taste is termed *flavour*. Flavour has been defined as an intermediate sensation between taste and smell.

Some races, such as Ethiopians and North-American Indians, are remarkable for the acuteness of their sense of smell, accounting for their wonderful power in tracking enemies. This perfection is in a great measure attributable to their mode of living; hunting and war are their chief pursuits, to which they are trained from their earliest childhood, and, although often subject to privation, their hardy vigorous life in the free air is healthful in the highest degree; their physical faculties are developed by constant practice. The senses of sight and hearing in these wanderers are as singularly perfect as their sense of scent. Such perfection is quite unknown to dwellers in cities, whose physical faculties are deteriorated by luxurious habits of civilization, idleness, sedentary toil, disease, and other causes and curses.

The organic compound *mercaptan* above referred to is intensely powerful in odour, otherwise it could not be perceived by such minute molecules of its vapour. Although a laboratory product, it occurs in nature in the plant *Allium ursinum*, a species of onion. In small quantities the *flavour* of onions is pleasant, but the *odour*, even in very small quantities, is to most people unpleasant.

The *Allium ursinum*, sometimes called "Ramson's garlic," is frequently found in shady meadows; it diffuses when in flower an odour of garlic, and imparts this flavour to the milk of the cows that feed upon it. This odour, occurring in different degrees of strength in most alliaceous plants, appears to be mainly due to the presence of a sulphide of allyl, the principal constituent of essential oil of garlic. The difference between oil of garlic and oil of asafœtida seems to be that the plant furnishing this last contains a larger proportion of sulphur, consequently develops the odorous principle in a more offensive degree. A number of plants belonging to the genus *Ferula* possess an alliaceous odour; it is most intense in the *Ferula fœtidissima* (Regel), which also secretes more asafœtida than any other asafœtida plant. The *Scorodosma fœtida*, a gigantic umbelliferous plant found in the sandy Steppes east of the Caspian, and closely allied to the *Ferula*, is also said to furnish a sort of gum asafœtida. On cutting into the upper part of the root a juice exudes, which hardens by exposure. Persians and other Asiatics use it as a condiment; they even call it " the food of the gods," in strange contrast to the appellation which its disgusting odour has merited for it amongst Europeans, viz. "Devil's dung" or "*Stercus Diaboli.*"

Vanillin has been prepared from asafœtida. The possibility of deriving it from this plant was suggested by Tiemann, who showed its connection with ferulic acid *.

The curious way in which very dissimilar odours are generated in the same plant is exemplified by the *Tritelia uniflora*, a handsome white-flowered species of Lily from Buenos Ayres; it has a delicate odour of violets, but when bruised this is quite overpowered by an odour of garlic. There is some obscure cause connecting these odours; it is stated that Cassie flowers (*Acacia Farnesiana*), possessing an odour analogous to that of violets, have the singular property of imparting to the breath of those who eat them a strong odour of garlic, imperceptible to the user, but intolerable to all near him; the root-bark of the same tree growing in the West Indies has the same alliaceous odour.

Several species of *Petiveria*, such as the *P. alliacea* (the "guinea hen-weed" of the West Indies), the *P. tetrandra* of Brazil, possess a strong odour of garlic. The root, wood, and leaves of the *Seguiera alliacea* have a powerful odour of garlic and asafœtida, as has also a Petiveria called *Ajo del monte*, found in the forests of Bolivia. Silver that has been in contact with iodoform, or which is even touched by the fingers after they have been in contact with iodoform, acquires a nauseous odour resembling that of garlic, which even becomes more perceptible upon rubbing the silver. A drop of saliva from a patient fully under the influence of iodoform is said to be sufficient to impart the odour to silver; also that the odour is evolved by the mere exposure of iodoform and silver in the neighbourhood of one another. The odour is not that of iodoform but is thought to be due to a decomposition product †. The *Mycena alliacea*, a fungoid plant, has a fœtid smell of onions. Amongst other

* Ph. Jul. vi. p. 813, and xvii. p. 83.
† Ibid. [3] xvii. p. 575.

curious fungi the *Clitophilus sinuata* smells of burnt sugar, the *Telamonia sublanatus* and *T. bulbosus* both smell of radishes, the *Dermocybe cucumis* of cucumbers, the *D. cinnamomeus* of cinnamon, and the *Micropus suaveolens* of anise. The *Tricholoma myomyces* is generally called the "Mouse-mushroom" on account of its mouse-like smell. The *Astragalus caprinus*, a perennial leguminous plant, native of Barbary, smells of goats. The *Orchis hircina* also smells of goats, and the *Orchis coriophora* possesses the disgusting odour of bugs. The *Psoralia bituminosa*, which is abundant on the mountains of Languedoc, recalls the odour of tar or asphalte, the leaves of the *Spiræa Ulmaria* that of carbolic acid.

The *Helleborus fœtidus*, or "Stinking Hellebore," used to be grown in gardens as an effective remedy for groundworms. The *Ballota nigra*, or "Black Stinking Horehound," an herbaceous labiate perennial, is often found in temperate climates near towns and villages, by the wayside, where it suffers little by being generally covered with dust; the whole plant is as offensive in odour as it is unattractive in appearance. The leaves of the *Comocladia dentata* ("the tooth-leaved maiden plum") of the West Indies are very sensitive to injury; when bruised they emit a sulphurous odour, and birds which happen to break them fall asphyxiated, it is only after a considerable time that they can fly away. The odour emitted by this tree when wounded has been compared to that of dung. It grows in Cuba, where the natives believe it is dangerous to sleep under its shade.

The *Anagyris fœtida*, a leguminous tree found in the South of France, Spain, Greece, and Cyprus, is perceptible at a distance by its strong odour of human excrements; stercorarious flies deceived by the stench congregate in great numbers about it. The leaves possess purgative properties, and are used by the peasantry in Greece as senna; they are then called "pseudo-sinamiko," false senna leaves. In Cyprus they are called Agriophaselo, or wild beans, by reason of the appearance of their fruit. Although the common name of this tree in English is "Bean-trefoil," the common names in French, German, and Dutch are all indicative of its offensive smell.

The flowers of *Sterculia fœtida* * and the leaves of another species of *Sterculia*, both natives of the East Indies, emit the same abominable odour. The word *Sterculia* is derived from *Stercus*, " excrement." The odour of the flowers of *Pandanus fœtidus*, Roxburgh, a native of Bengal, is similarly offensive. The wood of *Celtis reticulosa* of Java also possesses a fæcal odour. This, and other Javan trees such as *Premna corymbosa*, *P. fœtida*, and *Saprosoma arboreum*, are all three distinguished by the same local name, *Ki-tari*, meaning " Stink-wood," because they all smell very bad.

The *Arum dracunculus*, producing a large liver-coloured flower, exhales a stink of carrion so strong that few persons can endure it; in fact the similarity is so striking that blow-flies, carrion-flies, and other nauseous insects which frequent slaughter-houses mistake this flower for putrefied meat and ˙come to it from all quarters to deposit their eggs. The appearance of this flower is hideous; it is not at all uncommon, and is often cata-logued by English nurserymen. There are other plants which from their abominable odour are called carrion-plants, yet whose blossom exhibits considerable beauty, and they are extensively cultivated on that account, such as the Stapelias, a genus of Asclepiadaceæ, natives of the Cape; their branches are mostly 4-angled, toothed, and spiny, resembling Euphorbias. One of the finest of the species is the *S. asterias*.

The *Stapelia gigantea* from South Africa has a singular resem-blance to a star-fish in appearance, the five fleshy lobes of the corolla being of a biscuit-colour, with close, wavy, red veins, and the centre of the flower crimson. It is known, from its evil odour, as the " great carrion-flower." The *Phallus fœtidus*, a fungus, is equally distinguished.

At a recent meeting of the Royal Horticultural Society an enormous flower of a species of *Aristolochia* was exhibited from Kew. The flower was 22 inches long by 18 inches broad, and had a tail 34 inches long; it was of a creamy tint flushed with rose,

* Rheede, Mal. iv. t. 30.

and was marbled with purpled veins and had a velvety-black throat. The plant has been identified by Mr. Hemsley as *Aristolochia gigas*, Lindley, and is said to be very nearly allied to *A. grandiflora*, Linn., a species described by Dr. L. Planchon as distinctly poisonous to animals, and evidently possessing very marked properties. This, like several of the allied species, possesses a foetid odour like that of decaying animal matter, which doubtless causes insects to visit the flowers and cross-fertilize them. An illustration of the flower of the *A. gigas* is given in the 'Gardeners' Chronicle,' Nov. 7, 1891, p. 553.

The "Bladder-headed Saussurea," an Alpine herbaceous plant belonging to the thistle group, has this smell of putrid meat. The *Arum muscivorum* not only gives off a similar effluvium, but it is furnished with hairs bending inwards in such a way as to offer no opposition to the entrance of flies but quite prevents their escape, so forming a trap in which they perish.

A chemical examination of the bark of the *Rhus aromatica*, which is called the "Fragrant Sumach" or "Stink-bark," was made by H. W. Harper and reported in the 'American Journal of Pharmacy' in May 1881. He obtained, by distilling it with water, an essential oil having a disgusting odour resembling very much that of bed-bugs, but on being mixed with ether and the ether allowed to evaporate, the remaining oil left in contact with air for 24 hours acquired a pleasant odour distinct in itself.

There is a fungus or Morel of the *Marchella* species which on being bruised smells of roast beef, and several varieties of "Cranesbill" smell like roast mutton.

The *Durio zibethinus*, native of the East Indies, is a very remarkable tree. Its fruit, which is about the size of a man's head, is said to be the most delicious of all the fruits of India; the edible part of it most resembles whipped cream or "blanc-mange" of our tables, but a considerable drawback from the extreme gratification it procures to the palate of the epicurean is its intolerable stench; even the rinds emit such an offensive effluvium that it used to be forbidden by law to throw them out near any public path. Some compare this smell to putrid animal substances, others to that of rotten onions, but all agree that, if the first repugnance be once overcome, no fruit is more enticing than the Durion. The tree grows to a height of 80 feet.

The *Chenopodium olidum*, or "Stinking Goosefoot," has long been known for its disagreeable odour, which is compared to that of putrid salt-fish. If a portion of the plant be distilled with a solution of common soda the distillate smells strongly of boiled crabs, herring-brine, or haddocks which have long been kept—due to the presence of trimethylamine. If herring-brine be distilled in the same way with soda the same volatile compound passes over, thus proving that the same chemical compound which imparts its offensive odour to dead or decaying fish is formed in the living plant of "Stinking Goosefoot." Propylamine, which is isomeric with trimethylamine, also possesses this fishy odour, and it has been found in the flowers of *Cratægus oxyacantha, C. monogyna, Pyrus communis*, and *Sambus Aucuparia*. The odour of these flowers has often been thought to resemble that of decaying fish. This fishy odour has also been observed in the aqueous distillate of English rosemary, and freshly-distilled oil of rosemary is tainted with it until the watery particles held in mechanical suspension are deposited by rest or dried out by calcic chloride.

The *Nepeta cataria*, or "Cat mint," possesses an odour so delightful to cats that it is almost impossible to cultivate the plant in town gardens, for as soon as the cats in the neighbourhood smell it they rush in numbers to roll on it, and after having well broken it down with their convulsive capers, they finish by tearing it to pieces with their teeth. These animals have also a great liking for the odour of melons, and especially so for valerian.

Dogs take great delight in smelling the *Chenopodium vulvaria*; they roll on it, and the fœtid odour exhaled by the plant excites them to such an extent as to provoke urinary excretion.

Toads are attracted by the odour of *Stachys palustris*.

Mention is made by Mr. Louis Piesse of a tree known in Central Australia as the "Stinking Acacia," by reason of the putrescent odour of its blossoms; but in singular contrast to this unpleasant smell is the agreeable fragrance of its wood, on account of which it is termed in Western Australia (where the wood is marketed) " Raspberry Jam " wood, from some resemblance in its fragrance to the odour of that well-known preserve; the wood is described as of a dark colour, very similar in appearance to rosewood, very heavy, sinking in water like a stone, and so hard when dry as to turn the edge of a saw or chisel. The leaves are said to yield no perceptible smell when fresh, but after 48 hours of being plucked to emit a strong unpleasant odour, something like rotten cabbage. A slab of the wood of this tree was exhibited at the Colonial Exhibition.

The *Pogostemon purpuricaulis*, a tall fruticose labiate, possesses in all its parts a strong odour of black currants; it is a perennial; its smooth leaves are often 6 inches long, broadly ovate, acuminate, and serrated. The plant is very common in Kokun.

The *Gaillardia scabiosoides*, a bushy plant about 15 inches high, growing in dense masses on low clay plains on the western side of temperate S. America, has blossoms which possess a strong odour of ripe apricots.

The alcoholic extract of the bark of *Colubrina reclinata* (" Snake-wood " of Martinique), when boiled with very dilute acid, gives off an odour exactly resembling ripe raspberries.

As a general principle, a larger proportion of white flowers are fragrant than those of any colour, yellow comes next, then red, blue, violet, green, orange, brown, and black. Orange and brown are frequently unpleasant in scent, white flowers very rarely so. It must be remembered, however, that odours are differently appreciated by different people, and what pleases one person may have a reverse effect on another; thus the strong odour of *Tagetes Patula* (French Marigold) and *T. erecta* (African Marigold) is not unpleasant to some, while others consider it very objectionable.

Many flowers have a strong odour of honey; it is very powerful in those of the *Asclepias Syriaca*; they are much visited by the bee. The odour of honey or brown sugar is noticeable in the " Sweet Scabious," also in the aquatic Ranunculi and in some varieties of Lotus. (The words Lotus or Water-Lily, the Latin *lotus* " washed," and the English *lotion* " a wash," are expressed in India by the word *lota*, which is there applied to a globular brass bowl, sometimes melon-shaped, with a long narrow neck, universally used in ceremonial and other ablutions.)

The flowers of the well-known yellow Water-Lily, *Nuphar lutea*, a plant which is common in most parts of Britain, and frequently grown in ornamental waters, have a curious alcoholic odour, hence the name " Brandy-bottle " which is applied in some counties to this plant; the flowers are used by the Turks in the preparation of cooling drinks. The same peculiar odour of brandy is also found in the yellow catkins of the *Salix caprea*, the " Goat Willow."

The *Hippocrepis comosa*, a sort of " Horse-shoe Vetch," common on chalky soils, recalls the smell of cheese, an odour which is also observable in the blossoms of *Genista Scoparia*, a thorny shrub, native of Spain ; to some persons the odour of this flower is more like that of the fruit of the cocoa-nut. The leaves of the *Philadelphus Coronarius* have an odour and flavour precisely resembling cucumbers.

Some odours are developed by desiccation, as Deer's-tongue leaves, Hedychium root, and Iris root; and some by partial fermentation, as Vanilla pods, Patchouli, tobacco, and tea leaves.

The leaves of *Scopolia luridus* (Dunal), a solonaceous plant of Nepaul and the Himalaya, emit a tobacco-like odour.

The *Hedyosmum nutans* (Swartz), called the " Tobacco bush " in Jamaica, is a common plant on the hills about Port Royal and on the Blue Mountains at an elevation of 5000 to 6000 feet above the sea. The aromatic oil distilled from it certainly has an odour somewhat like that of cake " honey-dew " tobacco. The *Critonea Dalea*, D.C., is another Jamaica plant locally called the " Cigar bush "* or " Cigar-maker's vanilla." Its odour recalls that of new-made hay and especially that of the *Liatris odoratissima* (" Deer's-tongue "), an herbaceous plant abundant in North Carolina and Florida, whose leaves are also used by tobacco-manufacturers for aromatising tobacco.

* Samples of Tobacco-bush oil and Cigar-bush oil were in the Colonial Exhibition. They have been recommended for perfuming toilet-soaps.

LIST OF ILLUSTRATIONS.

FIG.   PAGE

1. APPARATUS USED IN THE MANUFACTURE OF BULGARIAN OTTO OF ROSE   30

2. INTERIOR OF STILL-ROOM AT GRASSE 32

   SKETCH-MAP OF THE BULGARIAN ROSE-FARMING DISTRICT .. *To face* 32

3. LEAF OF PELARGONIUM CULTIVATED AT GRASSE 43

4. PIVER'S APPARATUS FOR " MACERATION " 53

5. PIVER'S APPARATUS FOR "ENFLEURAGE " 57

6. APPARATUS FOR EXTRACTION OF OILS FROM CITRUS FRUITS 69

7. ILLUSTRATING THE PROCESS OF ARTIFICIAL FECUNDATION OF THE
   VANILLA FLOWER 151

8. APPARATUS FOR ESTIMATING RECTIFICATION RESIDUE OF CASSIA OIL .. 211

9. FLOWER OF *Pogostemon Patchouli* AND OF *P. Heyneanum* 294

10. LEAF OF *Pogostemon Patchouli*................................. 295

11. MICROSCOPIC SECTIONS OF PATCHOULI LEAF 303

12. LAVENDER (*Lavandula vera*); APPEARANCE IN JULY 357

13. DITTO; APPEARANCE IN MARCH 358

LIST OF PRINCIPAL WORKS REFERRED TO, AND ABBREVIATIONS.

—◇◇◇—

American Journal of Pharmacy.	*Am. Journ. Pharm.*
Annalen der Chemie und Pharmacie.	*Ann. Chem. Pharm.*
Annales de Chimie et de Physique.	*Ann. Chim. Phys.*
Apotheker Zeitung.	*Apot. Zeit.*
Archives der Pharmacie.	*Archiv. der Pharm.*
Asiatic Researches.	*As. Res.*
Aublet, Histoire des Plantes de la Guiane Française.	*Aubl. Guian.*
Baillon, Dictionnaire de Botanique.	
Bentham & Hooker, Genera Plantarum.	
Bentley & Trimen, Medicinal Plants.	
Bericht von Schimmel & Co. in Leipzig.	*Schimmel.*
Berichte der Deutschen chemischen Gesellschaft.	*Ber. Deutsch. chem. Ges.*
Blanco, Flora de Filipinas.	
Blume, Bijdragen tot de Flora van Nederlandsch Indië.	*Blume, Bijdr.*
Ditto, Floræ Javæ.	*Blume, Flor. Jav.*
Bulletin de la Société Chimique.	*Bull. Soc. Chim.*
Cavanilles, Monadelphiæ classis dissertationes.	*Cav. Dis.*
Chemist & Druggist.	
Comptes Rendus des Séances de l'Académie des Sciences.	*Comptes Rendus.*
Consular Reports—various.	
Cooke, Report on the Gum-resins in the India Museum.	
Curtis, Botanical Magazine.	*Bot. Mag.*
DeCandolle, Prodromus Systematis Naturalis.	*DC. Prodr.*
Don, Prodromus floræ nepalensis.	*Don, Prod.*
Drury's Useful Plants of India.	
Dunal, Monographie de la Famille des Anonacées.	*Dunal, Mon. Anon.*
Dupetit Thouars, Orchidées d'Afrique.	
Dymock, Materia Medica of Western India.	*Dymock, Mat. Med. Ind.*
Edwards, Botanical Register.	*Bot. Reg.*
Flückiger & Hanbury, Pharmacographia.	
Ditto, ditto. French edition. "Histoire des Drogues."	
Forskål, Descriptiones plantarum floræ Ægyptico-Arabicæ.	*Forsk. Descr.*
Gærtner, De fructibus et seminibus plantarum.	*Gærtner, Fruct.*
Gallesio, Traité du Citrus.	*Gallesio.*

Gardeners' Chronicle.

Gleditsch, Mémoires de l'Académie des Sciences de   *Gled. Act. Soc. cur. Nat.*
Berlin.   *Berl.*

Guibourt, Histoire Naturelle des drogues simples.

Hall's Dictionary of the Economic Plants of India.

Hooker's ' Journal of Botany ' and ' Kew Miscellany.'

Hooker & Thomson, Flora Indica.

Hooker, Flora of British India.

Indian Agriculturist.

Jacquin, Hortus botanicus vindobonensis.   *Jacq. Hort. Vind.*

Journal of the Indian Archipelago.

Journal of the Linnean Society.   *Journ. Lin. Soc.*

Journal of the Royal Horticultural Society.

Journal de Pharmacie.   *Journ. de Pharm.*

Journal of the Chemical Society.   *Journ. Chem. Soc.*

Kæmpfer, Amœnitates exoticæ.   *Kæmpf. Amœn.*

Kew Bulletin.

Kunth, Genera terebenacearum.   *Kunth, Gen. Tereb.*

Lamarck, Illustrations des genres.   *Lam. Ill.*

Ditto, Encyclopédie Méthodique, botanique.   *Lam. Encyc.*
   Lam. Dic.

Lindley's Flora Medica.

Linnæus fil., Supplementum plantarum.   *Lin. Sup.*

Linnæus, Amœnitates academicæ.   *Lin. Amœn.*

Loureiro, Flora Cochinchinensis.   *Lour. Flor. Cochin.*
Ditto, ditto, 2nd ed. Edited by Willdenow.

Oil and Drug News.

Pharmaceutical Era.

Pharmaceutical Journal and Transactions.   *Pharm. Journ.*

Pharmaceutishe Centralhalle.   *Pharm. Centralh.*

Poggendorff's Annalen.   *Pogg. Ann.*

Pomet, Histoire des drogues.

Ray, Historia plantarum.   *Ray, Hist. plant.*

Répertoire de Chimie.

Retzius, Observationes botanicæ.   *Retz. Obs.*

Risso, Histoire Naturelle des Orangers.   *Risso.*

Roscoe & Schorlemmer, Organic Chemistry.

Roxburgh, Plants of the Coast of Coromandel.   *Rox. Cor.*

Ditto, Hortus Bergalensis.   *Rox. Hort. Beng.*

Ditto, Flora Indica.   *Rox. Fl. Ind.*

Royle, Illustrations of the Botany of the Himalayan   *Royle, Ill. Bot. Him.*
Mountains.

Rumphius, Herbarium Amboinense.   *Rumph. Amb.*

Schlechtendal, Linnæa, ein Journal für die Botanik.   *Linnæa.*

Silliman's Journal.

Sprengel, Systema Vegetabilium.   *Spreng. Syst.*

Swartz, Icones plantarum Indiæ occidentalis.   *Swartz, Icones.*

Sweet, Hortus Britannicus.

Sweet, Geraniaceæ or Natural order of Geraniums.

Tenore, Flora Napolitana.   *Teno. Fl. Nap.*

Transactions of Bombay Geographical Society.

Transactions of Linnean Society of London.   *Trans. Lin. Soc.*

Transactions of Medical and Physical Society of Calcutta.

Trommsdorf's Journal of Pharmacy.

Vahl, Symbolæ botanicæ.

Valmont de Bomare, Dictionnaire d'Histoire Naturelle.

Van Rheede, Hortus Indicus Malabaricus.   *Rheed. Hort. Mal.*

Ventenat, Description des plantes nouvelles du jardin de J. M. Cels.   *Vent. Jard. Cels.* / *Vent. Cels.*

Wallich, Catalogue of dried specimens in Botanic Garden, Calcutta.   *Wall. Cat.*

Ditto, Plantæ rariores Asiaticæ.   *Wall. Plant. As. Rar.*

Wight & Arnott, Prodromus floræ peninsulæ Indiæ orientalis.   *Wight & Arnott, Prodr.*

Wight, Illustrations of Indian Botany.   *Wight, Ill.*

Willdenow, Species plantarum.   *Willd. Spec.*

Woodville, Medical Botany.

Zeitschrift für physiologische Chemie.   *Zeitschrift.*

ODOROGRAPHIA.

CHAPTER I.

THE ODOUR OF MUSK.

THE excessive love of perfumes generally, and of strong perfumes especially, is only manifested by Orientals, but yet in Northern Europe there are often found people with an inordinate partiality for the odour of musk—pure, or in combination. This odour is very widely distributed in nature, being found developed in birds, beasts, fishes, insects, reptiles, and plants, yet its principle has not been isolated and is not understood. It is even produced artificially by chemical combinations, but the reason of its development is not apparent.

The Musk of commerce, which may be taken as the type of this odour, is the dried secretion of the preputial follicles of the male Musk Deer (*Moschus moschiferus*, Linn.). The Musk Deer is essentially a forest animal, inhabiting cold, mountainous districts on the Atlas and Himalayan ranges, at elevations above 8000 feet where coniferous plants abound. Although an inhabitant of the forest it is partial to woody ravines, and is frequently found on the spurs or projecting points jutting from the eternal snow-ranges at an altitude of from 10,000 to 14,000 feet. It is found in Thibet, in Yun-nan, Sze-tchuen, and occasionally in Petsche-li (sometimes spelt Chih-li).

The bag or pod containing the musk is situate near the navel, and is composed of several layers of thin skin. The pod varies in

B

size and shape according to the age of the animal and the time of year at which it is procured; the shape may be described as conical, oval, or pear-shaped. There is an orifice through the skin into which, by a slight pressure, the little finger will pass, but it has no connexion whatever with the body. It is probable that musk is at times discharged through this orifice, as the pod is often found not half full, and sometimes even nearly empty. The hairs are brownish yellow, or greyish, or whitish, bristle-like and stiff, arranged in a concentric manner around the orifice of the pod. The quality of the musk secreted in the pod varies considerably, the older the animal the more valuable the musk. Musk is only found in adult males, although the pouch destined to contain it is well-formed at birth. For the first two years of the animal's life the pod only contains a soft milky substance with a disagreeable smell. When it first becomes musk there is not much more than an eighth of an ounce, and as the animal grows it increases in quantity. In some individuals as much as two ounces are found. An ounce may be considered the average from a full-grown animal, but as many of the deer are killed young, the pods in the market should contain on an average half an ounce. Though not so strong, the musk of young animals has a much pleasanter smell than that of old ones. The secretion is known to have a much stronger odour in animals that inhabit Thibet and China than in those which are found farther north in Siberia.

Musk is also secreted by the *Moschus Altaicus*, another Musk Deer, inhabiting the mountainous Kirgesian and Sangorian steppes of the Altai on the river Irtysch.

The principal depots of musk produced in Thibet and Annam, as far as Tonquin, are Ta-tsien-fu, in about 30° N. lat., west of the province of Sze-tchuen, Silung-chow in Kwang-si, and Wuting-chow in Yun-nan. The greater portion is carried to Chang-hai by river, where the pods are opened, emptied, and the contents, after being carefully adulterated, are replaced in the pods and sewn up.

A description of the native method of drying the pod was given some years ago by Mr. Peake to the Pharmaceutical Society; some extracts from that paper are here given:—" The pod is cut from the deer with a portion of the outer skin, then pressed and dried on a hot stone to prevent putrefaction; but by this heating process much of the odour is driven off, consequently its value to the perfumer greatly diminished." It is further added :—" It would be a difficult matter for a native to resist the temptation of adulterating and falsifying these pods. They cut the young pods containing no musk at all, and fill them with the liver and blood of the animal mixed with the yellow fluid (which age would have matured into musk), add a small portion of genuine musk, then sew up the skin and dry; or those which yield half a drachm to a drachm they mix and dry in like manner."

Pereira says :—" The great sophisticators of musk are the Chinese; they actually export artificial pods from Canton. The hairy portion of the sacs is formed of a piece of the skin of the musk animal (readily distinguishable by its remarkable hairs) coarsely sewn at the edges to a piece of membrane, which represents the smooth or hairless portion of the sacs. These pods are distinguished from the genuine ones by the following characters : the absence of any aperture in the middle of the hairy coat; the hair not being arranged in a circular manner; and the absence of the remains of the penis (found in every genuine musk-sac). The odour of the false sacs is ammoniacal."

The grains of musk contained in the pods should be unctuous to the feel and of a dark reddish-brown colour. An infusion of genuine grain musk gives *no* precipitate with a solution of bichloride of mercury, but does with tincture of nut-galls and acetate of lead. By incineration genuine musk leaves behind a greyish-white ash, whereas blood yields a reddish one. An imitation musk is prepared by rubbing in a mortar dried bullocks' blood with caustic ammonia, and mixing the half-dried product with genuine musk. The substances generally used for adulteration, or to fill the counterfeit pods are said to be :—blood, boiled or baked on the fire, then beaten to powder, kneaded into a paste, and made into grains and coarse powder to resemble genuine musk ; a piece of the liver or spleen prepared in the same manner ; dried gall and a particular part of the bark of the apricot tree, pounded and kneaded as above. Lentils, peas, pieces of leather are also common adulterants ; it is sometimes found mixed with particles of dark-coloured earth and pieces of lead, to increase the weight.

The microscope is very useful for detecting adulteration in musk. The colour of the individual grains should not be reddish or reddish brown, but, even under the microscope, should appear deep brown or blackish brown. If isolated particles are glassy, jelly-like, and transparent, they indicate adulteration with coagulated blood.

The spirituous extract of musk and the infusion should not be reddish brown, but deep blackish brown.

The Siberian or Russian musk (sometimes called Carbadine), and that coming from the Altai Mountains, is rarely adulterated to any extent, but its odour is much less powerful, being more nauseous and somewhat empyreumatic. The Assam musk occurs in very large pods, and is very strong, but considerably lower in value than the Tonquin or Chinese musk on the London Market, a result of the heavy adulteration to which it is subjected.

The value of musk is subject to considerable fluctuation. At the beginning of 1883 Chinese pod-musk of poor quality realized 105s. per oz. at public sale in London. The position of this market in 1891 may be gathered by the following extracts from Trade Reports :—

"*April* 24, 1891.—MOSCHUS. Messrs. Gehe observe that confidence in Tonquin musk has gradually lessened, as the increased shipments of the last few months from Shanghai show. It is noteworthy that Paris, after a long pause, has again appeared in the *rôle* of a large buyer. During 1890 the total shipments from Shanghai were 1072 catties below those of 1889, when the figure was 2266 catties. The firm do not think, however, that artificial musk had the least to do with bringing this about. In fact, the purchase of musk in China by Paris is taken as a proof that the artificial product is not suitable for use in perfumery. It is further stated that it does not answer even in the manufacture of soaps, as the odour is very unstable, and very soon entirely disappears."

"*April* 23, 1892.—The price quoted at this date by wholesale London houses is 135s. to 145s. per oz. for Chinese grain-musk."

The analysis of musk does not disclose any clue to the nature of its odoriferous principle. The various musky-scented substances derived from the Animal Kingdom are strongly suggestive of a condition of peculiar putrefaction or decay. There are instances in the Vegetable Kingdom of odorous principles being generated by similar causes—for instance, Oriental Lign-Aloes (or "Eagle-wood"), *Aquilaria Agallocha*. The wood of this tree is impregnated with a resinous matter often found collected in clots here and there throughout the stem; a fact which is in all probability due to a diseased condition of the tree, a condition which is in fact induced by *wounding* the tree in order to increase the formation or collection in nuclei of the resin. This condition is also brought about by burying pieces of the wood in damp soil. The substance called ambergris (afterwards described) is considered to result from a disease of the whale. In nature nothing dies—it simply changes; it takes another form; additions cannot be made to, and nothing can be subtracted from, the original total of the contents of this globe, otherwise the equilibrium would be instantly upset.

The molecular particles of the odour of musk are so infinitely small, that for a long time loss of weight by exposure is inappreciable. A little musk will impart a durable scent to everything in its immediate neighbourhood. This odour is so persistent, and has such power to resist decomposition, that when musk is taken internally as a medicine (as it very frequently is in the East), it passes through the pores of the skin and impregnates the perspiration with its strong odour. This odour of musk can be disguised by keeping it in capsules of wax, or in contact with lime, milk of sulphur, sulphuret of gold, or syrup of almonds, all of which have the power of *concealing* it: but it is at once restored by being moistened with a little liquid ammonia *.

It is asserted that the odour is very powerfully increased by mixing the musk with alkaline salts, especially with carbonate of potash. Liquid ammonia has also been mentioned as exerting a revivifying effect on musk which has become partially exhausted; but some observers refute the possibility of alkalies having the power of restoring or strengthening its odour, and assert that ammonia simply increases the *volatility* of musk. It is, however, an established fact that combination with soap intensifies it, and as soap is an oleate of soda or potash, this result is naturally attributed to alkaline reaction.

When musk is moistened with water, the odour is more powerful than when in a dry state. Grain-musk is soluble in water to the extent of 90 per cent., and in alcohol to only 50 per cent. As a basis for toilet perfumes, musk is of great value by reason of its stability. By its great diffusibility it deserves the appellation of the " wings " of the perfume to which it is added.

For many years attempts have been made to imitate artificially the odour of musk. Experiments, successful to a certain extent,

* Repert. für die Pharm. Band xxix. Heft 1, p. 51.

were made by Margraff and by Elsner *. Coarse pieces of amber, reduced to powder and mixed with sand, are distilled in an iron retort; the oil which distils over is separated from the fetid liquor and succinic acid which accompanies it, and after being rectified at a gentle heat with about six times its volume of water, is gradually added to and digested with $3\frac{1}{4}$ parts by weight of fuming nitric acid, artificial cold being employed to prevent any portion of the oil being carbonized. An orange-yellow resinous matter forms, which, after being carefully dried, is the required product. It is also said to be formed by digesting for ten days one ounce of fetid animal oil, obtained by distillation, and half an ounce of nitric acid, then adding rectified spirit one pint, and digesting for a month.

The crystalline substance now met with in commerce under the name of " artificial musk," or " Musk Baur," is a trinitro deri-vative of butyl-toluene, produced by the action of nitric and sulphuric acids upon the hydrocarbon meta-butyl-toluene obtained from isobutyl bromide and toluene, and also found in resin-spirit†. A great number of homologues and isomerides have been prepared, but though many possess a musk-like odour, none have so great a technical value; it has therefore been found advisable to carefully purify the butyl-toluene for technical purposes. According to the English specification of Baur's Patent (No. 4963, 21 Mar. 1889), this " substitute " for musk consists of a nitrated hydrocarbon of the $C_{11}H_{16}$ group, for the formation of which five parts of toluene are mixed with one part of butyl bromide, or butyl chloride, or butyl iodide. To these may be added gradually, whilst boiling, one fifth part of aluminium chloride or aluminium bromide; this results in the development of hydrobromic or hydrochloric or hy-driodic acid respectively, and a product of reaction is obtained from which, by the action of steam, the hydrocarbon $C_{11}H_{16}$ and unchanged toluene are distilled. By the admission of steam the hydrocarbon is carried along, and may be obtained in a condenser as a colourless oil floating on water. The oil, removed and dried by means of chloride of calcium, is fractionated, and in this manner the necessary hydrocarbon for the production of artificial musk is

---

* Journ. für praktische Chemie, 1842.

† 'Comptes Rendus de l'Académie,' cxi. pp. 238-240; Berichte Chem. Ges. xxiv. p. 2832; Pharm. J. li. p. 266.

obtained. One hundred parts of the former give a like quantity of musk preparation. Three parts of fuming nitric acid of 1·52 sp. gr. and six parts of fuming sulphuric acid are mixed together, and to this mixture is carefully added, whilst cooling, one part of the hydrocarbon aforesaid. Each drop causes a violent reaction. As soon as all the hydrocarbon is added, the whole mixture is heated up to a temperature of about 100° C. After cooling, the nitro-product is precipitated by pouring into cold water of about five or six times the volume, and is separated from superfluous acid by washing with cold water. The nitro-product separates first as a heavy viscid oil, which after some time hardens into a firm crystalline substance. The raw nitro-product is then purified by re-crystallization from alcohol of 90 %. The pure product crystallizes out in yellowish-white needles possessing a strong smell of musk. It is insoluble in water, easily soluble in alcohol and ether, and slightly volatile with water-vapour.

Another "substitute" for musk has been patented in England by Emil Schnaufer and Heinrich Hupfeld of Frankfort (No. 18521, 18 Dec. 1888), according to the specification of which "three parts of metaxylol, two parts of isobutyl alcohol, and nine parts of chloride of zinc are heated in a digestor to from 220° to 240° until the pressure, which at the commencement is from 25 to 29 atmospheres, sinks to below 6 atmospheres. The resulting hydrocarbon, corresponding to the formula $C_{12}H_{18}$, is collected, and the fraction which distils over at from 190° to 230° is nitrated with HNO_3 or with HNO_3 and H_2SO_4, whilst being cooled. The product of the reaction is poured into water, whereupon a reddish-brown oil separates which is washed several times with alkaline water. The formula of this oil is $C_{12}H_{17}NO_2$, and in a concentrated condition it possesses a sweet smell, whilst in a dilute solution it gives off a penetrating and enduring musk-like odour."

The Complete Specification states that "aromatic hydrocarbons containing the iso-propyl, iso-butyl, or iso-amyl group, on treatment with fuming nitric acid or a mixture of strong nitric acid (40° to 44° B.) and sulphuric acid (66° B.), produce derivatives which, in very dilute alcoholic solution, furnish a liquid possessing an odour resembling tincture of musk in the highest degree." In the Provisional Specification only one example of the process is given by way of illustration, but of course the process may be carried out with the other well-known homologues. "The hydrocarbons may be produced in the ordinary way, but we produce them by the following operation :—Toluene or xylol is heated in a digestor with iso-propyl, or iso-butyl, or iso-amyl alcohol in molecular quantities with the addition of from four to five times the quantity of chloride of zinc to the boiling-point of the hydrocarbon, or to about 40° or 50° above the boiling-point of alcohol, until the pressure, which at the commencement was equal to about 26 atmospheres, sinks to a little above 2 or 3 atmospheres. The product of the reaction is subjected to fractional distillation.

"By the above process the following hydrocarbons are obtained:—

    1. From Toluene—Methylisopropyl-benzene.
                  Methylisobutyl-   „
                  Methylisoamyl-   „
    2. From Xylol — Dimethylisopropyl-benzene.
                  Dimethylisobutyl-   „
                  Dimethylisoamyl-   „

"To produce the 'Musk-substitute' we add to the above-mentioned hydrocarbons, which during the operation should be kept thoroughly cool, a little more than the molecular quantity of fuming nitric acid or nitro-sulphuric acid. The acid should be gradually run in and the whole then allowed to stand undisturbed for from one to two hours, the resulting mass being then poured into water in order to get rid of the excess of acid. The well-washed substances thus obtained are then subjected to distillation by means of steam, whereupon simultaneously-formed bodies which smell like nitro-benzol and overpower the musk-odour readily distil over, whilst the pure substances remain behind."

The musk-substitute obtained by Baur's process (the trinitro derivative of isobutyl-toluene above-mentioned) is insoluble in water. A process has been devised by Valentiner for obtaining a product which he thinks may be more useful in perfumery inasmuch as it is soluble in water. This is effected by nitrating a sulpho-acid of butyl-xylene. A mixture of isobutyl alcohol and aceto-xylene in equivalent proportions is gradually mixed with five parts concentrated sulphuric acid without being allowed to become hot. After some time the mixture is diluted with a four-fold quantity of water, and the oily layer of unaltered material thus separated is removed. The clear, rose-coloured, watery solution is mixed with saturated sodium-chloride solution until the para-isobutyl-xylene-sulphonic acid is deposited in white crystals, which are collected by filtration, recrystallized, and dried *. It is commercially known as "Tonquinol." Valentiner's *English* Patent is dated 3rd October, 1890 (No. 15687), and is abstracted in the 'Official Journal of Patents' as follows :—The formation of "artificial Musk consists in condensing molecular proportions of iso-butyl alcohol and xylene by means of sulphuric acid at a temperature not exceeding 45° C., and introducing the mixture into fuming nitric acid, whereby a dinitro derivative of the condensation products is produced. This is separated by addition of water and purified by crystallization from alcohol. Instead of xylene, oil of turpentine or cymene may be employed, and iso-propyl or iso-amyl alcohol instead of iso-butyl alcohol. The product of condensation consists of a new hydrocarbon and a sulphonic acid. The separation and nitration of the latter is also described. In order to prepare a soluble artificial Musk for perfuming soaps, the product of condensation is sulphonized with fuming sulphuric acid before nitration." This Patent has been opposed. The German Patent gave rise to a law-suit on the ground that it is an infringement of Baur's Patent, and the case is not yet decided. Baur's second English Patent is dated 11th August, 1891 (No. 13613).

Two Patents have been taken out in London by Link and Avenarius jointly, for the manufacture of products of a very similar nature, and both these Patents are "opposed." The first one, dated 1st January, 1891 (No. 48), consists in first producing tertiary amyl-toluene or its homologues by boiling tertiary amyl chloride with toluene or xylene in presence of ferric chloride. The hydrocarbon produced is converted into its tri-nitro derivative by heating it on a water-bath with a mixture of nitric acid sp. gr. 1·5 and fuming sulphuric acid. After separating excess of acid and purifying by re-crystallization from alcohol, light yellow crystals of musk-like odour are obtained. The second Patent, dated 3rd January, 1891 (No. 115), consists in first heating combinations of isodibutylene and halogen acids together with toluene or metaxylene or ethyl-benzene in the presence of ferric chloride, whereby tertiary butyl-methane, polymeric propylenes, and iso-propyl toluene and homologues of these bodies are produced. The hydrocarbons are distilled off with steam, dried, and fractionally distilled to separate the tertiary butyl-propyl-methane

* Pharm. Centr. xxxiii. p. 80.

and the hydrocarbons of the composition $C_{15}H_{21}$ or their homologues. These products are next converted into their tri-nitro derivatives and purified as specified in the first Patent.

At the Meeting of the Society of Chemical Industry held on the 4th April, 1892, a paper was read entitled "Studies on Musk Baur" by Dr. Baur, the inventor of that product. A study of the cresols in their relation to butyl enabled the author to find that a body with an intense musk-like odour is obtained by boiling pure meta-cresol ether with iso-butyl bromide and aluminium chloride, subsequently nitrating the product. If meta-cresol be treated with butyl alcohol, then with zinc chloride, and the mixture boiled in contact, with a reflux condenser attached, an unsymmetrical butyl-cresol is obtained which, when etherified and nitrated, yields a substance with an odour very like civet.

It does not appear that artificial musk blends well or "fixes" perfumes well, as does the natural article. It is not stable, and under certain conditions its odour is destroyed. Also, it is but little understood, and there exists an idea that such nitrated compounds are explosive.

CIVET.

Next in importance as an example of the musk type is Civet. This is secreted by the *Viverra Civetta* and the *Viverra Zibetha* in a pouch divided into two bags and situated beneath the tail. It is so powerful that it infects every part of the creature. This secretion is increased when the animal is irritated, a fact which is sometimes cruelly taken advantage of by enclosing the animal in a cage in which it cannot turn round, and then tormenting it. The cage being opened by a door from behind, a spoon is introduced through the orifice of the pouch and the contents carefully scraped out; the operation being repeated two or three times a week. The yield is said to be increased by feeding up the animal on foods which it is fond of. The secretion is a thick, unctuous, pale yellow matter about the consistence of honey, repulsive both in appearance and odour; the object of its formation is not obvious: it may be intended for purposes of defence, as is the case with the skunk and the polecat. The secretions are in relation to the habits of life and conditions of existence of various creatures, such as the poison of the viper for attack, and the fetid exhalations of some insects for defence. A poison-bag would be useless to the serpent if it fed on vegetables. There is a little insect termed the " Bombardier " (*Carabus crepitans*) which, when pursued, emits with an explosive noise a bluish, acrid vapour, very highly irritating to the senses of its enemy, which is an insect of the same tribe but of three or four times its size and strength. The inky secretion of the cuttle-fish, which that animal employs as a means of baffling its enemies and escaping pursuit, derives its utility from the circumstance of its being diffusible through water.

The odour of civet is much more powerful than musk, although its diffusiveness is not so great. On being much diluted its odour becomes bearable and even fragrant. It is very useful to assist other perfumes in the same way as musk. The first of the above-mentioned Civettæ is a native of the hottest countries in Africa; the second is an inhabitant of India, the Moluccas, and Philippine Islands. A third species is found in Java, called the *Viverra rasse*.

The *Viverra Zibetha* is sometimes entrapped with the fruit of the *Durio Zibethinus* tree, a delicacy which the animal is extremely fond of, so much so that the tree is specifically named after it. The fetid odour of this fruit is already mentioned.

The Canadian Musk Rat, *Ondrata Zibethica* *, is an amphibious animal related to the beaver. It abounds on the margins of rivers and lakes of the United States and Canada, inhabiting mud huts, which it constructs. It lives on aquatic plants, principally the roots of the *Nymphæa* and the *Acorus*, which last, as a food, may have some influence on the production of the characteristic perfume of this animal ; but its voracity is such that when unable to find vegetable food it will eat flesh, and, failing that, these animals will even eat one another †. This Musk Rat is frequently mentioned by early writers on America, on account of its odour, which is due to a whitish fluid deposited in certain glands near the base of the tail. It is particularly strong in spring. The skins and tails, which long retain their odour, are used in Russia to preserve clothes from moth. Being cut up and macerated with spirit, a very powerful tincture is obtained : to one pint of spirit two drachms of slaked lime are generally added, the idea being that calcium hydrate or potassium hydrate softens the perfume and

* Buffon, Hist. Nat. x. t. 1.　　　† Guibourt, Hist. Nat. iv. p. 37.

helps the solvent powers of the menstruum *. The bags are only properly developed in the male animal. The evident purpose of the odour is, so far as the animal is concerned, that of attracting the opposite sex.

There are other small animals going by the name of " Musk Rat," having the upper lip elongated into a snout or short proboscis, such as the Russian Musk Rat or " Desman " (*Mygale moscovita*), figured in Buffon's Nat. Hist. tab. x. This is common on the borders of rivers and lakes in the South of Russia; it feeds on worms, larvæ, and leeches, which it extracts from the mud with its flexible proboscis; its odorous principle is secreted in small follicles beneath the tail. Its odour is so powerful as to be communicated to the pike which feed on it. The Musk Rat of the Antilles (*Mus pilorides*) is a true rat, and a very voracious and noxious little animal.

There is another Musk Rat, native of India, called the " *Sondeli*," which often utterly spoils provisions by the persistency and strength of its odour. It is called the " *Ondrata* " by Rimmel, but it may be the *Sorex Indicus*. It is common in the lower and central regions of Nepal, also in Spain. In some parts of Ceylon this rat is a great pest. It is asserted that wine-merchants have to carefully seal with wax every bottle of wine in a cellar, to prevent the powerful secretion of these rats from penetrating the corks, and so spoiling the wine.

The " Musk Ox " (*Bos moschatus*), found in the coldest parts of North America, has many striking peculiarities which appear to give it an alliance to the goat, rather than to the ox, yet the general figure and size will warrant the naturalist in placing it in the bovine tribe. A singular secretion of musk strongly pervades and taints its flesh, particularly the heart and kidneys; this is said to be much more manifest in the lean than in the fat kine.

The natives living in the vicinity of the Sahara Desert collect the droppings of a small Antelope (*Antelope Dorcas*), which, when dried, is quite as fragrant as musk. Analysis proves this product to contain 63 per cent. of undigested vegetable matter, 26 per cent. of insoluble mineral matter, and only 10 per cent. of matter soluble in water and spirit; this consists of a musk-like resin, benzoic acid, biliary acid, and biliary colouring-matters. This antelope is very common in the Desert, and is called by the Arabs

* Ph. J. [3] xv. p. 87.

the " Retsal." The aroma is said to be due to the product of the secretion of some sebacic glands situated in the inguinal region of the animal.

The Florida Alligator has four glands which secrete a whitish-yellow fluid possessing the exact odour of ordinary musk. Two of these glands are situate on the lower part of the head immediately under the throat, and one on each side of the vent. A similar alligator is found in British Honduras. There are two marine Turtle which have a strong smell of musk : the *Chelonia caouana* and *Chelonia caretta*.

The *Cerambyx moschata*, a coleopterous insect, owes its specific name to the same cause.

AMBERGRIS.

It has been already remarked that some perfumes seem to be a result of decay or disease ; as an instance of this amongst fishes may be cited Ambergris, which is a biliary concretion of the Spermaceti Whale (*Physeter macrocephalus*), and is, according to several authorities, an undoubted product of disease ; its odour recalls that of musk, but is much more delicate ; it gives a permanency to floral odours which are in themselves evanescent. For fixity and permanence the animal odours are unrivalled, and with careful blending in bouquet perfumes their identity is not predominant. A handkerchief scented with ambergris will retain the odour even after being washed.

It has been repeatedly asserted that the odour of ambergris can be evolved from cow-dung by careful distillation of that unsavoury material, taken fresh, in the months of May and June. This assertion, originally made by an ancient writer on Chemistry (and Alchemy), does not appear to have been contradicted by modern scientists, and although rather a dirty experiment, it is worth testing—especially as the material is so easily obtainable, and competition has reduced the price of soap. Experiments made with the urine of the horse led to satisfactory results as regards the production of hippuric acid and its convertibility into benzoic acid *.

Some early writers even go so far as to say that night-soil,

* Pogg. Ann. xvii. p. 339 ; Ann. Chem. Pharm. xxvi. p. 60, xxviii. p. 40 ; Journ. Prakt. Chem. xxxvii. p. 244.

under certain treatment, evolves an odour of ambergris. The early writers were evidently daring experimenters, and had stronger stomachs than modern manufacturing perfumers.

Ambergris is found floating on the sea near the coasts of, and thrown up on the shores of, various tropical countries. As it has not been found in any whales but such as were dead or sick, its production is generally supposed to be owing to disease. Most specimens of ambergris, especially the large ones, are found to contain embedded in them the beak-like nasal bones of a species of sepia, *Sepia octopodia* or *Sepia moschata*, which is the common food of this whale, and to which food some observers attribute the odour of ambergris.

Ambergris is found in pieces of various size, generally in small fragments, but sometimes in pieces so large as to weigh nearly 200 lbs. The very high price which fine ambergris has lately realized on the London market is the best proof of the indispensability of the drug in the preparation of high-class perfumes. During the past year the price of the best ambergris has risen from 180s. to 215s. per oz., at which price it is now quoted by wholesale London houses (23 April, 1892).

The small compass within which a very valuable quantity of the drug may be imported without attracting attention, and the ease with which the requirements of the Customs regulations, that all goods shall be entered under their proper name and at their full value, may be circumvented, render it exceedingly difficult to follow closely the imports of the drug, where it is advisable to keep secret any important consignment of ambergris. It is stated, for instance, that although for many months fine ambergris has been thought to be exceedingly scarce in our market (and the visible supply has in reality been so), there has been a far greater supply available than has appeared on the surface, in fact, that a piece weighing 136 lbs. has been recently imported from Melbourne, and that the consignees have, as far as possible, kept the matter secret *. The greater part of the ambergris sold in London during the last few years has been obtained by the New Zealand and Tasmanian whalers, who ply their trade in the Antarctic Ocean. Whale-fishing was once an important industry in Tasmania. Now, the Tasmanian industry has practically ceased to

---

* Chemist & Druggist, 17 Oct. 1891.

exist, and there is no hope of its revival. New Zealand still possesses fisheries of some importance, and will probably continue to supply our market with much of its ambergris for many years to come. Meanwhile, spermaceti whales are getting scarcer year by year, and the time may come when the scarcity of ambergris will be chronic instead of spasmodic. It is to be hoped that before that date science will have taught us how to supplant nature in the production of ambergris; but at present there are no indications whatever of an efficient synthetic substitute " *.

When taken from the whale it is not so hard as it afterwards becomes on exposure to the air. Its sp. gr. ranges from 0·780 to 0·926. If good, it adheres like wax to the edge of a knife with which it is scraped, retains the impression of the teeth or nails, and emits a fat odoriferous liquid on being penetrated with a hot needle. On rubbing it with the nail it becomes smooth like soap, but is not so tenacious, and more easily broken than soap. Its colour varies, being white, ash-coloured, yellow, brownish black, or the colour of ochre. It is sometimes variegated or mottled, grey with black or with yellow spots or streaks. It is inflammable. Its smell is peculiar, and not easily to be counterfeited. At $62^{\circ} \cdot 2$ C. it melts, and at 100° C. it is volatilized in the form of a white vapour; on a red-hot coal it burns, and is entirely dissipated. Water has no action on it; acids, except nitric acid, act feebly on it; alkalies combine with it and form soap; ether and the volatile oils dissolve it, also ammonia when assisted by heat; alcohol dissolves a portion of it. The principal constituent of ambergris is *ambrein*. Succinic and benzoic acids are said to be sometimes found among the products of its destructive distillation. Its inorganic constituents are carbonate and phosphate of calcium, with traces of ferric oxide and alkaline chlorides. The principal market for ambergris is London, and its high price leads to many adulterations; these consist of various mixtures of benzoin, labdanum, meal, etc., scented with musk. The greasy appearance and smell which heated ambergris exhibits afford good criteria, joined to its solubility in hot ether and alcohol.

By digesting ambergris in hot alcohol, sp. gr. 0·827, the peculiar substance called ambrein is obtained. The alcohol on cooling deposits the ambrein in very bulky and irregular white crystals,

* Chemist & Druggist, 17 Oct. 1891.

which still retain a very considerable portion of alcohol. Ambrein thus obtained possesses an agreeable odour, but by repeated solution and crystallization it loses this. It is destitute of taste, and does not act on vegetable blues. It is insoluble in water, but dissolves readily in alcohol and ether, and in much greater quantity in those liquids when hot than when cold. It melts at 30° C., softening at 25° C. When heated above 100° C., it is partly volatilized and decomposed, giving off a white smoke. It does not seem capable of combining with an alkali or being saponified. When heated with nitric acid it becomes green and then yellow, eliminates nitrous gas, and is coverted into an acid which has been called ambreic acid. Ambrein is perhaps impure cholesterin, which sub-stance it greatly resembles in its properties. Pelletier * found it to contain very nearly the same proportion of elements in com-bination.

Whilst on the subject of fishes and insects, it may be opportune to remark that the odour emitted by the flesh of the Grayling has been likened to that of thyme; this is attributed to a habit of this fish of feeding on the *Gyrinus natator,* an insect of so strong an odour that when several of them are collected together they may be scented at a distance of 500 paces. Many insects are aromatic; there are ants in Bahia which, when squeezed, give off a strong smell of lemons.

The food of animals undoubtedly affects the odour of their secretions and excretions. It has been remarked that the Musk Deer only frequents districts in which the birch-tree is found; the reason of this is not apparent, but the animal certainly fre-quents localities where certain plants of the larkspur species thrive, species which possess such a strong odour of musk that the peasants of the locality believe the odour of this animal to be due to feeding on this plant; a belief which may be wrongly conceived, because the Musk Deer is found in other localities where the plant does not exist. Still, it is a curious coincidence.

To quote from the 'Flora Indica' of Hooker and Thomson : writing on the botany of the Himalayas :—

"Owing to the great power of the sun there is scarcely any vegetation even at 15000 feet; above that, though plants may be

---

* Ann. Ch. Pharm. vi. p. 24.

gathered up to 19,000 feet, the vegetation is excessively scanty and only found on the margins of rills by the melting snow. The flora of these arid regions includes some plants of great interest. . . . amongst others the *Delphinium Brunonianum.*" The species of this genus generally smell of musk, but the authors discredit the fact of the plant furnishing food to the Musk Deer, which is quite believed by the mountaineers. The *Delphinium moschatum* grows at an elevation of 14,000 feet; its flowers are pale blue. *D. glaciale* is found at an altitude of 18,000 feet; its flowers, which are pale blue, appear in August and September. *D. Brunonianum* is found on the mountains of Eastern Thibet at an altitude of 18,000 feet; its flowers are also pale blue and appear in August and September.

Delphinium Brunonianum is apparently abundant, as the juice of the plant is used in Afghanistan to destroy ticks in animals, especially in sheep.

Hooker's 'Flora of British India' says :—" *D. glaciale* grows on the Eastern Himalaya at an elevation of 16,000 to 18,000 feet, the whole plant has a *rare musky odour*, and *D. Brunonianum* in Western Thibet at 14,000 feet; it is synonymous with *D. moschatum* of Munro " * .

If these plants were more accessible they could doubtless be turned to some commercial practical use, especially as some plants which have appeared even more difficult to obtain are now successfully grown. As an instance of this may be mentioned the "Sumbul" root, *Euryangium Sumbul,* a plant which was jealously guarded, and only obtained after a reward had been offered for a root by the Russian Government. It was discovered in 1869 by a Russian traveller in the Maghian mountains near Pianjkent, a small village eastward of Samarkand, whence a living plant was forwarded to the Botanic Garden, Moscow, and it flowered there in 1871. It is a perennial umbellifer, growing to the height of 9 or 10 feet, and has a branched fleshy root about 11 inches in circumference at the base, with numerous rootlets. In 1876 it was reported by Wittmann that the plant was found in large quantities in the extreme Eastern regions of Siberia which border on the Amoor river †. The word "Sumbul" seems to be employed in Arabic to designate various substances, especially the

* Jacq. Voy. Bot. viii. t. 7, and Bot. Mag. t. 5461.
† Pharm. Journ. 1876, p. 329.

Indian "Nard" or root of *Nardostachys Jatamansi* (the true
" Spikenard "); but when or why it was first applied to this root
remains an unsolved problem (the word " Sumbul " appears to be
an incomplete name, or rather an abbreviated name).   It is known,
however, that the Sumbul was first introduced (imported as a
drug) into Russia about the year 1835, *as a substitute for musk*
(which was at that time recommended as a remedy for cholera);
it began to be known in Germany about 1840, and in England ten
years later.   It was admitted into the English pharmacopœia in
1867.   The root, as we know it in commerce, is usually cut
transversely into slices of from 3 to 5 centimetres, and sometimes
12 centimetres, in length, by about 3 or more centimetres in
thickness, sometimes mixed with small inferior shoots not thicker
than a goose-quill.   It is covered with a dark papyraceous bark.
The internal surface of the slices is pale brown, marbled with white
streaks; examined with a glass, an exudation of a large number of
resinous drops is noticeable, especially near the circumference.
The internal structure has a spongy, fibrous, farinaceous aspect.
It exhales an agreeable odour of musk, and possesses an aromatic,
bitter taste.   Prof. Flückiger remarks that the Indian Sumbul
root mentioned by Pereira is unknown to him, and that the root
imported from China mentioned in Pereira's ' Materia Medica '
appears to be quite a different root to Sumbul ; a fact confirmed
by Dr. Dymock of Bombay, who states that in China the root of
Dorema Ammoniacum is perfumed with musk and sent to Europe as
Sumbul.   Microscopically examined, the internal structure of
Sumbul root is very irregularly formed of wood and medullary rays,
and the bark consists of a soft spongy parenchyma.   The ana-
tomical structure of the root becomes very apparent when a thin
slice is moistened with a solution of iodine, the medullary rays
acquiring an intense blue colour by reason of the starch contained
in them.   The irregularity of the structure resembles that of
rhubarb root, but this last has not the large resinous cavities
observable in Sumbul root and in many other umbelliferous
plants.   Sumbul root contains about 9 per cent. of soft balsamic
resin, soluble in ether, and a small proportion of volatile oil (about
0·3 per cent.).   When the resin is brought into contact with
water it develops a musky odour.   A solution of potash is said to
convert this resin into a salt of potassium and sumbulamic acid,
smelling very strongly of musk.

Continuing the list of musk odours in the vegetable kingdom, may be cited " Mal-oil " or oil of apples. This is produced by *cellulostasis*, a disease of the apple which imparts a musky odour to this fruit. It is obtained from the diseased apples by distillation with water. It is a yellowish-grey oil, lighter than water; it boils at 109° C. It smells strongly of musk and has a rough sharp taste. It volatilizes completely when heated. It dissolves readily in alcohol and ether, and imparts a musk-like odour to water *.

The " white Musk Mallow " (*Malva moschata alba*), a British hardy perennial, is found growing abundantly in some localities by the roadsides in dry gravelly soil. This white variety is an attractive plant, and forms a branching pyramidal bush about 2 feet high, clothed with dark green, deeply-divided foliage. The flowers are pure white, from 1 to 1½ inches in diameter; the whole plant is slightly musk-scented.

The *Mimulus moschatus*, also called *Erodium moschatum*, a native of North America, usually known in England as the " Musk plant," and cultivated in pots for window decoration, is too well known to require description. The old-fashioned variety is more strongly scented than the large-flowered sort.

The *Hibiscus Abelmoschus*, an herbaceous plant attaining about 3 feet in height, a native of the hottest parts of India, of which two varieties are cultivated in tropical countries, is a somewhat important plant commercially. Its large yellow flowers are succeeded by greyish-coloured seeds which possess a very pronounced odour of musk; these seeds, known as " Ambrette " seeds, are distilled for their fragrant oil, the yield of which is estimated at 0·2 per cent.

The *Eurybia argophylla* or *Guarea Swartzei*, the " Silver-leaved Musk-tree " of Jamaica, New South Wales, and Tasmania, is a meliaceous tree attaining a height of 25 feet. In Jamaica it is called the " Musk-wood." It is often cultivated in greenhouses as a shrub, and valued for the musky odour of its leaves. A sample of " Musk-wood " was exhibited at the Paris Exhibition, 1878, from Queensland, said to be derived from *Marlea Vitiensis*, Benth.

The *Carduus nutans* (Musk Thistle) is not uncommon on waste land, fallow fields, and barren pastures where the soil is gravelly,

---

* Pharm. Journ. [3] xxvii. p. 158.

or more especially calcareous. It is an annual plant, flowering (in England) in July and August. The flowers are not ornamental, but smell strongly of musk in warm weather. Their drooping posture distinguishes them at sight from our other thistles. The stem rises from a spindle-shaped root to the height of 2 or 3 feet, and is striated, slightly invested with cottony down ; its flowers are drooping.

The *Adoxa moschatellina,* a small tuberous plant, 4 to 6 inches high, may be found in flower in April (if the weather be genial), in woods or on shady banks in many parts of England; its flowers are musk-scented.

The *Hyacinthus muscari,* a bulbous plant with dull purple flowers having a strong musky odour.

Gnaphalium odoratissimum is similarly scented.

The *Cyanus orientalis (major) moschatus* or *Amberboa moschata* (" Purple Sultan ") is considered by some to have a musky fragrance, as its name implies, but by others the odour is more comparable to that of honey (it may be comparable to the " Sweet Scabious ").

The odour of musk has been observed in the root of the common Beet (*Beta vulgaris*).

The musky odour noticeable in some grapes may be due to the presence of succinic acid.

The *Achillea moschata* (Musk-scented Milfoil), a little plant found on the Swiss Alps at elevations of 5000 to 10,000 feet, yields by distillation of its flowers and leaves an essential oil of musk-like odour. It is used to perfume the liqueur known in Switzerland as " Iva."

It has been remarked that the dried flowers of the *Canella alba* (" Wild Cinnamon ") when softened in warm water, have an odour nearly approaching that of musk.

The " Musk-Cranesbill," *Geranium cicutæ folio moschatum,* a British perennial, has a faintly agreeable odour of musk, which is destroyed by bruising the plant.

The greenish-yellow flowers of the *Cestrum nocturnum,* a bastard Jasmine, native of Cuba, and there called " Dama de Noche " or Lady of the Night, give off a strong musky odour after sunset, but as soon as the sun rises this odour is replaced by one of a very nauseous kind.

The term " Noctu-olens " or night-smelling has been applied to

many flowers which are scentless by day and smell powerfully at
night. Linnæus calls them "*Flores tristes*," melancholy flowers,
belonging to various tribes as discordant as possible ; the colour of
such flowers is generally white, pale yellow, greenish yellow, dull
brown, or faded blue-tint.   Amongst them may be noticed :—

Hesperis tristis, the double-flowered night-scented Rocket.
Cheiranthus tristis, the night-flowering Stock.
Daphne pontica.
Crassula odoratissima.
Mirabilis jalapa.
    ,,   *dichotoma.*
    ,,   *longiflora.*
Datura ceratocaula.
Cereus nicticalus.
    ,,   *serpentinus.*
    ,,   *grandiflora,* which puts forth a bloom as large as a
    cauliflower, smelling powerfully of vanilla.
Rivea Bona-nox (*Lettsomia Bona-nox*).
Ipomœa grandiflora.
Nyctanthes arbor-tristis.
Geranium triste.
Sanseviera Pumila.

CHAPTER II.

THE ODOUR OF ROSE.

THE organs of the sense of smell can be trained to the *appreciation* of perfumes, especially by young persons, as easily as the palate can be trained for business purposes to the tasting of the flavours of wines, tea, or coffee. Of course a taster, sampler, or evaluer of such beverages must naturally be possessed of a finely developed nervous susceptibility to the slight variations occurring in every sample which comes under his notice. Such natural perfection of susceptibility is not common, although many possess the gift without being quite aware of it : they may not be in the tea-trade or the wine-trade, or in any trade at all; and so the gift is not trained, or even appreciated. To those who are not constantly occupied in the culture of the rose, it may seem that one rose is very much the same as another, and excepting a few variations in colour and habit of growth there is very little difference distinguishable. Probably many persons would never believe that there are not only roses perfectly devoid of odour, but there are some which stink. There are experienced gardeners who can *name* many varieties of rose in the dark : this means that the perfume of roses is very varied, and that no two varieties possess the same odour. What is called the *pure* odour of rose is unique, undefinable, incomparable. It is in fact a *type*, and no imitation can approach it. It may be best represented by the *Rosa centifolia* cultivated in Provence, or by the *R. Damascena* cultivated in Bulgaria. The odour of tea is not perceptible in the so-called " Tea-roses," indeed many " Tea-roses " are odourless, such as the Mélanie Souppert, Marie Guillot, Marie Caroline de Sertoux, Triomphe de Milan, &c. Some " Tea-roses " possess very delicate fruity odours, somewhat approaching that of raspberry, such as the Maréchal Niel and the Madame Bravy. The odours of some

Tea-roses, such as Gloire de Dijon, are so soft and undefinable that any comparison is impossible. The *Rosa Socrates* has an odour of peach; the *R. Elizabeth Barbenzien* and *R. Souveraine* an odour of melon; *R. Isabelle Narbonnaud* and *R. Banksia alba* an odour of violets; *R. Safrano* an odour of pinks; *R. bracteata* and *R. Macartnea* an odour of apricot.

As a rule the red roses are more odoriferous than the white. Cut roses placed in a vase diffuse their fragrance more powerfully than when growing on the plant. The majority of Noisette roses are inodorous, but the variety known as " Unique jaune " recalls the odour of hyacinth, and the " Desprez " that of fruit. Some sorts of *Rosa canina* and *R. arvensis*, also *R. sæpium* and *R. alpina*, exhale a fine odour of mignonette. The *R. moschata* (Miller) possesses a fine odour of pinks, but none whatever of musk. The odour of pinks is also disengaged from the oil-glands in the peduncles and sepals of *R. Brunonii* (Lindl.). In estimating the quality of the odour of a rose, care should be taken in handling the stalk, the calyx, or any green part, as a very slight friction breaks the glands which the green parts mostly contain in quantity, and in which is secreted an oil or oleo-resin totally different in character to the oil developed in the microscopic glands of the petals. In the green parts of some roses the odour is sometimes rank and terebenaceous, as in *R. pomifera*, *R. mollis*, and *R. tomentosa*; sometimes it is balsamic, as in *R. centifolia*; sometimes fruity, as in *R. sæpium*, *R. micrantha*, *R. rubiginosa*, *R. graveolens*, and *R. glutinosa*, the lower part of whose leaves contains innumerable oil-glands, which being broken by friction exhale an agreeable odour, which has been very rightly compared to that of an apple called the " Pomme Reinnette," or " Pippin "; this is very distinctly developed in *R. rubiginosa* (the " Sweetbriar "), and to such an extent as to be disengaged spontaneously, especially on a warm day. (The composition of the body contained in these glands is apparently unstudied and little understood, but its nature is suggestive of valerianate of amyl *.) The *leaves* of *R. lutea*, Dalech.

* A similar odour has been noticed in *phenylnitro-ethylene chloride*, which can be prepared by passing chlorine into a cooled solution of phenylnitro-ethylene in chloroform. On the evaporation of the latter it remains as a thick oil, which has a penetrating odour, resembling, when dilute, that of *pippins*. On standing for some time, large lustrous crystals are deposited, which are extremely soluble in ether and chloroform, and are again left on evaporation as an oil, which

(*R. eylanteria*, L.), possess a still finer odour, recalling that of Jasmine. The *flowers* of this group (*Rubiginosæ*) are generally quite devoid of odour, but those of *R. lutea* are said to develop an odour resembling a mixture of bugs and coriander; the same is said of *R. platyacantha* and *R. capucine*, especially the variety *bicolor* (Jacq.). It has been noticed that roses flowering under glass give off a greater amount of perfume than those cultivated in the open air; the reason of this is obscure, but it is perfectly certain that under no conditions is the odour fully developed except in very hot climates, where the power of the sun affords the maximum benefit of light and heat.

The flowers of *Rosa gallica* (which are used officinally) are but feebly odoriferous when freshly gathered, their perfume develops gradually in the process of desiccation, while that of the Damask rose is almost destroyed by drying.

In Bulgaria the flowers grown for the distillation of the otto are gathered before they commence to open, and a little before sunrise. Were they gathered later in the day, when fully expanded by the heat, the perfume would be stronger, but not so sweet, and the resulting essence would be of less value. It has been noticed that previous to a storm, or atmospheric disturbance, the odour of the rose seems strangely increased; this may be by reason of the oxidizing influence of the ozone in the atmosphere, or it may be that our perceptive faculties are sharpened at such moments. In further illustration of the capricious nature of this perfume, and the extraordinary complexity of its forms, it is stated that not only in the whole list of roses are there no two which develop precisely the same odour, but that in the same species, and even on the same plant, there are not found two flowers absolutely identical in odour,—even yet further, that it is a well known fact amongst rose-growers that at different times in the day an individual flower will emit a different perfume.

The essential oil of rose can rarely be obtained pure. In India the natives seem to *prefer* it adulterated, especially with oil of santal-wood. The word *Atar* in India is used like the word *Abir*

---

solidifies, when placed in contact with a fragment of the original crystals, to a mass, which melts at 30°. This odour of pippins, akin to that of Sweetbriar, is noticeable in the flowers of *Agrimonia eupatorium*, and in all parts of the *Agrimonia odorata*.

for many mixed perfumes ; the *Abir* of Bombay is compounded of santal, violets, orange-flower, rosewater, musk, and spikenard. Apart from the systematic adulteration of Otto of Rose, which gives us a false idea of the true perfume, the quality would undoubtedly be finer if the rough apparatus now used in the East for its extraction were replaced by modern appliances, and if greater care were taken in the process of the distillation. The quality would especially be improved by removing all the calyces, seed-receptacles, bits of stalk, and in fact carefully rejecting every green particle of the plant, as such contain, as above explained, oils and oleo-resins of very different and deteriorating odours.

The rose cultivated in Bulgaria for the otto has been clearly identified by botanists as the *R. Damascena,* Miller, the Red Damask rose. It is a native of Syria, and is distinguished from the *R. centifolia* by the greater size of its spines, green bark, elongated fruit, and longer reflexed sepals. It forms a small branching shrub, reaching the height of 5 or 6 feet ; the branches are spreading and rise from the bottom of the stem ; until they become old they are covered with brown, straight, very closely set spines, sometimes a centimetre in length. The leaves are about 10 to 15 centimetres long, composed of seven folioles which are unequal, sessile, elliptical, non-acuminate and sharply serrate ; their upper surfaces are bright green and glabrous ; the under surfaces are of a dull glaucous colour, the margins and nerves being finely pubescent. The petiole is furnished with recurved spines and is covered with short glandulous brown hairs.

The flowers are grouped in 2- or 3-flowered cymes. The branches bear on the average seven flowers, and in good years as many as thirteen have been counted. The peduncles are slender and about 4 centimetres long, bristling with numerous very fine spines intermixed with glandulous hairs which render the stalk very sticky to the touch. (The workmen who gather the flowers find that their fingers become hardened so that they do not feel the pricking of the thorns, but they become covered with a dark resinous substance emanating from the glands of the flower-stalks ; the odour of this substance is strongly terebenaceous, and at the end of the day is scraped off the fingers, rolled into balls, and kept for mixing with tobacco in cigarettes.)

The small receptacle (seed-vessel) is almost conical and gradually diminishing in size to the stalk ; this also is full of resinous glands. The sepals are very pointed and sometimes 3 centimetres in length. The margins of the two exterior sepals and the exterior margin of the partly exposed inner sepal of the bud are provided with several long thin tongue-like growths, covered with hispid glands on the outer surface. The two sepals and the sepal partly concealed in the bud are simply hairy. The internal surface of the five sepals is covered, towards the broad concave base, with a fine pale down. The petals are orbicular, pink, almost red in the bud, and becoming paler as the flower expands; they are thin, not shiny, but not velvety. The stamens are few in number. The styles are free in their entire length. The exquisite odour of the flower is very analogous to that of the *R. centifolia.* The colour of the berry is cherry-red.

A microscopic examination of the transverse section of a rose petal reveals that the otto is secreted in cells on both its surfaces, those of the upper epidermis being of a papillary form and those of the lower of an elongated cubic form. The presence of the oil in these cells is clearly demonstrated by moistening the section with a dilute aqueous solution (1 in 200) of osmium tetroxide (OsO_4, sometimes called osmic acid); a reagent of great sensitiveness in detecting the presence of both essential and fixed oils. The section becomes almost instantaneously bordered by two bluish-black lines of cells, resulting from the reduction of the acid and deposition of the osmium [*]. The section must be then washed in distilled water and mounted in glycerin.

Information respecting the cultivation &c. of the rose in Bulgaria was published by D. Pappazoglou of the firm of Pappazoglou Brothers of Kézanlik, and issued in the form of a pamphlet at the Philadelphia Exhibition, where the firm exhibited specimens of their otto. Later information on the subject has also been published by Christo Christoff, also a merchant of Kézanlik, and an exhibitor of otto at the Paris Exhibition, 1889. These two pamphlets contained much useful information which is to a great extent embodied in the following details of the culture of the rose in Bulgaria:—

The bushes are planted close together, so as to form hedges, in long parallel rows, with a space of about six feet between the rows. The bushes grow about six feet high.

The ground selected is preferably of a sandy porous nature

[*] Blondel, in Bull. de la Soc. Bot. de France, Feb. 1889.

allowing of free percolation of moisture, and sloping towards the south so as to be sheltered from the cold winds of winter. The situation should be in the vicinity of running water, which is necessary for the distillation. The propagation is not effected by cuttings or layers, but by cutting down or digging up entire bushes, such as appear to leave blanks in the lines of flower in the hedgerows by reason of the plants being worn out. These bushes are pulled completely up and chopped into pieces. As the branches spring from the bottom of the stem, the stump with as much of the root as possible left attached is torn up and divided with a spade or hatchet, and is placed with the boughs, leaves and all, in pieces four or five abreast horizontally in long straight trenches about 40 centimetres broad and the same in depth; part of the mould taken from the trench is then thrown lightly on, and a thin dressing of manure is thrown over all. If possible it is then watered. The planting is done in October or November; the young shoots make their appearance about the following April, being of a deep red colour until they attain the height of about 20 centimetres. The ground is then weeded and raked with great care; it is again weeded at the beginning of June, and in November, when the rest of the mould which was not put back into the trench the year before is carefully heaped up to the base of the young shoots (which by this time are about 30 centimetres high), to protect them against the cold of winter. The following May the plants will have attained the height of about 60 centimetres, and produce a few flowers which will be harvested with the rest, and be sufficiently productive to cover the expenses incurred in keeping the ground clean. The next year they will be in full bearing. Its maximum production is in the fifth year. When the bush has attained its tenth year, many cultivators prune it right down to the ground to strengthen it; new branches and even flowers will appear the following year. The harvest commences about the third week in May, according to the season, and lasts about a month; there is sometimes another small gathering of flowers in November. After the harvest the ground is cleaned of weeds, and in October it is slightly manured, the earth being hoed up to the roots of the bushes to protect them in the winter. The pruning takes place in March, when all withered branches are carefully removed. In April the ground is again cleaned, and the mould which was heaped up at the roots of the bushes before the winter is removed. Watering is very rarely necessary. As the bushes grow older, the branches, which are more or less spreading and spring from the bottom of the stem, interlace and form a very close thicket. The life of the bush exceeds twenty years.

The rose-tree requires unceasing care; the ground must be hoed at least four times during the year, and kept scrupulously clean of weeds. Pappazoglou states that the plant must be manured every other year, but admits that "such benefits the *quantity* but harms the *quality.*" The rose is very susceptible to climatic changes. In very cold winters the branches die. Frost and fog are very dangerous to the tree, especially if occurring when the sap is rising. The quality of the crop depends greatly on the temperature during the harvest; if during that time the weather be cold and wet, the flowers will develop very slowly, and a very great heat expands them too quickly.

The borders of the Bulgarian plantations are defined by hedges of a white rose, the *Rosa alba,* L. It is a bush of more vigorous growth than the *R. Damascena,* and flowers about a fortnight later. Its odour is agreeable, but much inferior to that of the red rose. The oil derived from it is of very poor quality, but it is rich in stereoptene, and unscrupulous manufacturers distil its flowers with those of the red, in order that the otto of the latter may bear adulteration with a larger proportion of "geranium oil."

The annexed sketch of the apparatus used in the manufacture of Bulgarian otto of rose is copied from one drawn by a Turkish engineer, and published in a Consular Report in 1872. It is very primitive in construction and capable of great improvement, but it appears to agree in every particular with a sketch recently published by Christo Christoff of Kézanlik of the apparatus in use at the present day in Bulgaria. The still is of copper, about five feet high, resting on a furnace built of bricks or stones. The condenser is simply a straight tube passing obliquely through a wooden vat. The fuel for heating the furnace consists of long pieces of wood or poles, which are lit at one end, and pushed into the furnace as fast as the end is consumed, and of course to lower the fire, or put it out altogether, it is only necessary to pull out the wood by the unburnt end. There is no door to the furnace, and the smoke escapes by a short piece of pipe stuck in the brick-work. The cold water for condensation is supplied by a wooden gutter suspended over the condenser. There is a hole at the bottom of the condensing tub, into which an upright pipe is fixed; the other end of the pipe reaches nearly to the top of the tub. This allows for the overflow of the hot water, which runs away in a trench in the floor of the building. 10 kilogrammes of flowers, just as they are gathered, including their green parts, are put into the body of the still, with 75 litres of water. The separate parts of the still are then adjusted together, and the joints luted with strips of cotton rags moistened with wet clay. A brisk fire is first made up and then allowed gradually to moderate. The operation lasts about an hour and a half, and the fire is drawn when 10 litres of liquid have distilled over. This has been received in two flasks of 5 litres each, and placed on a shelf to cool. The parts of the still are then disconnected; the spent petals are separated on an osier sieve, and the dirty hot water is put back in the still to do duty for a fresh charge of flower, so economising fuel. The operation is repeated all day and sometimes all night, as long as there are flowers to be distilled. Petals kept twenty-four hours from the time of gathering lose much of their fragrance and afford an unsatisfactory yield, but yet they are sometimes allowed to accumulate in large quantities, and while waiting for the still to be vacant to receive them are spread out on the ground in cool shady places, or in low-built sheds constructed for the purpose.

The two flasks of rose-water above-mentioned contain all the otto obtainable from the 10 kilos of flowers, yet some distillers continue the operation until they have obtained three flasks of distillate, making 15 litres in all. The result is not improved in quality, but it contains more stereoptene, and so, by reason of its higher congealing-point, permits of the otto being adulterated with a larger quantity of "geranium oil" (*Andropogon Schœnanthus*, L.) without the fraud being detected. When 40 litres, or 8 flasks, of rose-water have been collected they are redistilled together. The distillation is conducted as before, but only 5 litres of distillate are collected; the water remaining in the still being reserved for fresh flowers. The flasks which receive the first distillate are bell-shaped, with short necks, A, fig. 1. The flask which receives the product of the second distillation is globe-shaped, with a long neck, like a laboratory boiling-flask, B, measuring 40 centimetres in height by 20 centimetres diameter, made in thin Hungarian glass. Its capacity is also 5 litres. The second distillate, which

is at first white and cloudy like an emulsion, gradually clears as it
becomes cold, and the oil rises to the surface in the narrow neck,
forming a yellowish stratum a few millimetres in thickness. It is
removed by a sort of funnel-shaped spoon, C, made of tin, about
2 centimetres broad, and pierced with a minute orifice at the apex.
This is dipped into the neck of the flask and repeatedly plunged
below the surface of the layer of oil, the oil gradually flowing over

Fig. 1.

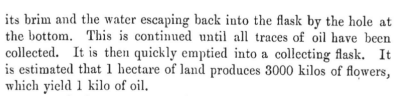

its brim and the water escaping back into the flask by the hole at
the bottom. This is continued until all traces of oil have been
collected. It is then quickly emptied into a collecting flask. It
is estimated that 1 hectare of land produces 3000 kilos of flowers,
which yield 1 kilo of oil.

A more barbarous process of treating a flower so delicate and
fine as the rose can scarcely be conceived. The oil so manufactured is, comparatively speaking, scorched and empyreumatic in
odour, and is very far from conveying to the mind a true idea of
the natural perfume; but, apart from this consideration, a recent

Report of the English Consul at Constantinople states that the Otto industry of Turkey suffers from the effect of the mistrust naturally aroused by the admitted wholesale adulteration of the product with geranium oil. It is therefore highly probable that the efforts made to grow the rose in the north of Europe, and prepare the otto with the most complete technical appliances, will shortly succeed to a great extent in competing with the Turkish product, especially as the firms who are now making these efforts are of a standing and respectability which guarantee the *purity* of their products. Certainly, with a little more sunlight and heat, they would entirely supersede the Turkish manufacturers. At the Paris Exhibition of 1878 a sample of about 3 oz. of pure otto of rose was exhibited by M. Hanart of Anzin, Département du Nord, as being the produce of about 24 cwt. of rose-petals grown in the Anzin district.

In about 1886 a rose plantation was established near Leipzig by Messrs. Schimmel and Co. of that city. In their Report for April 1891, they state that these rose-fields extend over about 180 Prussian acres, the oil from which is introduced into commerce in a liquid state, *i. e.* practically free from stereoptene, which has been mechanically extracted from it. Such oil will therefore remain liquid at ordinary temperatures, will readily dissolve in spirit, and, used in the preparation of compound bouquets, will not deposit stereoptene. (A very great advantage to perfumers.) Undoubtedly a reliable product of this sort, which must have cost much material sacrifice and perseverance to establish, will be duly appreciated and supported by the buyers, but it remains to be seen whether this otto will compete in fragrance with that distilled in the south of France, where the rose culture has attained an enormous development. The French oil has a greater consistence than the Turkish, and is more green in colour. At Grasse 8000 to 10,000 kilos of rose-petals yield about 1 kilo of oil. It is collected as a by-product in the distillation of rose-water. It does not appear that this oil has been offered for sale "*sine* stereoptene."

In contrast with the rough distillatory apparatus in use in Bulgaria, the annexed woodcut illustrates the more rational system of using steam-jacketed stills in France, and represents the actual view of a still-room in the factory of Messrs. Sozio and Andrioli at Grasse.

Fig. 2.

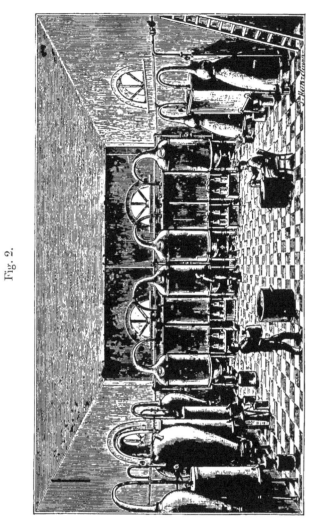

INTERIOR OF STILL-ROOM AT GRASSE.

The average production of Bulgaria since 1871 is estimated as follows :—

		kilogrammes.
1871	... .	3000
1872	....	1500
1873–76	....	2000 to 2400
1877	....	2000

DISTRICT.

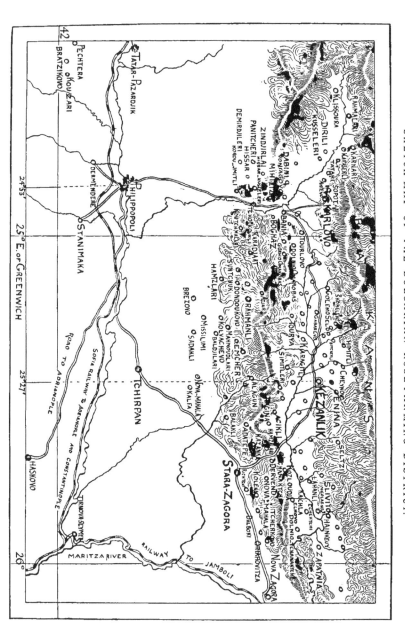

SKETCH-MAP OF THE BULGARIAN ROSE-FARMING DISTRICT.

	kilogrammes.
1878	1800
1879–85	1800 to 2000
1886	2500
1887	2800
1888	2600
1889	3000
1890	2500
1891	2200

The importance of the yield from the different villages in each rose-growing district in Bulgaria may be estimated by the following approximate totals of each village in each Canton for the year 1888, which was considered a fairly good year. (The figures are taken from the Report of Christo Christoff of Kézanlik.)

DEPARTMENT OF STARA-ZAGORA.

CANTON OF KÉZANLIK.

	kilos.		kilos.
Doïmouchlari	15	Kachla	10
Malko-Sélo	15	Karaghitli	15
Golémo-Sélo	100	Pavel	18
Gornio-Sarhané	8	Bania	20
Skobélevo	3	Tourya	45
Himitli	30	Alexandrovo	30
Chipca	35	Oftchilari	15
Enina	50	Tarnilchani	12
Gussovo	15	Gabarevo	30
Seltzi	2	Eurenli	8
Magliche	60	Dolnio Sarhané	5
Popovtzi	30	Bitcherli	6
Jergovetz	15	Saltacovo	10
Slivito	12	Sekiretchovo	2
Kozloudja	60	Cheynovo	12
Tchanatchi	12	Hass	8
Ragina	15	Kézanlik (town)	35
Achikli	10	Emichléri	5
Horozovo	15	Tcherganovo	5
Armaganovo	5	Toulovo	13
Karalajdéri	25	Gornio-Sofilari	20
Simitléri	30	Dolnio-Sofilari	3
Ikanli	40	Lahanli	50

D

CANTON OF NOVA-ZAGORA.

	kilos.		kilos.
Terziobas	1	Hainkeny	5
Tvarditza	6	Dolnio-Tchanaktchi	8
Couzosmadé	7	Zapalnia	12
Kounari	10	Nicolaevo	4

CANTON OF STARA-ZAGORA.

	kilos.		kilos.
Tcherkovo	4	Hamzalari	6
Oriahovitza	10	Nova Mahala	7
Dalboki	10	Kazanka	12
Koléna	10	Balakli	5
Richteni	7	Ada-Tépé	6
Dervent	10		

DEPARTMENT OF PAZARDJIK.

CANTON OF PECHTERA.

	kilos.		kilos.
Bratzicovo	60	Pechtera (town)	10
Kouzari	10		

DEPARTMENT OF PHILIPPOPOLIS.

CANTON OF KARLOVO.

	kilos.		kilos.
Kalofer	30	Dirili	50
Mitirizovo	30	Kusséléri	50
Arapovo	15	Ghilléri	20
Karlovo (town)	50	Obchilari	20
Sopot	25	Voniagovo	25
Ahiévo	5	Dabini	40
Aïganovo	10	Doughandji	3
Tatari	4	Bania	5
Karatchévo	2	Mihiltzi	20
Achiklari	5	Tchoukourli	20
Karnari	15	Kourou-Himitli	10
Téké	25	Boghaz	30
Rahmalari	100	Dolnio-Omarobas	40
Klissoura	90	Gornio-Omarobas	5
Slatina	25	Kourtovo	5
Karaserli	20	Madjeri	15

CANTON OF TCHIRPAN.

	kilos.		kilos.
Gornio-Enicher	25	Konouktchi	5
Baldjilari	15	Nova-Mahala	5
Aldjioglari	6	Kalfa	3
Karodjovren	2	Karaptcha	2
Alagun	40		

CANTON OF NOVO-SÉLO.

	kilos.		kilos.
Hissar-Kusséléri	15	Koprivchtiza	4
Zindjirlai	30	Panagurichté	1
Panitchéri	20	Doudini	2
Aïréni	2	Daoudjioglou	3
Novo-Sélo (town)	20	Hissar	2

CANTON OF BRÉZOVO.

	kilos.		kilos.
Mratchinik	25	Tchaalari	40
Adjar	25	Kolvàtchevo	25
Rahmanli	100	Missilimi	5
Babek	10	Sadakli	25
Hamzalari	50	Karadjaat	10
Brézovo (town)	5	Suutchuk	10
Sari-Démirdjileri	8	Mousloutchali	7
Dondoukovo	5	Koutchmalari	10
Mabmoudlari	3		

Pure Turkish otto of rose is at the ordinary temperature a pale yellow liquid of a sp. gr. of 0·87 to 0·90. A reduction of temperature causes it to assume the appearance of a concrete mass, by reason of the formation of clear, brilliant, laminar (or plate-shaped) crystals, the proportion of which vary according to the geographical position of the country in which the flowers producing it were cultivated, the degree of temperature at the time the flowers were gathered, and other causes more obscure. According to Bauer it solidifies at between 11° and 16° C. In some experiments made by Hanbury in 1859 *, the point of solidification of pure Turkish otto varied between 16° and 18° C., that of an Indian sample was 20° C., one distilled in the south of France was between 21° and 23° C., one distilled in Paris was 29° C., and one obtained in London by distilling rose-water was between 30°

* Pharm. Journ. [2] xviii. p. 504.

and 32° C. These results indicate that the cold climates of the North are unfavourable to the production of a very odoriferous oil, the high points of solidification resulting from a larger proportion of stereoptene, which is an inodorous constituent of no value to the consumer, but valuable to the fraudulent dealer, who takes advantage of its high proportion, and can thereby add a proportionately larger quantity of geranium oil.

Pure otto of rose is, therefore, a mixture of a liquid oxygenated body to which it owes its perfume and a solid hydrocarbon or stereoptene which is absolutely odourless. The proportion of each of these constituents is certainly very variable; the Turkish otto may contain as much as 18 per cent., in the French and the English as much as 35, 42, 60, and even 68 per cent. have been estimated.

The fluid constituent of otto of rose, according to Markownikoff*, is composed of two bodies, $C_{10}H_{20}O$ and $C_{10}H_{18}O$. According to Poleck †, only the body $C_{10}H_{18}O$ is present. Investigations made by Eckart at the Breslau University, and published in the 'Archives der Pharmacie,' 1891, pp. 355–389, show that both Turkish and German otto contain 5 per cent. of ethyl alcohol, which can be distilled off below 100° C., and a body which he names Rhodinol, $C_{10}H_{18}O$, constituting the remainder or bulk of the liquid portion. The stereoptene is separated by dissolving the otto in 75 per cent. alcohol at 70°–80°, and cooling to 0° C. The alcoholic solution is then evaporated in a vacuum to obtain the rhodinol. The physical constants of this body vary, according to its source, between the following limits :—boiling-point, 216°–217° C.; refractive index, 1·4710–1·4725 ; refraction equivalent, 48·97–49·28 ; dispersion, 11·1–12·5 ; specific rotation, −2·7 to −2·8. The vapour density corresponds with the molecular weight 142·43 ($C_{10}H_{18}O$ =154). Rhodinol shows all the reactions of an alcohol. Oxidation with potassium bichromate and sulphuric acid converts it into an aldehyde, Rhodinal, believed to be an isomer of, and closely resembling, citral, which can be obtained from geraniol by similar treatment. By phosphoric anhydride rhodinal is transformed into dipentene, losing one molecule of water.

Citral, $C_{10}H_{16}O$, which exists to the extent of $7\frac{1}{2}$ per cent. in oil

---

* Bericht d. Deutsch. chem. Ges. xxiii. 1890, p. 3191.

† Ibid. xxiii. 1890, p. 3554.

of lemon, has been studied by Semmler [*], and is believed to be identical with the aldehyde obtained by the oxidation of geraniol.

The nature of stereoptene was studied by Fluckiger [†] in 1869, and by Gladstone in 1872 [‡].

Recent experiments made by Schimmel & Co., of Leipzig, showed that by digesting stereoptene at a temperature near to its melting-point, it was possible to obtain, by crystallizing, fractions of different melting-points, and by repeatedly putting the substance through these operations, two constituents were finally obtained, one of which melted at 41° C. and the other at 22° C. The result of this experiment on Turkish stereoptene was confirmed by repeating it on stereoptene of known purity resulting from the distillation of roses grown by themselves. These investigations tend to prove that, contrary to the views hitherto held, stereoptene is not a simple body, but a mixture of two or more homologous hydrocarbons.

A drop of stereoptene let fall on paper is not dissipated by the heat of a stove, even after several days. When it is carefully melted at the temperature of sun-heat and then allowed to cool, it sets in microscopic crystals of a peculiar form, most of them being truncated hexagonal pyramids, which, nevertheless, do not belong to the rhomboidal system, as their angles are manifestly unequal; many of them are curved into the form of an S or §. Examined under the microscope by polarized light, they present, by reason of their refractive power, a very brilliant aspect. Stereoptene is a very stable, unalterable body, but on boiling it for several days in fuming nitric acid it slowly dissolves and decomposes into various acids, homologues of the fatty acids, and possibly also fumaric acid. Amongst the former, butyric and valerianic acids are recognizable; the principal product, however, is succinic acid, which was found in pure crystals, giving the well-known reactions. According to experiments of the same observers, there are many points of resemblance in the physical characteristics of stereoptene and paraffin.

Besides the crystallizing test, which is applied to estimate the purity of otto of rose, it is necessary to examine the *manner of crystallization* in order to ascertain if that crystallization be caused

---

[*] Bericht d. Deutsch. chem. Ges. xxiii. p. 3556, and xxiv. p. 203 (1891).

[†] Pharm. Journ. [2] x. p. 147.   [‡] Journ. Ch. Soc. x. p. 12.

by stereoptene or by spermaceti. The crystals should form in brilliant plates and in aigrettes reflecting the prismatic colours, *in all parts of the liquid.* Spermaceti is precipitated in a solid mass, easily recognizable, besides which its melting-point is 50° C., as is that of most varieties of paraffin; the microscopic crystals of the last, although somewhat resembling those of stereoptene, are easily distinguishable by careful comparative examination.

A test for the presence of the oils of rose-geranium, palma rosa, etc., is described by Ganswindt *:—" On mixing a few drops of pure oil of rose with an equal bulk of sulphuric acid, the rose odour is not changed, but oils used for adulteration change their odour; or 5 drops of the oil are mixed in a dry test-tube with 20 drops of pure concentrated sulphuric acid; when the mixture is cool it is agitated with 20 grams of absolute alcohol, when a nearly clear solution should be obtained, which, heated to boiling and then allowed to cool, remains clear yellowish brown. In the presence of the oils of rose-geranium, etc., the alcoholic mixture is turbid, and on standing separates a deposit without becoming clear."

The above-mentioned test has been confirmed ten years later by Panajotow †, who mentions that the brownish-red fluid resulting from the mixture of equal parts of oil of rose and concentrated sulphuric acid dissolves completely in 95 per cent. alcohol to an almost colourless solution, while the similar product resulting from oil of geranium is rendered turbid by the addition of alcohol, and a yellow, fatty, flocculent mass separates.

O. F. Müller has recently observed that a number of resins, oils, and lacs yield colour reactions with fuchsin solution de-colorized by sulphurous acid, the so-called "Schiff's reagent." Applying this test, Panajotow found ‡ that if two or three drops of "Indian geranium oil" be shaken in the cold with 2 c. c. of the reagent, it gives at first a blue-violet, and after two hours a beautiful blue coloration. Under the same conditions *pure* otto of rose only gives a red coloration after twenty-four hours, and hence the slightest admixture of geranium oil is recognizable, because the bluish coloration is always formed at once.

This test is, however, condemned by certain distillers of otto of rose as perfectly inefficient and useless, by reason that the reaction

---

* Am. Journ. Pharm. 1881, p. 250.

† Berichte d. Deutsch. chem. Ges. xxiv. p. 2700.　　　　‡ *Loc. cit.*

is occasioned by the *presence of an aldehyde*, citral or geranium aldehyde, formed by oxidation of geraniol. This is coloured by the reagent at first blue-violet, after some time a turbid greenish-blue; whilst the aqueous liquid assumes an intense blue-violet colour. Otto of rose (the liquid portion of it) consists, according to the investigations of Eckart [*], principally of an alcohol, Rhodinol, $C_{10}H_{18}O$, which is isomeric with, and nearly related to geraniol. Geraniol yields on oxidation Citral (geranium aldehyde) according to Semmler [†], whilst rhodinol yields, on similar treatment, an aldehyde *Rhodinal*, closely resembling citral, and this oxidation is partially effected when the otto is exposed to the air.

On this logic, the aldehyde rhodinal producing the same colour reaction as the aldehyde citral, the fuchsin-sulphurous acid test can hardly be depended upon.

The addition of "Indian geranium" oil to otto of rose was formerly made in Constantinople, but now the mixing takes place at the seat of the manufacture of the otto. It is stated that in many places the roses are sprinkled with it before being placed in the still; this probably makes a more perfect "blend." Although the introduction of geranium oil into Bulgaria is now forbidden by the Government, it is still brought in secretly by Jews and Greeks. It is said that *pure* or "Virgin" otto never leaves Kézanlik; that is probably true, and the assertion that Turkish "Virgin" otto never arrives in England at all (even to the Turkish Ambassador) is almost certainly true. The manufacturers avow it. The method in use in Bulgaria to detect the amount of adulteration with oil of geranium is very simple, and very defective; it is called the congelation test, and is based on the fact that the addition of geranium oil lowers the temperature of congelation of the otto in proportion to the quantity added. A perfectly pure Bulgarian otto congeals at from 14° to 16° Réaumur (63°·5 to 68° Fahr.). If geranium is added to it, the same otto only congeals at 13° R. (61°·2 Fahr.), 12° R. (59° Fahr.), 11° R. (56°·7 Fahr.), and even lower according to the amount of "geranium" added. The purchaser takes a 20-gramme flask containing 15 grammes of the otto to be tested, and plunges this bottle into a basin of water, the temperature of which is regulated

---

[*] Inaugural Dissertation, Breslau, 1891.
[†] Ber. Deutsch. chem. Ges. xxiii. p. 2965.

by the addition of hot and cold water, and read off on a Réaumur thermometer. In a pure oil the congelation commences in three minutes, and at the end of ten minutes should be so complete that on turning the flask upside down the contents should not run out. Payment is made according to the degree marked by the thermometer when this congelation is arrived at; the degree of congelation indicating the degree of adulteration.

Mr. Christo Christoff, of Kézanlik, states that "formerly paraffin was added to the otto, in which it dissolves very well, and in spite of the presence of geranium the congelation takes place at 65° to 68° F., but the crystals are opaque, of a dirty yellow, and break up, forming a sort of muddy substance which collects at the top of the flask. The simplest method of adulteration consists in adding to the roses to be distilled some white roses, the product from which, though less fragrant, is much richer in stereoptene than that of the red rose; furthermore, this otto, which normally would congeal at 68° F., can, by geranium, be brought down to 65° F., and remain still within the prescribed limits." The otto of rose manufactured in Germany may appear finer than the Turkish from the fact that it is *pure*, distilled with greater care; doubtless the green parts of the flower (calyx, receptacle, and stalk) are separated, and a still capable of distilling the oil at reduced pressure is used, and many precautions taken; but to develop the true fragrance of the rose in perfection requires the full power of the sun and greater heat than the climate of Germany can ever furnish. No doubt localities might be found in Persia or in Tunis where the climate and soil are suitable to the growth of the rose, and with good stills and trustworthy European managers a pure otto could be produced at a very remunerative price, and the Oriental fraud and jugglery with "geranium oil" put an end to —as far as the London market is concerned. Further, the stereoptene might be separated, as it is a perfectly odourless product, and otto so sold "*sine* stereoptene" would be more difficult to adulterate and yield no deposit in bouquet preparations.

Rose Culture in France.

As before observed, the cultivation of the rose in the south of France, especially in the neighbourhood of Grasse, has enormously increased. Otto of rose and rose-water are made, but the flowers are principally used for the manufacture of rose pommade and the spirituous extract therefrom, the demand for both of which for France and abroad is very great.

The variety of rose there grown is the *R. centifolia,* Linn. The bushes are set in rows, but not so close together as to form thick hedges as in Bulgaria, nor do the plants attain such a height. Ground gently sloping to the south-east is preferred. Young shoots are taken from a five-year old tree, and are planted in ground which has been well broken up to a depth of three or four feet. When the young plant begins to branch out, the top of it is cut off about a foot from the ground. During the first year the farmer picks off the buds that appear, in order to develop and strengthen the plant. In the fourth or fifth year the tree is in its full yielding condition. A rose-tree will live to a good age, but does not yield much after its seventh year; at that period it is dug up and burned, and the land planted with some other plant for one year.

The flowering begins about mid-April, and lasts through May to early June, a time so short that difficulty is experienced in dealing with the enormous quantity of blossom produced, for of course loss of perfume occurs if the flowers are not immediately used. The buds on the point of opening are picked in the early morning. On some days as many as 150 tons of roses are gathered in the province of the Alpes Maritimes. At the factory the petals are first completely separated from the green parts; this work is done by women, seated at long benches under a shed. The separated petals, of which there are sometimes as much as four tons accumulated on the floor at one time in one factory, are then either distilled with water for the production of rose-water, or for the otto, or they are subjected to the process of maceration in warm fat or in olive-oil for the purpose of obtaining the perfumed pommade or oil, and these products afterwards finished off by the process of "enfleurage." From such pommade and oil the "extrait de rose" is afterwards obtained. These several processes are hereafter described.

PERFUMES SOMEWHAT RESEMBLING THE ROSE.

Geranium.

African, Spanish, French, and Réunion *geranium oil* is derived from three species of *Pelargonium*, the *P. odoratissimum* (Willdenow)*, *P. capitatum* (Aiton)†, and *P. roseum* (Willd.)‡, a variety of *P. radula* (Ait.) § ; but the plants in actual cultivation are *varieties* of these pelargoniums, and are not true in character to specimens of the above-named as grown in England, nor do they exactly accord with descriptions and plates of the same.

The plants are cultivated in open fields in many parts of Algeria, notably at La Trappe de Staoüeli near the Bay of Sidi Ferruch, at Castiglione, at Sahel in the good red soil consisting of a decomposition of micaceous schists, at Boufarik, at Blidah, at Grand Chérakas and at Guyoville, in the environs of Constantine, and in the plains of Métidja close to Algiers ‖.

Originally the plants were cultivated on dry arid slopes, where they were stunted in growth but yielded a perfume of great delicacy ; now, on the contrary, the plantations are established on low-lying and rather humid soil, which yields three crops annually instead of one. By a system of irrigation which floods the plantations, the proprietors force the growth of the plant to a height of about 30 inches, and nearly an inch in thickness in the stem. Under these conditions the oil is produced in much greater abundance, but the quality is sensibly inferior. (This observation on the immediate effect of a moist soil on the secretions of a plant which naturally prefers a dry soil is in accordance with the observations of Linnæus.) The irrigation process is now so general that for 1 hectare of land cultivated " dry," 200 hectares will be found " irrigated." The very superior product of the " dry " method is rarely sold separately, but is generally mixed with common oil (called " géranium irrigué ") to ameliorate the quality.

Ordinary stills are used for the distillation, which is carried on during the whole time of each harvest. It is estimated that 300 kilos of the plant yield 1 kilo of oil. The plant is gathered a

---

* Cavanilles, Monadelphiæ Diss. iv. t. 103. fig. 1.
† Andrews, Coloured engs. of Geraniums.
‡ Botanist's Repository, p. 173.   § Botanical Mag. t. 95.
‖ Exp. de Paris, 1878, Cat. Spéc. de l'Algérie.

little before the opening of its flowers, when the lemon-like odour
which it at first possesses gives place to the odour of rose. The
critical point is recognizable by the leaves beginning to turn
yellow. The oil is formed entirely in the leaves and all the green
parts of the plant, the petals yielding no odorous product whatever, but in order to waste no time in detaching the flowers they
are put in with the branches. The odour which may be thought
to be perceived in the flower is simply due to the secreting organs
in the calyx and peduncle.

The Pelargonium is also cultivated and distilled in other countries,
as in Spain (near Valencia, and recently in the province of Almeria),
Italy, Corsica, the Island of Bourbon, and in Provence. The
Spanish oil is considered the finest (probably owing to the fact
that the plantations are not " irrigated "); the plant which pro-

Fig. 3.

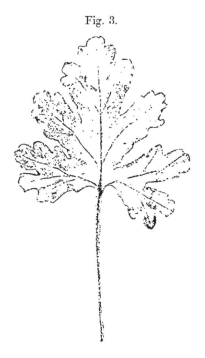

duces it is said to be the same variety of pelargonium as the
Algerian plant. The oil from Provence ranks equally as regards
quality with the Spanish ; a " superfine " oil is also manufactured
in Provence by adding rose-petals to the still. The Corsican oil

is only exported in small quantities, but the Bourbon production annually increases in importance.

The variety of Pelargonium cultivated at Grasse is represented by the annexed illustration, which is a facsimile of a living leaf of a specimen of the plant kindly supplied to the writer by Messrs. Warrick frères, of Grasse. The plant is propagated by cuttings which are set in well-sheltered beds in October. During the frosts they are covered over with straw matting. In April they are taken up and planted in rows in fields or upon easily irrigated terraces. They soon form bushes three or four feet high. At Nice they generally flower at the end of August. At Grasse and cooler places they flower about the end of October. The whole flowering plant is put into the still.

The green colour of the Réunion oil is not the result of the presence of copper, but is probably due to the same cause as the colour in Cajeput oil.

Oil of Pelargonium should dissolve perfectly in from 2 to 3 volumes of 70 vol. per cent. alcohol at 20° C.*, forming a clear solution. If any considerable proportion of fatty oil be present, the mixture appears milky, but if a very little oil is admixed it appears only turbid, and after standing some time a greasy coating forms on the sides of the vessel.

It has been remarked that if geranium oil be kept in tins it acquires a very repulsive odour by prolonged contact with the metallic surface. It is, therefore, advisable to pour the oil into glass bottles as soon as received, and should the smell not disappear, the oil must be poured out and exposed to the air for several hours in an open basin.

The sp. gr. of the Algerian oil is 0·899, and that of Réunion 0·891 (Schimmel).

"Indian Geranium," or Ginger-grass Oil.

This oil is distilled in India from the leaves of *Andropogon Schœnanthus*, Linnæus ; synonymous with *A. Martini*, Roxburgh, *A. nardoides*, Nees, *A. Pachnodes*, Trinnius†, *A. Calamus aromaticus*, Royle‡, *A. Iwarancusa*, Schultes §, and *Cymbopogon Mar-*

* Jaillard, Journ. de Pharm. xxvii. p. 205.
† Spec. Gram. iii. t. 327. •
‡ Illust. Bot. Himalayan Mountains, i. p. 425, t. 97.
§ Phil. Trans. lxxx. p. 284, t. 16.

tini, Munro; but *not* the *A. Schœnanthus* of Wallich [*], which is identical with *A. citratus*, De Cand., and yields Lemon-grass oil. In Modeen Sheriff's 'Supplement to the Pharmacopœia of India,' the vernacular names of Lemon-grass are given under *A. Schœnanthus* instead of under *A. citratus*.

The various vernacular names for the plant in India are :—*Agya-ghás* and *Ganda-bena*, Bengal; *Bujina* and *Pala-Khari*, N.W. Provinces; *Mirchia-gard*, Siwaliks; *Rose-gavat* and *Rohisha*, Bombay; *Raúns*, Punjáb.

The oil is known in commerce under a variety of names, such as, in England, " Ginger-grass," " Turkish Oil of Geranium," " Rusa-grass Oil," " Oil of Nimar " or " Nemaur." In the otto-producing districts of the Balkan it is known to Europeans as " Essence of Geranium " and oil of " Palma-Rosa ; " in India it is called " Rusa Oil," " Roshel," " Rùsa-ka-tel ; " in Egypt, Arabia, and Constantinople it appears under the names of " Idris-Yaghi " and " Entreshah," names which may mislead to the belief in a variety of oils produced from several plants. These names seem to be mostly of modern origin. The name " Rusa-grass " is certainly more appropriate than " Ginger-grass." There is a grass found about old wells near Bombay which really has an odour of ginger, but it is rather a rare plant. The name " geranium oil " has caused much confusion with the true geranium oil derived from various species of *Pelargonium*, and has apparently come into existence from the fact that the so-called " ginger-grass oil " is used to adulterate the true geranium oil, which, in its turn, is used to adulterate the otto of rose. The grass is found growing wild in large tracts in the Northern and Eastern provinces, and particularly in the North-west Provinces and the Punjáb; it is abundant everywhere in the Deccan, in Central India, and is cultivated in Kashmir in localities formerly devoted to the rose. It has recently been discovered in British Baluchistan, by J. H. Lace, Deputy Commissioner of Forests in India. In his valuable paper on the "Vegetation of the Hurnai Railway Route," recently read before the Linnean Society [†], he mentions having found it on the lower hills.

Apparently the first mention made of this oil was by Maxwell in 1825 [‡], but it is only within comparatively recent times that

* Plant. As. Rar. iii. t. 280.
† Journ. Linn. Soc. xxviii. p. 293.
‡ Calcutta Med. and Phys. Trans. i. p. 367.

it has had any commercial value. From the fact of one of the largest supplies of Rusa oil being from the Nimar district of Khandésh, Bombay Presidency, the oil has come to bear the commercial name of Nimar, Nimaur, and Namar. Dr. Dymock, describing the manufacture in this district, states that an iron still is used, and only a very small quantity of water added to the grass; when the still is carelessly worked the grass burns, and communicates a dark colour to the oil, which should be of a pale sherry colour when good. The odour at first recalls that of the rose, but this sensation is almost immediately followed by a strong odour of lemon or citron. By rectification the oil is rendered colourless, and the odour of lemon is then much less marked.

The grass flowers in October and November, and is then fit for cutting. Dr. Dymock states that 373 lbs. of grass received from Khandésh and distilled under his own superintendence in Bombay, yielded 1 lb. $5\frac{1}{2}$ ozs. of oil.

The oil is largely adulterated in the districts where it is distilled, the distillers being regularly supplied with turpentine from Bombay; the oils of gurjon, coker-nut, ground-nut, rape, linseed, and cotton-seed being also used. With turpentine and ground-nut the resulting turbidity passes off in a day or two, hence they are preferred; especially turpentine, as it is not at once detected by the evaporation test. The oil is exported from Bombay to the Red Sea ports (chiefly to Yedda), to Constantinople, Trieste, and London. Before being sent to Turkey, which absorbs the great bulk of it, large quantities are sent to Paris for rectification. In Turkey it is subjected to a special treatment, which appears to render it more fit to mix with otto of rose without betraying its odour; this process consists in shaking it with water acidulated with lemon-juice, and then exposing it to the sun and air. By this process it loses its penetrating after-smell and acquires a pale straw-colour. This process was described by Mr. Baur of Constantinople*.

As found on the London market it varies greatly in quality. A distinction is often made commercially between Oil of Palma Rosa and Essence of Indian Geranium, although both are identical products of the same plant; the first is probably only a superior quality, or contains a small addition of oil of pelargonium. For

---

* Neues Jahrbuch für pharm. Jan. 1867.

some years past an " essence of geranium " has been received from Java, possessing all the characters of Palma Rosa, but its exact botanical origin and method of production are unknown. Some samples of Palma Rosa have a decided odour of Lily of the Valley. An oil termed " Huille essentielle de Pataque Malgache " has been introduced from the Island of Réunion, described as distilled from "*Andropogon fragrans*" (evidently a fancy name), with an odour identical with Indian ginger-grass oil.

The sp. gr. of " Indian geranium oil," according to Semmler *, is 0·8868 to 0·8871 at 16° C. (two specimens being examined), and the optical rotation 20' to the left, in a column of 100 mm. (Dr. Dymock found the oil to be dextrogyre, deviating the ray 39° to the right in a 100 mm. tube, and 78° in a 200 mm. tube ; he appends the remark that some samples of " commercial " oil rotate the ray about 13° to the right, and some have little or no effect upon it. As he further remarks that all " commercial " oil is more or less adulterated, it is inferred that the optical rotation is more or less active proportionately.)

This oil, when pure, forms a clear solution in dilute alcohol in very nearly the same proportions as does oil of pelargonium, so that when fixed oils are present they can be easily detected. A process for the quantitative estimation of fixed oil by saponification with caustic potash has been recommended †.

Geraniol, $C_{10}H_{18}O$.—This alcohol constitutes 92 per cent. of pure oil of " Indian geranium." It is a colourless, strongly re-fractive liquid boiling at 120°·5 to 122°·5 C., under a pressure of 17 mm.; by oxidation with potassium permanganate it yields valerianic acid, and by oxidation with chromic acid mixture it is converted into geranium aldehyde, $C_{10}H_{16}O$, which is identical with citral. By further oxidation with argentic oxide, Semmler prepared geranic acid, $C_{10}H_{16}O_2$, a limpid oil, and by treating citral with acid sulphate of potassium Cymol was formed, a molecule of water being abstracted ‡. Geraniol is nearly related to, and iso-meric with rhodinol, $C_{10}H_{18}O$, the alcohol which forms the prin-cipal constituent of otto of rose, and which, by oxidation, yields the aldehyde rhodinal, closely resembling citral as above observed.

Geranyl chloride, $C_{10}H_{17}Cl$, is obtained by the action of hydro-chloric acid on geraniol. It is an oily liquid having an odour

* Ber. Deutsch. chem. Ges. xxiii. p. 1098.   † Chem. News, xxx. p. 293.

‡ Ber. Deutsch. chem. Ges. xxiii. (1890) p. 3556, and xxiv. (1891) p. 203.

resembling that of camphor, and decomposes on heating. Geranyl bromide and iodide may be obtained from this by the action of potassium bromide and iodide.

Geranyl valerate, benzoate, and cinnamate have been prepared. They are pleasant-smelling liquids, but cannot be distilled without decomposition.

Geranyl ether, $(C_{10}H_{17})_2O$, is formed when the chloride is heated with water, or to between 100° and 200° with geraniol. It has a characteristic smell of peppermint, and boils at 187°–190°.

Many plants contain in their leaves, wood, and roots volatile oils which are likened in fragrance to that of the rose, but the likeness is very distant, although the rose is their type of odour.

The *Rhus aromatica* (Aiton) or Fragrant Sumach has been described by Harper, in the 'American Journal of Pharmacy,' as possessing an odour similar to Rose-Geranium.

"Rose-wood oil" is distilled from the wood of the *Convolvulus scoparius*; this oil is also called Oil of Rhodium, and before the rose-geranium was much cultivated used to be in great demand. The odour, which is very weak, is not perceptible until the wood is rasped, and the yield of oil is very small.

The root of the *Rhodiola rosea*, or "Rose-root," when bruised, and even when dried, yields a rose-like scent; it is a species of *Sedum* growing on damp rocks on the high mountains of Scotland, Ireland, and the north of England, also on sea-cliffs.

The "Tulip-wood" of Brazil, *Physocalymna floribundum*, sometimes called Brazilian Rose-wood, emits a slight odour of rose when rasped; a precisely similar wood has been imported from China.

The wood of "*Dysoxylon Fraserianum*," a tree of New South Wales, has the same perfume.

The wood of the *Colliguaja odorifera* when burned exhales an agreeable rose-like smell*.

In the French Guiana collection of woods at the Paris Exhibition, 1878, a specimen of rose-scented wood was exhibited, said to be derived from the *Licaria Kanali*, a species of *Acrodiclidium*, but the essential oil is identical with that of Mexican Lign-aloe.

The volatile oil of *Asarum Canadense* is said to be now used in the United States to a considerable extent in perfumery for strengthening the odour of other perfumes. Its odour is compared to that of rhodium or santal-wood.

* Molina's 'History of Chili,' i. p. 129.

The odour of rose has been noticed in the double rose-coloured flowers of the *Pæonia albiflora* var. *fragrans,* a handsome shrub of about three feet in height, native of the north of China [*]. The flowers of *Pæonia albiflora* var. *Whitlèji,* when on the point of falling from the tree, emit a scent somewhat like that of elder-flowers [†].

The flowers of *Chamædorea fragrans,* a native of Bolivia, have the odour of the Maréchal Niel rose.

There are several artificial preparations known in Chemistry possessing an odour somewhat like Rose and Rose-Geranium, for instance :—

Ammonium salicylite.—By agitating salicylol (oil of *Spiræa Ulmaria*) with strong aqueous ammonia at a gentle heat it crystallizes on cooling in yellow needles. It is slightly soluble in cold water, still less in alcohol, melts at 115°, and volatilizes without alteration at a higher temperature. When kept in a moist state in a closed vessel, it gradually decomposes, blackens, becomes semifluid, gives off ammonia, and acquires a very penetrating odour of roses [‡]. Calcium salicylite, after being kept for a time and then distilled with water, is said to give the same result; and an aqueous distillate faintly approaching it results from treating methyl salicylate (oil of *Gaultheria procumbens*) with caustic potash.

The name Methylbenzylenic ether was applied by Wicke [§] to a compound having a geranium odour prepared as follows :—" A mixture of one atom of chloride of benzylene with a solution of two atoms of sodium in absolute methylic alcohol is heated for some hours, when sodium chloride separates in abundance, the methylic alcohol is distilled off, and the residue mixed with water, when the ether rises to the surface and is removed with a pipette, dried, and rectified. It is a transparent colourless liquid, heavier than water, soluble in alcohol or ether. It boils at 208° C., leaving a brown residue arising from decomposition.

Phenyl paratoluate.—Paratoluic acid is prepared by the oxidation of cimene with nitric acid. The cymene which is contained in Roman cumin oil, and which can easily be obtained from camphor, is heated for a considerable time in an apparatus connected with an inverted condenser with a mixture of one volume of nitric

---

[*] Anderson, in Linn. Trans. xii. p. 260; Bot. Reg. p. 485; Hort. Trans. ii. t. 18.
[†] Anderson, Linn. Trans. xii. p. 259, and Bot. Rep. p. 612.
[‡] Ann. Chem. Pharm. xxix. p. 309.   [§] Ibid. cii. p. 356.

E

acid, sp. gr. 1·38, and four volumes of water, then neutralized with caustic soda and boiled in order to remove unattacked cymene and nitro-products. It is then precipitated with hydrochloric acid and the precipitate freed from nitroparatoluic acid &c. by boiling with tin and hydrochloric acid. The product always contains terephthalic acid, which remains behind on treatment with water. The paratoluic acid is finally purified by distillation with steam *. Paratoluyl chloride is next prepared, and this is heated with sodium salicylate. A viscid mass of paratoluylsalicylic acid, resembling turpentine, is formed, and decomposes into phenyl paratoluate and carbon dioxide on distillation with lime. It forms white plates, which have a nacreous lustre, smell like geranium, and melt at 71°-72° †. The methyl and ethyl paratoluates are also fragrant.

Phenyl benzoate.—Benzoyl chloride is first obtained by the action of dry chlorine on pure oil of bitter almonds. It is a transparent liquid, the vapour of which violently attacks the eyes, and has a peculiar, very penetrating odour resembling the sharp smell of horseradish. When benzoyl chloride is heated with phenol until hydrochloric acid ceases to be evolved, phenylbenzoate is formed. It is readily soluble in alcohol and ether, and crystallizes from a mixture of these in lustrous, monoclinic prisms, which melt at 71° and sublime at a higher temperature. In odour it resembles that of geranium ‡.

Methyl benzoate ("Niobe essence") is afterwards described.

The investigations made on "grass oils" by F. D. Dodge at the Organic Laboratory, School of Mines, New York (1889), are interesting as regards the rose and geranium odours, especially as the grass oils are procurable at their place of production in unlimited quantities. The following abstract of his paper on oil of Citronella refers to the *Rose odour*.

Citronellic aldehyde.—Oil of *Andropogon Nardus* is shaken for ten minutes with a saturated solution of sodium bisulphite. The liquid solidifies into a white magma with considerable evolution of heat; the vessel should therefore be kept cool with ice. The precipitate is wrapped in flannel, and after draining on a large funnel is carefully pressed in a filter press, thinned with ether or chloro-

---

* Ann. Chem. Pharm. clxii. p. 339, and ibid. ccv. p. 113.
† Chem. Centralbl. 1859, p. 84.
‡ Jahresber. 1879, p. 675.

form, filtered, pressed, and washed with the same solvents (in which it is practically insoluble). The washed bisulphite compound, now free from residual oil, is freed from ether by exposure to the air for a few hours. At this stage it has the appearance of wax or soap. To liberate the aldehyde, the dry mass is mixed with crystallized sodic carbonate, in the proportion of 450 grammes of the former to 350 grammes of the latter, in a large flask. Steam is then admitted into the mixture, which now liquefies, and yields a distillate of about 250 grammes.

On fractionating this oil, the first portion, amounting to 75 c. c., is limpid and colourless, boils between 177° and 180°, and has a lemon-like odour. The second portion, a somewhat thicker oil amounting to 120 c. c., and boiling between 222° and 224°, is of a slightly greenish colour, and has a pleasant *rose-like* odour [*].

Another observer, Kremers (Proc. Amer. Pharm. Assn. 1887), found that the sodium bisulphite compound could be decomposed by dilute sulphuric acid at a gentle heat, and the sulphur dioxide removed by dry potassium carbonate. The aldehyde thus obtained being a viscid yellow liquid, possessing a *geranium-like* odour.

As before observed, otto of rose and rose-water, obtained by distillation, do not give an idea of the true odour of rose, the most delicate fragrance being destroyed by contact with boiling water. In a manufacturing way the fragrance of the rose in its integrity is more nearly obtainable by subjecting the separated petals to the process of " maceration " in pure grease, and finishing off the product by the process of " enfleurage ; " the perfume being ultimately extracted from the grease by a solvent. Many flowers whose perfumes are destroyed by heat are submitted to these processes.

In the first process, when oil is used, it is the very finest olive-oil produced by the trees in the neighbourhood of Grasse, or that part of the coast. This is put into copper vats holding about 50 gallons ; 1 cwt. of flowers is added and well stirred in. After an immersion of some hours, and in some cases a whole day, the flowers are strained out by means of a large tin sieve. The oil is treated with several successive charges of flower until sufficiently impregnated. It is then clarified by filtration through paper or a

[*] American Chemical Journal, xi. p. 457.

hair-cloth sieve. When fat is employed as the macerating agent, the fat used is beef-suet mixed with the lard of the common hog; that kind known commercially as " corn-fed lard " being preferred, as it is superior to other grades in many respects. It melts at 81° F., and is softer and more fusible than suet. The success of the operation depends on the absolute purity of the grease. To bring it into a suitable condition it is " pan-rendered" by dry steam, and after being repeatedly melted with alum, salt, and nitre, it is washed over and over again with plain water and then with rose-water. Finally, it is again melted with a little gum-benzoin. The grease thus purified has lost all trace of animal origin, and is carefully stowed away until required for use in the basements of the factories, which are cool and dry and where it is not subject to change.

Considerable efforts have been made to introduce vaseline, petrolin, paraffin, and other such purified mineral products in the place of the purified animal grease, and practical experiments on a large scale have been made to test the efficiency of these products, with the result that although they were found to have a *very absorbent* property, they have not the power of *retention,* and that after a few months they lose about half the perfume imparted to them ; consequently for pommades and enflowered oils for export they cannot be used. This is a recent opinion, given in answer to my inquiry, by one of the largest firms at Grasse ; my informant adding that purified grease and olive-oil are the best materials known at present for absorbing the perfume of flowers successfully.

In the process of maceration, one hundredweight of the purified grease is placed in a tinned copper vessel capable of holding 5 cwt. and melted at as low a temperature as possible by means of a water-bath or steam-jacket. About 1 cwt. of flowers are then added and well stirred in with a wooden spatula. The flowers are wholly immersed and left for a day, care being taken to stir the mixture occasionally. Every day the flowers are strained off, and fresh ones put into the same pommade until it is at full concentration. As the flowers which are strained off take up a quantity of grease, this is separated by re-melting, straining through a hair-sieve, and submitting the marc, wrapped in cloths, to hydraulic pressure (the same is done with the marc from olive-oil). The fat squeezed out is accompanied by the moisture of the flowers, from which it is separated by skimming. All the grease is finally cooled very slowly so as to allow the impurities to settle. The clear upper portion is then poured into tin canisters and is ready for sale, or for immediate use in making the spirituous extract.

The annexed diagram represents an improved apparatus invented by Piver for maceration and described by Turgan, in the 'Grandes Usines de France.' The tank on the left supplies the liquid grease heated to the proper temperature, which circulates slowly through the macerating tank, in which a constant temperature is maintained by means of a steam-pipe. The macerating tank is divided into seven compartments, in which perforated metal baskets containing the flowers to be extracted are suspended. The basket on the left contains the flowers which have successively

Fig. 4.

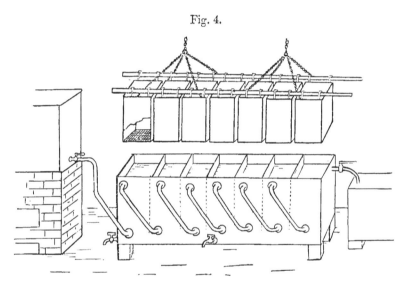

passed through all the compartments; it is removed from time to time, filled with fresh flowers and then attached on the extreme right; the other baskets being each moved a compartment to the left. In this way the fresh flowers have to traverse each compartment from right to left, while the grease flows slowly from left to right, and, saturated with the perfume of the flowers, collects in its greatest strength on the extreme right. By this method, the inventor states that all the perfume is extracted from the flowers; the grease in the first compartment being quite odourless as it proceeds from the cistern readily absorbs the last traces of it, while the grease in the seventh compartment, already almost saturated, takes up the superabundance of perfume from the fresh flowers, whereas it would not be able to absorb the last traces from the flowers which have been nearly exhausted by passing on to the first basket.

The method of making the spirituous extract consists in beating up the perfumed pommade with alcohol. The pommade is first granulated or passed through a macaroni press (what is called in London a " Piping press "), so as to bring it into a fine state of division and offer a large surface to be acted upon by the strong spirit (80 or 90 per cent. alcohol) with which it is agitated. This takes place in large drum-shaped copper cylinders, standing upright, and provided with powerful stirrers which revolve in opposite directions, the motion being given by cog-wheels connected by a band with the main shafting. About 12 gallons of alcohol is first poured in, the pommade is then introduced, and the lid bolted on; this is rendered air-tight by an india-rubber washer. The stirrers are then set in motion, and in a little time the contents are thoroughly blended and form a creamy mass. The stirring is continued for many hours, at the end of which time the contents are allowed to remain quiescent. The fat subsides and the alcohol is drawn off and passed through tubes surrounded with iced water, which solidifies and separates the remainder of the fat mechanically held in solution in the alcohol. The same pommade is subsequently washed with one or two more charges of spirit, by which time all the perfume is absorbed. The grease is now free from all scent, but by this process has acquired a great tendency to become rancid. It would appear that decomposition of the fat commences during the long kneading with alcohol in contact with air, although the drums in which the extracts are made are kept well closed. The fat is lastly placed in a steam-jacketed still to recover the alcohol mechanically mixed with it; but it is not again employed in any part of the factory. It is used by toilet-soap manufacturers, to whom it is very valuable on account of its great purity. Another sort of "agitator" for this purpose was invented by Piver; it seems particularly suitable for use with the perfumed oil; the contents of the vessels being subjected to a duplicated movement by reason of the vessels not being fixed axially in the same plane of rotation as that of the machine; by this arrangement the extremities of each cylinder are alternately high and low as the axle revolves, which of course adds a "shaking-up" motion to the revolving motion, and so effects constant admixture or friction of the particles. This peculiar oblique rotatory movement produces a more intimate mixture than an ordinary churn-movement. It is, however, doubtful whether Piver's "agitator" can bring the particles of grease into such a fine state of division, or mix them so thoroughly with the spirit, as does the upright cylinder enclosing agitators revolving both ways like an "egg-whisk."

The cold process of "Enfleurage" may be described as follows:—Grease, purified as before described, is spread to the thickness of about a quarter of an inch on both sides of panes of glass enclosed in wooden frames like window sashes; these frames are called "*châssis*," and are about two feet wide and three feet long. The flowers are then sprinkled on the upper surface of the greased surfaces and the châssis piled up in a stack. As the edges of these frames are quite flat, they superpose one on the other in such a way as to form a series of almost air-tight compartments, the upper and lower surfaces of which are composed of grease. Every day, during the flowering season, these sashes are taken down, and the flowers changed. Each time that fresh flowers are put on the grease it is worked about with a palate-knife, and the surface serrated or furrowed, so as to present a new surface to the flowers. This is repeated until the fat is sufficiently impregnated.

Olive-oil is also enflowered by a similar process, but in place of glass the frames are furnished with coarse wire gauze stretched between its four sides; these frames are called "*châssis en fer.*" Upon the wire gauze is laid a very thick soft cotton cloth with a fluffy surface like a bath-towel. This cloth is called a "*molleton*," and is saturated with pure olive-oil. The flowers are placed on these cloths, and, after frequent renewal with fresh flowers, the cloths are folded together and put under a powerful press.

The perfume is finally extracted from the solid grease, or from the oil, by agitation with strong spirit (90 per cent. alcohol); the same process as above described.

Other processes have been invented for the extraction of perfumes without the aid of heat. The superiority of any process over the old-fashioned one above described must consist in the facility of thoroughly extracting the flowers while quite fresh, before any decay has taken place in the petals, also without causing any change by oxidation of the molecules of aroma *during* the process. The process must work rapidly, so as to dispose of a great bulk of flower at once, and so keep pace with the harvest during the very short time which it lasts, in fact to use up every day the whole of the flower which can in that day be cut.

In 1856 an invention was patented in France by Piver* called a pneumatic process. This consisted in spreading the flowers on perforated metallic plates enclosed in an air-tight chamber, and passing through them a current of carbonic acid gas, which was afterwards conducted into a vessel containing liquid grease kept fluid by heat; this vessel was fitted with revolving plates so as to continually present a fresh surface to the gas, and so absorb the particles of perfume with which the gas was impregnated; the gas was then passed through a second vessel fitted in the same way, and was then forced again through the chamber containing the flowers, thus being made to circulate several times, in order to abstract all the particles of aroma and deposit them on the absorbent grease.

Another apparatus was patented in France by Piver in 1859†, and described by Turgan in his 'Grandes Usines de France,' 1865, p. 129. It is described as follows:—The apparatus consists of two chambers communicating with each other at the basement, and fitted with moveable perforated metal trays on which the flowers are strewn; between these trays are plates of glass on which is placed the grease, not spread as a paste but in a subdivided condition obtained by forcing it through a piping press and cutting it into small cylinders. The air is forced to and fro between the two chambers by means of a bellows arrangement on the top of each chamber, and the plates of glass, which are a trifle narrower than the width of the chamber, are arranged in an alternate manner so that the current of air travels between each plate, as explained in the annexed illustration. The doors of the apparatus fit air-tight, and are closed during the operation, so that the same air constantly travels from one chamber to the other. The inventor of this apparatus claims that flowers can be exhausted of their perfume with great rapidity, and it can be absorbed either by finely-divided grease or by olive-oil soaked into cotton cloths.

* Brevet 15950. † No. 41090.

In both cases the extraction by alcohol is effected in the usual way (as described for rose-pommade). Another patent was taken out in France by Piver in 1872* for " Improvements in processes for the ' enfleurage ' of fatty substances."

Fig. 5.

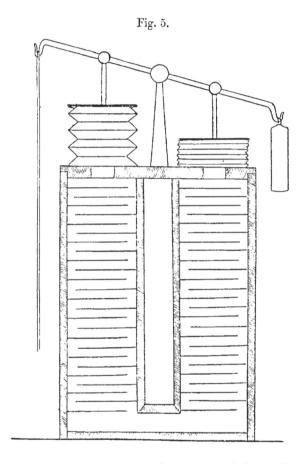

In 1888 a process was patented in England by Nellenstein, a chemist of Amsterdam†, to extract the perfume of flowers by means of a vacuum and refrigeration. The flowers being placed in a vessel of sufficient strength to resist external atmospheric pressure, the air was exhausted and caused to pass through a vessel surrounded by a freezing-mixture, and then into a third

* No. 95326. † Patent No. 15299.

vessel containing an absorbent such as vaseline or purified grease placed on perforated shelves. In this patent no arrangement is made for passing the same air a second time over the flowers, the inventor claiming to abstract the perfume in a condensed form in the refrigerator and in the grease at one operation.

Another process, very similar to Piver's inventions, was patented in England in 1888* by Hagemann of London. According to the Specification, carbonic acid or nitrogen gas is passed through a vessel containing the flowers, and becoming charged with the volatile particles is caused to pass through another receptacle into a substance suitable for fixing them, such substance to be one having little or no retaining action on the gas, such as purified fat, glycerine, or a strong solution of sugar. The gas, after being thus freed from the volatile bodies, is then caused to pass again through the flowers, and to circulate in this way until all the perfume is eliminated. The inventor claims that, as this opera-tion can be performed at a low temperature and as the circulating gas is practically innoxious to the flowers, the products obtained are in a high state of perfection.

The reason why some strongly scented flowers yield no volatile oil by distillation, even by repeated cohobation with water, is partly because the quantity of oil contained in them is very minute or is very soluble in water or (which is very likely) it is apt to decompose by the combined action of heat, air, and water. Of such plants are the Narcissus, Hyacinth, Jonquille, Violet, Tubereuse, &c. From some of them it has been found possible to isolate the odoriferous principle. As far back as 1835 Robiquet† exhausted the fresh corollas of the Jonquille with ether in a com-pression-filter, separated the upper yellow ethereal stratum of liquid from the lower watery one, distilled the ether from the liquid at a gentle heat, and obtained a residue consisting of crys-talline nodules together with a mother-liquid, which, when eva-porated in the air, gave off a strong and agreeable odour of the flowers. The crystalline nodules, when purified, formed an in-odorous camphor and appeared to be the odoriferous oil converted into camphor by exposure to the air. Buchner obtained similar results by applying Robiquet's process to the flowers of Lilac, Lime, and Mignonette‡. In 1855 further experiments were made by

* No. 6851. † Journ. de Pharm. et de Chim. xxi. p. 334.
‡ N. Br. Arch. viii. p. 70.

Millon and patented by Ferrand, his co-worker (Brevet 22404) *. The flowers were packed in a displacement-apparatus and exhausted by disulphide of carbon or ether : the odorous principle was then isolated by evaporating the solvent. This process was found impracticable on a large scale owing to the danger of manipulating large quantities of the solvent in open vessels ; it was also found that the residuary resinous matter retained the odour of the disulphide of carbon or of the ether, which could not be removed without destroying the odorous principle.

In 1862 the experiments were repeated by Piver, who endeavoured to remove the last traces of the solvent by washing the product with a weak alkaline solution and so leave the perfume pure † ; but that idea had already been patented by Deiss of Marseilles ‡, and does not appear to have proved successful, probably owing to the great susceptibility of such delicate perfumes to deterioration or alteration by contact with alkalies.

Patents were also taken out by Lemettais & Bonnière of Rouen for the extraction of the odours of spices and living plants by disulphide of carbon, and rectification by alkalies and salts of lead §. The processes of Ville ‖ and of Hirtzel ¶, which were modifications of Millon's process, the former employing chloroform and the latter petroleum ether, could not stand the test of practical working for similar reasons.

A method has been recommended by Millon for the purification of disulphide of carbon ; it is as follows :—The disulphide is first washed several times with water and then introduced into a large retort with a quantity of quicklime. After 24 hours' contact the disulphide is distilled from the lime and received in a flask containing a large quantity of copper turnings which have been previously calcined to remove organic matters and then reduced by heating in a current of hydrogen. The lime from which the disulphide has been distilled is deeply coloured and resembles crude soda-ash in appearance. The disulphide of carbon thus purified has an ethereal odour, which, if not actually agreeable, is quite different from the offensive smell of commercial disulphide. It was with disulphide of carbon thus purified that Millon and Commaille

* Journ. de Pharm. et de Chim. xxx. pp. 281 & 407.
† Rep. de Chimie, iv. p. 286.   ‡ Brevet 54126.
§ Brevet 35499, 1858.   ‖ Brevet 47285, 1860.
¶ Brevet 61486, 1864.

separated the perfume of the most delicate flowers, and from cow's milk were able to recognize the odours of the plants eaten by the animals.

From some plants which contain the odorous principle in exceedingly minute quantity, this principle has been isolated by Chardin and Massignon by distilling with a small quantity of water the pommades or oils saturated by the process of enfleurage, saturating the distillate with salt and agitating it with ether. On evaporating the ether, the essence remains. In this way they succeeded in eliminating small quantities of the essence of strawberries, of raspberries, of tubereuse, jasmine, and pinks. These rarities were very troublesome and expensive to make, and on a large scale, would be attended with many difficulties, such as the danger of the explosion of the ether vapour. This difficulty, on a small scale, could be overcome by substituting chloroform, because chloroform and its vapour are not inflammable and it evaporates most completely and rapidly, leaving no trace of unpleasant odour as ether does ; chloroform may be expensive, but it could be recovered by condensation of its vapour. As regards the first part of the process, it would certainly be necessary to employ as little heat as possible in the distillation of the grease, to avoid decomposition both of the grease and of the odorous principle. This might be accomplished by using a vacuum-still, such as are made by Berjot of Caen and Brinjes and Goodwin of London.

In 1879 a patent was taken out by Massignon * for the extraction of perfumes by making use of the solvent power of methyl chloride. The economic production of this liquid on a large scale from the molasses left in the manufacture of beet-sugar was invented by Camille Vincent †, for which he was awarded a gold medal by the Société d'Encouragement. Methyl chloride in the gaseous state is dried, purified, and condensed by pressure into strong metallic reservoirs. In this state it is a colourless, very mobile fluid, having a sweet ethereal odour. It boils at about $-23°$ C. under the normal atmospheric pressure of 0·76 m. It has the property of dissolving fatty bodies, resins, and essential oils. The apparatus used for the extraction of perfumes by the aid of this liquid is described ‡ as :—1, a digestor, in which the flowers are

---

* Brevet 130967, May 30. † 'La Nature,' June 21, 1879.
‡ 'Bulletin de la Société d'Encouragement,' Dec. 1879.

placed; 2, a closed vessel containing purified methyl chloride; 3, a closed vessel for the reception of the solvent after it is charged with the odorous principle of the flowers; 4, a pump for creating a vacuum in vessel no. 3, by which means the chloride is vaporized and the vapour compressed in a refrigerating condensing-worm, from whence it is conducted back in a liquefied state into the vessel no. 2 which originally contained it. This latter portion of the apparatus (and it is a very important portion) is the refrigerator which was illustrated in 'La Nature,' June 21, 1879. The action of the apparatus is as follows:—The digestor, being filled with flowers, is closed; then, by means of a valve, the liquid chloride is allowed to flow from vessel no. 2 into the digestor and to cover the flowers. After two minutes' digestion the liquid, then charged with the perfume, is allowed to flow into vessel no. 3. A fresh charge of methyl chloride is then introduced over the same flowers, and is afterwards passed into vessel no. 3; this is repeated until the flowers are exhausted of their aroma. Finally, a vacuum is created in the digestor to remove the remnant of the chloride mechanically contained by the flowers, and the vapour is forced into the refrigerator; a jet of steam is then passed through the spent flowers, all resulting methyl-chloride vapour sucked out by a pump, and, after desiccation, liquefied by pressure. The liquid containing the perfume in solution in the third vessel is evaporated by exhaustion; to effect this, the vessel is surrounded with water kept at a temperature of about 30° C. while the pump is working; the evaporation of the chloride is then rapid. When the pressure-gauge attached to the vessel registers a vacuum of half an atmosphere (it having originally marked a pressure of 3 or 4 atmospheres), the operation is finished. On opening the vaporizer, the perfume is found mixed with fatty and waxy matters. This mixture treated with cold alcohol yields up the perfume in all the fragrance and sweetness which it originally possessed in the plant.

At the meeting of the Société d'Encouragement, December 26, 1879, Messrs. Schneider and Naudin, chemists, of Montreuil-sous-bois, make a claim of priority as regards Massignon and Vincent's process. Schneider and Naudin's patents are dated April 12, 1879 *, May 21 and 26 †, May 30 ‡, and additions to these patents were made on June 14, September 19, and October 14, 1879.

* No. 130137.   † No. 130873.   ‡ No. 130967.

The patent of Naudin and Schneider certainly exhibits points of similarity to that of Massignon and Vincent, but a glance at the illustration accompanying the abstract of the specification of the former, given in the ' Pharmaceutical Journal ' [3] xiv. p. 44, suffices to show that with such complicated apparatus difficulty might be experienced in working off several tons of flowers daily if any part of it got out of order. The consequences of the least derangement in harvest-time, when perishable flowers are brought in quantity, would be serious.

Another patent for extracting perfumes by means of a solvent was taken out in London in 1890 * by the " Société Anonyme des Parfums Naturels de Cannes." The solvent employed, by preference, is stated to be a light petroleum boiling at about 80° C., the flowers being washed in a succession of " extractors " called " batteries." The specification of this patent is an elaborate study, consisting of ten pages of letterpress describing its intricacies, crowded with letters and numbered letters which are expected to be explanatory of twenty-eight figures illustrating the machinery, which is composed of a formidable array of batteries, extractors, boilers, gasometers, vacuum-pumps, purifiers, evaporators, reservoirs for compressed gas, gauges, purging-pipes, automatic valves, cocks and working cocks, and other things; to understand the technique of which would require workmen educated up to it. A comprehensive abstract of this patent might entice the reader on to the verge of insanity, so, as the specification is procurable at the London Patent Office for the modest sum of fifteen pence, the reader can purchase it and read it for himself on his own responsibility.

Whatever may be the merits or demerits of all these improvements and patents, it is stated by one of the largest firms of perfume manufacturers at Grasse, in answer to an enquiry for information about them, that, after having made trials of various pneumatic processes and solvent processes, they have been abandoned as not giving satisfactory results; and the old method of " enfleurage-sur-châssis " is always returned to after such trials. This statement, made in August 1891, does not say much for the value in actual work in a wholesale way of the various patented processes, however effectual they may be in working on small quantities of flower.

* No. 5940.

It may, however, be here recorded that the use of methyl chloride as a solvent is recognized in the Colony of Victoria, and " concrete oils " are there obtained by a simple process regardless of the disputes between French patentees. Mr. J. K. Blogg, manufacturing chemist of Melbourne, gave evidence before the Royal Commission of Enquiry into the Vegetable Products of Victoria * to the following effect :—" Concrete oils are by far the choicest and most valuable of all perfumes, and though the manufacture of them requires a little skill, yet the process is not beyond the reach of average Victorian intelligence. The plant required is not very expensive. A jacketed vacuum-pan (still) fitted with an air-pump and condenser, a close macerating vessel, and a receiver is all that is required. Fresh flowers, free from stalks, are put into the macerating-vessel, which is then filled up with deodorized methyl chloride. The flowers are allowed to remain in contact with this fluid for about 10 or 15 minutes. The solvent has during this time extracted the essence and holds it in solution. This solution is now transferred without exposure to the vacuum-pan, the pump is set in motion, quickly at first, and afterwards at a steady rate, to maintain a vacuum of half an atmosphere. The temperature in the vacuum-pan is at the same time raised to 98° F. The vacuum in front of the methyl chloride causes it to evaporate very quickly at a low temperature, the vapour passing from the pump into the condenser flows therefrom into the closed receiver, and is reserved for the treatment of fresh flowers. The concrete oil will be found in the vacuum-pan, and may be collected after each distillation or left in the pan to accumulate with other subsequent operations. As the methyl chloride is very volatile, care should be taken to prevent its exposure in transferring from one vessel to another.

" The methyl chloride may be deodorized by passing it, while in a gaseous state, through pure sulphuric acid. This might be done by evaporating from the vacuum-pan by means of the air-pump, the pipe from which might be made to pass to the bottom of a two-necked or Woulf's bottle containing the acid, and thence on to the condenser from the other neck of the bottle. It would be necessary to keep the sulphuric acid warm in a sand-bath, so that the methyl chloride might not condense in the acid and fill up the bottle."

* No. 2, publ. by the Gov. Printer, Melbourne, p. 45.

CHAPTER III.

THE CITRINE ODOURS.

THE *Citrine odour* is a very distinct type. It is represented by the products of the very numerous family of the *Citrus*,—trees which flourish in all parts of the Tropics and semi-tropical countries, furnishing a great variety of essential oils, varying considerably in perfume, from their flowers, leaves, and fruit. The word " orange " is derived from the Sanskrit " Narunga " and the Arabic " Narung ; " these names apply to the different varieties of the *Citrus aurantium*.

The Sweet or Portugal orange is the *Citrus aurantium* of Risso, who enumerates nineteen varieties. It is a native of Asia, and cultivated in all warm countries on account of its delicious fruit; but in every other respect the Bitter orange is superior to it.

The Bitter or Seville orange is the *Citrus Bigaradia*, Duhamel, *C. aurantium*, var. *amara*, Linnæus, and the *C. vulgaris* of Risso, who describes twelve varieties. It is a native of India, and is cultivated in most of the warm countries of the world.

The Bitter orange is propagated from seed, and is considered by most cultivators to be quite distinct from the Sweet orange, although not differing from it in any important botanical point; but it is at once distinguished from it by the appearance of the rind of its fruit, which is rougher, of a deeper reddish yellow, and by the very bitter taste of its fruit. The rind, the flowers, and the leaves are all more odorous than those of the Sweet orange. In the south of France the orange-trees are grown from pips, and the young plants are grafted when about three years old, generally with the Bitter orange. They are also propagated by grafting the Bitter orange on to seedlings of the Sour orange and on to rooted cuttings of the Citron, which is a rapid method of propagation and makes a strong stock for grafting on to. When a seedling is about 4 feet high it is transplanted, and allowed a year to gain strength before being grafted. It requires much care the first few years; it must be well watered in the summer, and if at all exposed must have its stem covered up with straw in winter. It is not expected to yield a crop of flowers before the fourth year after transplantation.

The Citron is the *Citrus Medica*, and is called in Arabic "Turnj" or "Utrej," and in Sanskrit "Beejapoora ;" it is the μῆλον μηδικὸν of Theophrastus, a native of the Himalayas, and cultivated apparently from the time of the earliest Aryan settlements in Media, whence it derives its Greek and specific scientific name. Theophrastus and Virgil both call it the *Malus Medica*, meaning "Apple of Media." It is quite by mistake that some writers assume the word *Medica* to refer to any medicinal properties it may possess, and it is erroneously translated into French as "citronnier médicinal." The tree is known in France as the "Cédratier," and the fruit as a "cédrat." The fruits which in France are called "citrons" are what we know as "lemons." The citron fruit sometimes attains an enormous size; according to Ferrari (in his work on the 'Hesperides') the Calabrian citron will weigh 4 or 5 lbs., the Genoese citron as much as 12 lbs. and sometimes even more.

The Lemon is "Nimbuka" in Sanskrit and "Limun" in Arabic. This is the *Citrus limonum*, a native of India. It was found by the Crusaders in Palestine in a cultivated state, and had previously been naturalized in Africa by the Arabs; also grown in the south of Spain, from whence it was introduced into Italy and the south of France. There are many varieties and hybrids of the lemon, the rind of the fruits of some of them being very fragrant, and the fruit containing abundance of acid juice—such as the Roman "Lustrata," the Genoese "Bugnetta," and the Spanish "Balotin." The ordinary Genoese lemon is cultivated on the borders of the Mediterranean between Nice and Genoa, in Calabria, Sicily, Spain, and Portugal, and on nearly the whole coast of Liguria, from Spezzia to Hyères, furnishing the bulk of the fruit met with in commerce, it being well adapted by the thickness and toughness of its rind to withstand pressure in transporting it to the North.

During the months of November to March the average yield of

F

1000 lemons in the factories at Palermo is 320 grammes of oil; in those of Messina, where a better quality of fruit is employed, the yield is about 400 grammes. The same number of lemons yield about 40 litres (10 gallons) of acid liquor, which, of course, is utilized for citric-acid manufacture.

Trees of the *Citrus* tribe do not often become diseased, but when a disease once manifests itself it spreads with great rapidity. The disease of the orange-tree was first discovered in the Azores in 1836, when it was found that the oldest and best trees, as much as 200 and 300 years old, producing each from 5000 to 20,000 oranges, were disappearing. It was observed that all the trees affected produced a very heavy crop the very year that the disease manifested itself, that the leaves became yellow and fell in great quantities, and on the trunks near the ground, and sometimes beneath the ground, the bark opened and drops of a kind of yellow gum exuded. The drops resembled tears—*lagrimas* in Portuguese; therefore the disease was named " Lagrima." Many orangeries were quite destroyed. The disease was remedied by cutting the bark across, to allow the exudation to run out; and if the disease was in a very advanced state the bark and the whole of the diseased wood was cut out, the roots being bared to a distance of a foot or two from the stem, and every portion of diseased root cut away. If the disease was taken at an early stage this process was successful. The disease is not the result of age, as there are trees in Spain now known to be 600 or 700 years old, and when the disease made its appearance in Australia trees of 22 to 25 years old, and even seedlings of one year old, were attacked.

A report by the late Dr. Landerer states that millions of lemon-trees grow on the islands of the Grecian Archipelago, on Chios, Paros, and in the Peloponnesus, but that the same disease appeared amongst the trees at Paros which had ravaged the plantations of Sicily. The methods of extracting the essential oil by the "éponge" and the " écuelle " were introduced into Paros by the Sicilians.

The oils from the peel or " zeste " of the citrine fruits are manu-factured in large quantities in the north of Italy and in the south of France, the fruit being taken when in a barely ripe state—the oil of the Bitter orange being by far more valuable than that of the Sweet. They are extracted by processes called the " Eponge " and the " Ecuelle-à-piquer," and are termed " Essence de Bigarade au Zeste " and " Essence de Portugal au

Zeste,"—the Bigarade referring to the Bitter, and the Portugal
to the Sweet orange. The oils obtained by distillation are very
inferior, and are termed "Essence distillée" of Bigarade or of
Portugal respectively. The same terms apply to Bergamotte,
Citron, Lemon, and to all the Citrine fruits. The process called
the "Eponge," as applied to the lemon in Sicily and Calabria,
is briefly as follows :—In the months of November and December
the small irregular-shaped fruits, which have but little value for
export, are selected, preferably whilst still green, as they are then
more rich in oil than when perfectly ripe ; the rind is removed by
making three incisions lengthways and three round, the fruit itself
being slightly cut into and left in a pyramidal shape, with a little
piece of rind at each extremity ; the fruit is then divided by a cut
across the middle, and put aside. The oil is then extracted by
holding in the right hand, between the finger and thumb, a segment
of the rind face downwards—that is to say, gripping it on the pith
side and pinching it so that the outer part of the rind, which originally was convex, now becomes concave in form ; the oil then
escaping from the fractured cells is received into a piece of sponge
held in contact with the rind by the fingers of the left hand. Each
segment is well pinched three or four times, the workman always
avoiding to squeeze any fragments of fruit adhering to the rind.
When the sponge is saturated with liquid the contents are squeezed
out into an earthenware bowl of about three pints capacity ; the
oil floating to the surface is afterwards decanted. The yield varies
from 9 to 14 ozs. from 400 fruits operated upon. The fragments
of rind and fruit are then pressed to extract the lemon-juice, and
the remnant is afterwards distilled to obtain the residual oil.

The "Ecuelle-à-piquer," which is in use at Mentone and Nice,
consists of a saucer-shaped vessel, about 20 centimetres diameter,
made of tinned copper. All over the bottom of this vessel are
numerous strong sharp spikes, projecting about 1 centimetre ; in
the centre of the hollow or lowest part of the cavity of this vessel
is an orifice about 2 centimetres diameter, leading into a tube or
hollow handle of about 12 centimetres in length. The workman
then places a fruit in the tool, and by a rapid rotatory motion of
the hand causes the oil-vessels of the rind to be pierced by the
spikes. The escaping oil flows into the hollow handle. When the
handle is full, the contents are emptied into another vessel to
clarify. A further small quantity of oil is obtained by immersing
the scarified fruit in warm water, and decanting the supernatant

oil. An inferior oil is sometimes obtained by roughly rasping the surface of fruit which has undergone the écuelle operation, and sometimes by rasping fresh fruit, and distilling the result with water, which yields an oil very inferior in perfume and in value ; it is called " Essence distillée," to distinguish it from the above, which is called " Essence au Zeste." Although the distilled oil is still considered inferior, experiments made in London by Moss* by distilling peel from fresh lemons, the cellular tissue of which was torn to free the oil, resulted in a product which was considered equal to or finer than the Italian oil made by the cold process. (Its specific gravity was estimated by Dr. Tilden at ·852 at 20° C.) This experiment led him to believe that the inferior fragrance of foreign distilled oils was partially attributable to the construction of the stills. Products distilled on a large scale are undoubtedly affected in quality by the form of the still and the method of manipulating it ; but probably in the experiment made by Moss every lemon was carefully peeled by hand, and no trace of acid juice of unripe fruit allowed to enter the still.

A sort of écuelle on a larger scale, capable of operating on six or eight fruit at a time, or about 7000 per day, is used for extracting Bergamotte oil. It consists of a metal vase perforated with small holes at the bottom, and is provided with a heavy rotatory lid communicating by cog-wheels with a handle. There is a channelled groove round the inner circumference, for the reception of the fruit ; the inner surface of the lid and the groove are fitted with short metallic blades projecting about 15 millimetres. The rapid rotation of the lid, which presses on the fruit, causes them to revolve and become lacerated at all points, and the liberated oil flows through the perforated bottom into a receiver. Oil thus obtained is greener than formerly obtained by the éponge process.

An apparatus called the *Strizzatore termo-pneumatico* (thermo-pneumatic extractor) was invented by Dominico Monfalcone for the extraction of oils from all varieties of Citrus fruits—the advantages stated by the inventor being that from a given weight of fruit double the quantity of oil is obtainable than by the older method of the éponge, also the economy of an immense deal of time and labour. The product is said to be equal in quality to that obtained by mechanical means alone. The apparatus is described as follows :—A is a hollow cylinder of sheet iron, the

* Pharm. Journ. 20th March, 1879.

interior surface of which is studded with a large number of
metallic knife-points. This cylinder revolves on two axles attached

Fig. 6.

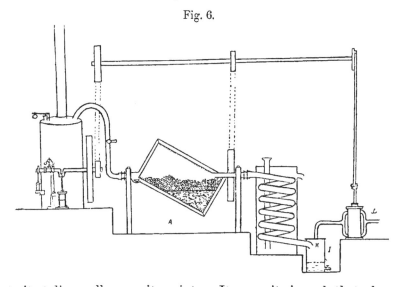

to it at diagonally opposite points. Its capacity is such that when
charged, as indicated in the illustration, it is capable of holding
1200 or 1500 lemons. The cylinder having been properly charged,
say with the above-mentioned number of lemons, together with a
small quantity of water, it is set in motion by aid of the shafting
and pulleys driven by the small steam-engine. The fruits are
thereby made to come in contact with the metallic lancets,
and their whole surface gradually becomes punctured sufficiently
to cut or rupture the cells containing the essential oil, which
escapes. The cylinder A has double walls, the space between them
being intended for the circulation of a current of steam, which is
admitted to heat the water and to facilitate the extraction of the
oil, while at the same time the vapours are rarefied or aspirated by
the vacuum-pump L. The axles of the cylinder on both sides are
hollow : that on the side looking towards the engine is imbedded
in such a manner that steam from the boiler may be admitted
at will either into the double walls of the cylinder or into the
interior of the cylinder itself; while the hollow passage in the other
axle communicates with a condensing-worm, the outlet of which
(H) descends into a cylindrical vessel I, intended to receive the
condensed products, consisting of water and essential oil. When

the apparatus is first set in motion, the pump L begins to produce a vacuum during the first revolutions of the cylinder. Steam being now cautiously admitted into the double walls of the cylinder, the water is raised to boiling at a comparatively low temperature, and the vapours charged with the essential oil rapidly pass over into the receptacle. With a 2-horse power engine and a boiler designed for a 5-horse power engine, five such apparatus may be driven at once, and the operation, including the time required for charging and emptying, is completed in three-quarters of an hour.

The expressed oil of lemon contains mucilage, which is apt to ferment and produce cloudiness. It has been suggested by Dr. Bond that this can be washed out by agitating the oil with water in the proportion of 2 ozs. of water to 1 lb. of oil, by which the mucilaginous matter is removed and sinks to the bottom with the water. The oil can then be decanted and dried.

Essential oils prepared by distillation with water often retain some of that liquid, even when they appear quite clear. The water may be detected by mixing the oil with several times its volume of petroleum-ether (the so-called benzin), whereby a turbidity is produced owing to the separation of globules of water.

Oil of lemon is very apt to oxidize and resinify by contact with air, developing a turpentine odour, which quite unfits it for use as a perfume. This change can be retarded by keeping it in full well-stoppered bottles in a dark cool place; also by mixing the oil with an equal bulk of alcohol or by pouring a little alcohol on the surface of the oil, which prevents its contact with air.

It is said that oils of lemon, lavender, and others, which tend to resinify and acquire a terebinthinous odour, may be kept indefinitely by the addition of sodium bisulphite in the proportion of 50 grains of the salt to each pound of oil.

Most of the commercial oil of lemon is adulterated with the cheap distilled oil and with oil of turpentine.

Oil of lemons has recently been examined by Oliveri[*], who fractionated it by distillation into three portions,—the first boiling between 170° and 175°·5 C., the second between 176° and 178°, and the third between 240° and 242°. The first was a colourless mobile liquid, sp. gr. 0·8867, with a very pure lemon oil odour, and consisted of limonen, $C_{10}H_{16}$, forming the characteristic crystalline tetrabromide, $C_{10}H_{16}Br_4$, melting at 31°, and a dihydrochloride, $C_{10}H_{18}Cl_2$, melting at 50°. The second fraction, amounting

[*] Pharm. Journ. 27th June, 1891, p. 1172.

to nine tenths of the whole distillate, had a sp. gr. of 0·899, and in respect to its formula and molecular weight was identical with limonen, but the rhomboid crystals of the tetrabromide melted first at 102° to 103°. The principal constituent of the third fraction was the sesquiterpene, $C_{15}H_{24}$, which is present in old oils in far larger proportion than in new oils ; it has a sp. gr. of 0·9847, and does not form a crystalline tetrabromide or dihydrochloride. The first two fractions are optically active, but the third is inactive. According to Oliveri * the rotatory power of oil of lemon in a column 20 centimetres long varies with age between + 117° and + 123°, and it is suggested that this character might be utilized to detect an adulteration with oil of turpentine. Investigations made by Schimmel and Co. indicate the sp. gr. of expressed oil of lemon to be 0·857 to 0·863 at 15° C., and the rotatory power in a 100-millim. tube + 40° 10′ to + 62°; the sp. gr. of the distilled oil being found 0·856, and the rotation of same + 66° 20′. It would appear that slight variations in the conditions under which oil of lemon is prepared are apt to cause considerable differences in the physical properties of the oil, and that a reliable commercial oil of uniform strength and free solubility in rectified spirit would be of great advantage.

A so-called "Terpene-free" oil of lemon has been prepared during the last ten years at Pirna-on-the-Elbe. Experiments recently made on this oil show that it more readily becomes altered in flavour by keeping and by exposure than when mixed in its natural state with its own terpene †.

In various essential oils of the Citrine series Messrs. Schimmel and Co. have discovered an aldehyde called *citral*, $C_{10}H_{16}O$, to which is ascribed the aroma. According to their researches, normal oil of lemon contains $7\frac{1}{2}$ per cent. of citral; but it is admitted that, " as it lacks some part of the freshness which characterizes good oil of lemon, it should be used mixed with the natural product. A mixture of 1 kilo of oil of lemon and 75 grammes of citral will be equal in strength and aroma to 2 kilos of oil of lemon (*i. e.* 75 grammes of citral possess the strength of aroma of 1 kilo of oil of lemon)." ‡ In a partial

---

* Apot. Zeit. 1891, p. 341.   † Ph. J. [3] xxii. p. 876.

‡ One ounce of citral in 15 ozs. of 95 per cent. alcohol is considered equivalent to 16 ozs. of lemon-oil. Even in liquids containing only 30 per cent. alcohol citral forms a clear solution.

vacuum under a pressure of 16 millim. Citral boils at 116° C., and under normal atmospheric pressure at 228° to 229° C. without decomposition, if quite pure. At 15° C. its sp. gr. is 0·899. Semmler ascertained * that the aldehyde $C_{10}H_{16}O$, obtained by the oxidation of geraniol with chromic-acid mixture, is identical with citral. According to Poleck † the fluid constituent of otto of roses is an alcohol, $C_{10}H_{18}O$, which by oxidizing agents is converted into the aldehyde citral, obtainable from geraniol by similar treatment. Messrs. Schimmel have ascertained the presence of citral in the following essential oils:—*Citrus limonum, Citrus limetta, Citrus Madurensis, Andropogon citratus, Eucalyptus Staigeriana, Backhausia citriodora, Tetranthera citratus,* and *Xanthoxylum piperitum.*

The Lime is the *Citrus limetta,* which grows wild in many tropical countries, but does not flourish even so far north as the Azores. It is found wild and cultivated in Jamaica, Dominica, and Tahiti; but the most important plantations are those established on the island of Montserrat, considered the most beautiful of the Antilles, situated in 16° 45′ N. lat. and 61° W. long. This island is only about eight miles in length from north to south, by a breadth of five miles from east to west, and is composed of a small cluster of volcanic mountain-tops rising out of the Caribbean Sea to the height of 3000 feet. The high mountains, whose steep sides are covered with virgin forests, seem to protect Montserrat from the hurricanes which desolate the neighbouring islands; and the forests ensure a rainfall when the other islands are parched with drought. The temperature of Montserrat is remarkably uniform, the thermometer at night seldom falling below 69°, or rising even at mid-day above 90° F., with an average of 78° to 80°. The average rainfall is 54 to 64 inches a year.

The Lime is a thorny, bushy evergreen tree, with handsome dark green leaves. The leaves are so fragrant that they are universally used in the West Indies to perfume the water in the finger-glasses at dessert. The small white flowers resemble orange-blossoms, and their scent is equally delicious. The very extensive plantations are owned by an English Company, which was formed for the main purpose of extracting the juice from the fruit; but a very fine essential oil is also obtained from the rinds of the fruit in the same way as the kindred ottos of orange, lemon, and

* Ber. Deutsch. chem. Ges. 1890, xxiii. p. 3556, and 1891, xxiv. p. 203.
† Ibid. 1890, xxiii. p. 3554.

bergamot, viz. by rasping the unripe fruits by rubbing them over a perforated concave metal rasp fixed over a basin, squeezing the pulp thus obtained, and purifying by filtration the oil which exudes. Thus obtained, it has the peculiar sweet sharp odour characteristic of the fruit, and very superior as regards delicacy of fragrance to the oil obtained by distillation. The flowers yield a "neroli," and the leaves and young twigs a "petit grain" oil by distillation.

The Lime harvest is heaviest from September to January, but the Montserrat plantations yield a considerable return all the year round. The trees require regular pruning, and to be freed from the mistletoe and other mischievous parasites, so that their cultivation during the years that elapse before they come into bearing involves a considerable outlay. Plants raised from seed come into full bearing in seven years. The Lime flourishes best in a light soil near the sea.

The physical qualities and properties of oil of limes have been carefully studied by Watts *. His investigations show that "pure 'écuelled' or hand-made oil is of a decidedly yellow colour, varying in intensity, being darker in new specimens. The sp. gr. also varies, being higher in newer samples,—the mean sp. gr. of seven samples, all under twelve months old, being 0·8734; the distilled oil of same age being 0·8554. Ecuelled oil may be regarded as an almost saturated solution of citroptene or lime-camphor, and this may be made a means of distinguishing it from the distilled oil. The difference in flavour and aroma, however, is so marked as to scarcely require any other means of distinguishing écuelled from distilled oil, the former having a decided and fragrant lemon-like smell, whilst the latter frequently possesses little more than the smell of turpentine. The distilled oil is usually almost colourless, is specifically lighter, and contains no citroptene. When the citroptene is treated with oxidizing agents, nitric-acid or chromic-acid mixture, a red resinous acid body is produced; so, if a sample of oil of limes be agitated with chromic-acid mixture for some few minutes, and the mixture filtered, the red resin will be left on the filter and sides of the test-tube if the oil be hand-made, but will not appear if a distilled oil."

As regards the production of oil of limes at Trinidad, the Director of the Botanic Gardens at that place stated †, in answer to an

* Pharm. Journ. [3] xv. p. 322.   † Ibid. xiv. p. 1005.

enquiry from the Pharmaceutical Society, to the following effect:—
The finest limes in the West Indies are grown at Trinidad, on
trees quite unequalled for size and exuberance by those of any part
of the Western tropics. Their odour more resembles that of
lemons produced in Europe than of the limes produced there. The
young shoots and all tender parts of these Trinidad limes have
the odour of *Aloysia citriodora* (the lemon-scented verbena), the
stronger lime-odour being developed in the older parts. In the
treatment of the fruit for obtaining the oil, the more rapid the
process the more pronounced is the lemon odour in the result, if
perfectly fresh fruits are used. The strong flavour of limes, more
or less tinged with that of turpentine, seems to be a result of
treating stale or decomposed fruit-tissue. The plan adopted on
economic grounds at Dominica and elsewhere, of crushing the
limes as received from day to day, and then, on the attainment of
a large quantity of pulp, proceeding to distil, seems completely
preventive of a fine flavour in the resulting oil. The sp. gr. of the
Trinidad oil was found to be 0·8741 and the boiling-point 177°·7 C.
It is soluble in five parts of alcohol of sp. gr. 0·838, as is ordinary
oil of limes. Commercial oil of lemon is barely soluble in 15 parts
of the same menstruum.

The Shaddock is the *Citrus decumanum*, a distinct species, remarkable for the large size of all its parts. It is a native of China
and Japan, and owes its name to Captain Shaddock, who introduced
it into the West Indies. Its fruit is spheroidal and greenish
yellow, the pulp red or white, the juice sweet or acid, the rind is
thick, fungous, and bitter. Thunberg says the fruit in Japan grows
to the size of a child's head, and Dr. Sickler states its weight at 14 lbs.

The " Mandarine," or Maltese orange, is the *Citrus nobilis*, a
very small orange of flattened shape, with a thin rind which
separates spontaneously from the pulp; so that when quite ripe
the latter may be shaken about inside. The perfume of the
flowers, leaves, and rind is delicately soft, and the taste of the
pulp very sweet. In China, where this delicious variety has been
raised, the fruit is generally presented to the Mandarins, hence its
name. It is now successfully cultivated in Malta and the Azores,
and essential oils obtained from it are quoted by manufacturers.
In 1857 the oil was not obtainable on the London Market; but
its properties were carefully studied by de Luca, who obtained
specimens of the fruit, expressed the oil, and communicated the

results of his investigation to the French Academy of Sciences *.
An abstract of these results, from unquestionably pure oil, may
here be worth repeating.   He says :—" The essence of mandarine,
prepared by expression, has a light golden-yellow tint; it is very
limpid and extremely mobile; its odour is very sweet and different
from that of citron or orange; it boils and distils exactly at 178° C.
(352°·4 F.), almost without leaving any residue, in which, however,
is found a small quantity of yellow colouring-matter.   The distilled
product is colourless, endowed with the same odour and taste as
the crude essence.   Its density at 10° C. (50° F.) is 0·852; the
same density was found with the first portions distilled, with the
last, and with average samples.   The density of the same essence,
determined at a previous period and with another sample, was
0·8517 at the temperature of 12° C. (57°·6 F.).   The essence does
not appear to contain oxygen.   It is soluble in about ten times its
volume of alcohol.

The Bergamot orange is yielded by the *Citrus bergamia*, a
small tree whose leaves and flowers much resemble in appearance
those of the Bitter orange.

It is stated in a small work, " Le Parfumeur François, par le
Sieur Barbe, parfumeur, 1693," that oil of bergamot is extracted
from the fruit of a lemon which has been grafted on a bergamot
pear.   The name of the latter is derived from the Turkish *Beg-
ârmûdt*, the " prince of pears."   Volkamer, in his ' Hesperides
Norimbergenses,' 1713, further describes *limon bergamotta* as
" gloria limonum et fructus inter omnes nobilissimus," and mentions that the Italians prepare one of the finest essences from it.
The name is derived from this fact, and is in no way connected
with the town of Bergamo in Lombardy as some writers on perfumery assert.   It is not even cultivated in that district.

The Bergamot is cultivated at Reggio and the adjacent villages
on low-lying lands near the sea, being frequently grown amongst
orange and lemon trees.   The oil is extracted from fully-developed
but unripe fruits, they being more or less green.   They are
gathered in the months of November and December, the amount
of oil obtained being from 45 to 60 grammes from 3 kilogrammes
of fruit.   The colour of the oil is a pale yellowish green, due
to traces of chlorophyl, which is proved by the spectroscope.

---

* Comptes Rendus, 23 Nov., 1857.

Its boiling-point varies from 183° to 195°. This oil is systematically adulterated on a large scale, and is rarely sent into the market pure. The adulterants are oil of turpentine, petroleum, essential oil obtained from the leaves, an inferior oil distilled from the residue of the fruit which has passed through the écuelle process, oil of sweet orange, and oil of lemon.

The admixture of oil of orange and of oil of lemon at once lowers the sp. gr., and augments the optical activity, as will be seen by the following figures :—

	Average sp. gr. at 15° C.	Optical rotation in 100 mm. tube.
Pure oil of Bergamot, expressed.	0·881 to 0·888 ⎫	$+4°·2$ to $+19°·30$
„ „ distilled .	0·873 ⎭	
Pure oil of Sweet Orange, distilled	0·849 to 0·855	$+97°·4$ to $+97°·32$
Pure oil of Lemon, expressed .	0·857 to 0·863	$+40°·1$ to $+62°$
„ „ distilled . .	0·856	$+66°·2$

One part of pure oil of bergamot forms a clear solution at 20° C. with half a part of 90 per cent. (vol.) alcohol, and is not rendered turbid by the further addition of alcohol of the same strength ; whereas the oils of sweet orange and lemon do not form clear solutions under those conditions. The addition of turpentine would of course decrease the solubility in alcohol. Additions of alcohol or petroleum would lower the sp. gr.

Additions of fatty oils are recognized by the higher sp. gr. of the specimens, and by the residue which they leave when volatilized at 100° C. If about half a drachm of oil of bergamot is warmed on a watch-glass to 100° C., until the odour has completely disappeared, there is left behind a green homogeneous residue of ointment-like consistency, which in genuine oil amounts to about 6 per cent. In the presence of fatty oil the residue on volatilization is increased, and has a different consistency ; it presents the appearance of a green thick layer on the inner surface of the glass with a supernatant oily yellow liquid. Upon the data of these experiments, the sp. gr. of oil of bergamot should not be under 0·873, and the optical rotation (in 100 mm.) not over $+20$. The oil must be clearly soluble in $\frac{1}{2}$ part 90 % (vol.) alcohol, and the solution should not become turbid on the further addition of alcohol of equal strength. The residue on volatilization should be a green homogeneous mass, and not exceed six per cent. (except in very recent specimens) *.

According to recent researches of Schimmel and Co.†, the most important constituent of bergamot oil (to the extent of 40 per cent.) is the acetic ester of Linalool, a body which is also contained in oil of lavender (which see). Linalool, discovered by Semmler ‡, forms the principal constituent of Mexican Linaloe oil (which see).

The crystalline constituent of Bergamot oil, Bergapten, $C_{12}H_8O_4$, has recently been studied by Pomeranz §. Its melting-point is 188° C.

The odour of Bergamot has been noticed in the *chloride of sylvestrene*:—

Sylvestrene.—Stockholm tar and Archangel tar are obtained in the North of Europe by the dry distillation of fir-wood (*Pinus sylvestris* and *P. Ledebourii*), and these are the source of both Swedish and Russian turpentine. The former was found by Atterberg to contain both *australene* and *sylvestrene*, which are stated by Tilden ‖ to be also present in Russian turpentine, while Wallach has observed the occurrence of *Dipentene* in addition to these.

Sylvestrene, $C_{10}H_{16}$, boils at 173–175°, and smells like fresh fir-wood. The chloride is formed by passing hydrochloric acid into its ethereal solution, and yields the pure hydrocarbon on heating with aniline, or, better, with sodium acetate and glacial

---

* Schimmel and Co., Report, April 1891.

To determine the solubility of an oil in alcohol, 2 c. c. of it are placed in a stoppered bottle, and the alcohol is gradually added from a burette, the mixture being well shaken after each addition. The termination of the experiment is known by the liquid becoming clear. A suspected sample being now similarly treated, it will be found that if turpentine be present a larger proportion of alcohol will be required. It must be borne in mind that some other oils, such as those of juniper, savin, eucalyptus, and copaiba, are but slightly soluble in dilute alcohol, and that they may, when mixed with other oils, give rise to appearances similar to those produced by oil of turpentine. Moreover, in some cases, changes take place in oils which have been kept for a long time, and these must be considered in applying the alcohol test for turpentine. A test for alcohol in essential oils is the red coloration produced, if alcohol be present, when a drop of the oil is let fall on a crystal of "magenta." To some oils, such as clove, however, this test does not apply.

† Ibid. April 1892.   ‡ Ber. Deutsche chem. Ges. 1891, xxiv. p. 207.

§ Monatshefte für Chemie, 1891, p. 379.

‖ Journ. Chem. Soc. xxxiii. p. 80.

acetic acid. In the pure condition it boils at 175–178°, and smells like *oil of bergamot*. The addition of a drop of concentrated sulphuric acid or fuming nitric acid to its solution in glacial acetic acid or acetic anhydride produces a splendid deep-blue coloration. It is dextrorotatory, and combines with the hydracids to form compounds from which it can be separated unaltered *.

The fruit of *Havenia dulcis* is eaten in China and Japan; its taste is said to resemble that of the Bergamot pear.

In the Southern States of America the orange, the lemon, and the lime are extensively cultivated; in Florida they grow wild and are found in great abundance, but on many very large tracts of land they are cultivated. In Louisiana and Mississippi they are grown from seed. The seeds are planted early in spring, or in hot-beds in January; when one year old they are transplanted in a nursery; at the age of two and a half years they are budded with fully-matured buds from bearing trees of the sweet orange (the seedlings being the sour variety). This renders the tree more hardy, seedlings of the sweet orange having been found subject to a root-disease called "heel," which does not attack seedlings of the sour orange. At the age of four years the trees are transplanted into orchards. At the age of six, flowers first appear, and at ten years the trees are called full bearers. Orange-trees were introduced into the States in 1816.

The cultivation of the *Citrus* fruits in Louisiana is confined to the Sweet orange (*C. aurantium*), and is restricted to the lowest parishes of the delta and of the Gulf Coast east of Vermillion Bay. In the parish of Plaquemine, the chief site of the orange orchards, groves from 10 to 200 acres in extent are found, yielding large incomes. The quality of the Louisiana or Creole orange is of the highest order. The crop produced in the State is scarcely sufficient to supply the demands of the home market. The Lemon (*C. limonum*) is raised only in a few sheltered localities on the coast.

A writer in the 'American Journal of Pharmacy' states that the humidity of the atmosphere materially affects the flowers— when too wet the pollen heads are injured and the secretions are imperfect; extreme dryness has a similar effect on the pollen, but does not affect the secretion of oil. When the temperature is low but few flowers are fructified. The most favourable temperature

---

* Ann. Chem. Pharm. ccxxx. p. 240; ccxxxix. p. 24.

is about 68° to 78° F. Under 60° F. the flowers are blighted. An ordinary tree in America is said to yield from 2 to 10 lbs. of flowers, generally about 7 lbs.; they are often collected on canvas cloths spread under the trees. The most fragrant flowers are those which fall in the early morning. Orange-flowers produced in the extreme southern borders are believed to possess a stronger odour and more oil; the difference is accounted for in this manner:—" In the tropics the trees do not begin to bear very much until about twenty years old, while in this country they begin at about seven; the development is more rapid, the tree more vigorous, and it is reasonable to suppose a better development of odour in the flower." The writer in the Journal above referred to states that "the flowers are more fragrant, and the fruit more juicy, but not so sweet as in some other countries;" further, that " collected flowers placed in the direct sunlight lose all their odour in the course of two days, in diffused daylight they retain it for at least three days, and, in a dark humid atmosphere, the odour is quite distinct after one week. When bruised they lose their odour in half the time stated. The flowers hermetically sealed up in tin canisters are known to have retained their odour unimpaired for nine months."

All the sorts of Citrus may be propagated by seeds, cuttings, layers, and grafting. The object of raising plants from seed is either to obtain new varieties or stocks for grafting. Shaddock stocks are the strongest, and next to these the citron. At Genoa and Florence citrus trees are grown in a strong yellow clay, which is richly manured, a soil which is considered by the first Italian gardeners as the most suitable. For growing citrus trees in boxes, the French gardeners recommend a fresh loam containing a third of clay, a third of sand, and a third of vegetable matter, which has lain a long while in a heap. An equal bulk of half-rotten cow-dung to be then added and the whole allowed to remain till the next year, when it is to be twice turned over. The succeeding year it is to be mixed with one half its bulk of decomposed horse-dung, and turned over two or three times. The winter before using it is to be further enriched with a twelfth part of sheep-dung, a twentieth of pigeon-dung, and a twentieth of dried night-soil.

The compost recommended by Henderson, an English culti-vator, consists of one part of light brown mould from ground that has not been cropped or manured for many years, one part of peat-earth such as is used for growing heaths, one part of rotted hotbed dung, and one part of rotted leaves of trees, mixed well together.

Though orange-trees grow exceedingly well in large boxes in a conservatory, yet to produce the finest crop of fruit they should be planted in the ground against the back wall of a narrow house, and treated like peach-trees. In this way they may be propagated quickest from cuttings of strong young shoots or pieces of two-year-old wood from 10 inches to 2 feet in length, covering with a hand-glass and giving a gentle bottom heat. They will strike in seven weeks or two months. The citron is the most easily struck and the freest grower. Budding and grafting seedlings or cuttings grown in England under glass or in warm climates in the open air may be performed at any time when the sap is in motion. Trees raised from seed and grafted in England are found to bear the cold better than imported trees.

Gallesio, in his 'Traité du genre Citrus,' has given a synopsis of the forty principal sorts cultivated in Italy. In the 'Histoire Naturelle des Orangers,' by Risso, of Nice, and Poiteau, of Versailles, 169 sorts are described, and 105 of them figured ; they detail 42 sorts of sweet orange, 32 sorts of bitter and sour orange, 5 bergamots, 8 limes, 6 shaddocks, 46 lemons, 17 citrons, and 12 other sorts. A great number of hybrids produced by accidental cross-fertilization are known, and details of very curious hybrids produced experimentally are described by Gallesio in his 'Storia della Riproduzione Vegetale' (Pisa, 1816).

The essential oils of each variety of Citrus vary in perfume, and the products of the flower, leaf, and fruit of the same tree varies. The oil distilled from the fresh flowers, termed Neroli or *Oleum Neroli*, is much esteemed for its delicate perfume. Oil of Neroli being so valuable when pure, and so difficult to obtain pure, the following extracts, from so high an authority as Flückiger and Hanbury's ' Histoire des Drogues,' may be translated with advantage :—" The Neroli of *Citrus aurantium*, var. *Bigaradier*, is slightly brown in colour, bitter in taste, and neutral to litmus. The specific gravity of a pure sample obtained from Mr. Warrick, of Nice, was found to be 0·889 at 11° C. Mixed with alcohol it presents a brilliant violet fluorescence, quite distinct from the blue fluorescence of a solution of quinine. This phenomenon is very evident when a little spirit of wine is poured on the surface of the oil, and the liquid gently agitated so as to cause a slight undulatory movement. (Fluorescence was noticed by de Luca in rectified oil of Mandarine, either pure or dissolved in spirit, but he did not observe the phenomenon in the crude or unrectified oil, attributing its absence to the yellow colouring-matter *.) Neroli agitated with a saturated solution of sodium bisulphite assumes a very pure, intense, and permanent crimson tint. Examined in a tube of 100 millim., it deviated the polarized ray 6 degrees to the right. The greater part distilled at 185° C., and the portion distilling at 195° being still colourless, of the same odour as the original oil, and still manifesting in a marked manner the violet inflorescence. The portion remaining in the retort was then mixed with rather more than its volume of 90 %, alcohol, and the addition of a few drops of water caused no turbidity. A small quantity of crystallized Neroli camphor was seen floating on the surface of the liquid, which was obtained pure by redissolution in boiling alcohol. No camphor could be extracted from the rectified oil. Neroli camphor is neutral, inodorous, insipid, fusible at 55° C., and assumes a crystalline form on cooling from its solution in hot alcohol. The quantity of this product found in the sample was small, being only 1 decigramme from 60 grammes of oil. The proportion of camphor in Neroli diminishes with the age of the oil." (The observers were unable to discover any substance of this nature in the oils of Petit-grain, of orange-peel, or of bergamot peel.)

The great variations in the prices quoted for Neroli indicate a great diversity of quality. The commercial oil is rarely pure, being adulterated with oil distilled from the leaves, and with a less odorous Neroli distilled from the sweet "Portugal" orange. Common oil of Neroli often consists of $\frac{3}{8}$ oil of Petit-grain (from the leaves), $\frac{1}{8}$ oil of Bergamot rind, and $\frac{1}{2}$ genuine Neroli.

The usual time for beginning the collection of the orange-flower crop in the south of France is the last week in April, and the gathering lasts about a month or five weeks. The quantity gathered is at first rather small, but gradually increases, and after May 10 reaches its full proportion. One of the principal centres of this industry is Vallaurie, the name being apparently derived from *Vallum aurantii* or its Provençal equivalent. There are fifteen distilleries of orange-flower in this town. The crop is said to

* Comptes Rendus, Nov. 23, 1857.

average a million kilogrammes (about 1000 tons). The yield of Neroli varies with the season when the flowers are collected. Those gathered at the beginning barely produce half a gramme to the kilo, while near the end of May they afford one gramme or more. The buds are picked when on the point of opening, by women, boys, and girls, who make use of a tripod ladder to reach them. These villagers carry the flowers to an agent, who weighs them and spreads them out in a cool place, where they remain until 1 or 2 A.M.; then he puts them into sacks and delivers them at the factory before the sun has risen. They are then taken in hand at once. On exceptional days as many as 160 tons are so treated in the whole province. At the factory the flowers are spread out on the stone floor of the receiving-room, in a layer some 6 or 8 inches deep. The sepals are then separated by girls, and such of the petals as are destined for the production of orange-flower water and Neroli are put into a still through a large canvas shoot, and are covered with water, which is measured by the filling of reservoirs on the same floor. The man-hole of the still is then closed, and the contents are brought to boiling-point by the passage of superheated steam through the coils of a surrounding worm. The water and oil pass over, are condensed, and fall into a florentine receiver, the oil floating on the surface thereof and the water flowing through the bent tube from below. A piece of wood or cork is placed in the receiver to break the force of the stream flowing from the condenser. This gives time for the small globules of oil to cohere and prevent them being carried away by the downward current. The first portions of the water coming from the still are put into large tinned-copper vats holding about 500 gallons, and there stored, to be drawn off as occasion may require into glass carboys or tinned-copper bottles.

After the flowering-season, or about the end of June, the farmers prune their trees; these prunings are carted to the factory, where the leaves are separated and distilled, the product being "oil of petit-grain." During the autumn the ground round about the trees is well weeded, dug about, and manured, for the old practice of planting violets under the orange-trees is being abandoned. The orange-trees produce a second crop of flowers in autumn, sometimes of sufficient importance to allow of their being taken to the factories, but always in request for the bouquet market. Late in the year those blossoms which escaped collection in the spring have developed into fruits. These, when destined for the production of the oil, are picked while green.

The Nerolis of the various Citri differ in odour from each other, and do not represent the true perfume of the flowers from which they are severally extracted, by reason of partial decomposition, modification or oxidation caused by heat, contact with water and hot aqueous vapour during the process of distillation; a finer perfume than Neroli being obtained by the process of maceration in pure warm grease. The same observations apply to the perfume of the rose. This process and that of absorption or " enfleurage à froid " are described in a previous chapter.

The oil of " Petit-grain " above referred to was mentioned as far back as 1692 by Pomet in his ' Histoire des Drogues '; it was then distilled from the small unripe fruit about the size of a cherry, which fall from the tree shortly after the flowers; they are called "orangettes." The name "petit-grain," or small seed, indicates this origin of the term. At present the oil of petit-grain is also made on a large scale from the leaves and young shoots of both the Bitter and the Sweet orange, the former being much more odorous than the latter, and worth twice the price. The leaves are gathered in districts of the Mediterranean where there are large plantations of Citrons. Citrons are generally grafted on to orange-stocks (seedling orange-trees), and these stocks during the summer put forth shoots which are allowed to attain the length of a few feet; they are then pruned off, tied up in bundles, and sent to the distiller. The strongest shoots are frequently reserved to make walking-sticks. The leaves of the Bitter orange are at once distinguishable from those of the Sweet by the odour given off on crushing them between the fingers. The oil of petit-grain is much used in perfumery, especially in the manufacture of Eau-de-Cologne.

Both the Bitter and the Sweet orange are abundant in Jamaica; the oils are obtained by distillation, and, on an average, 580 oranges will weigh 180 lbs., and yield 12 ozs. of oil. The Government Chemist at Kingston kindly supplied me with samples of both oils, which, although very fine for distilled products, are not equal in fragrance to those obtained by the cold process in Europe.

" *Orange-Flower Water.*"—Referring to the fact that Neroli

does not represent the true odour of the orange-flower, and that the watery distillate which comes over with the oil is identical in perfume with the blossom, Soubiran is of opinion that Neroli is a modified isomer of the natural oil of the flower, and not so soluble in water as the unaltered portion of the latter which remains dissolved in it. In confirmation of this view, he states that by agitating the watery distillate with ether and then leaving the decanted ether to spontaneous evaporation, a small quantity of an oil possessing absolutely the same perfume as the flower remains, which is capable of being easily re-dissolved in water [*].

In Paris, orange-flower water " double " is made by distilling the flowers of the Bitter orange with water and drawing over double the weight of the flower put into the still; the " single " orange-flower water being simply the addition to the " double " of an equal volume of distilled water. The preparation known commercially as " orange-flower water quadruple," as made in the South of France, is a distillate equal in weight to the weight of the flower put into the still (Soubiran). This indicates that the finest or the largest part of the essence comes over first.

According to Xavier Landerer, orange-flower water is distilled in large quantities in the Island of Chios; it is known in Greece as " Anthoneron," and is sophisticated with " Mythoneron," which is a water distilled from the small and highly aromatic leaves. The Jews of Thessalonica add to this " Mythoneron" some drops of oil of neroli and some salt, and then bring it into the market as " Anthoneron." It is sold by retailers and travelling merchants in small straw-covered flasks resembling those used for salad oil, but their necks are twice as long.

Orange-flower pommade is prepared in the same way as rose pommade by " maceration," or, more properly speaking, digestion of the flowers in warm inodorous grease or oil. The exquisitely fragrant " orange-flower extract " is made by washing the pommade or oil with inodorous spirit, which absorbs the fragrant molecules. This extract is quite distinct in perfume from the solution of Neroli in rectified spirit.

Dried Orange-peel.—The finest quality of the dried Bitter orange-peel of commerce comes from Barbados and Curaçao, under the

---

[*] ' Traité de Pharmacie,' p. 654.

name of " Curaçao des Iles " and " Curaçao de Hollande." The
first, which is derived from unripe fruits, appears in the form of
thick, hard solid sections, of a strong, persistent pleasant odour
and bitter taste. The second description is derived from ripe
fruits from which the white pith has been removed in Holland,
and is in the form of extremely thin yellowish-red rinds, externally
wrinkled and very aromatic. The dried rind is also exported
from Italy and Provence, made either from the small young green
fruit or from the more mature yellow fruit, but from both of
which the inner white pith has not been removed. All of these
descriptions of rinds are used for the manufacture of Curaçao
liqueur, alcoholic tinctures, and for flavouring syrups.

The " Curaçao " liqueur is made in Holland, mainly from the
orange-peel imported from the island of Curaçao, and it is named
after the island as a sort of guarantee of its quality, for un-
deniably the finest orange-peel in the world comes from there. It
is said that this island produces the bitterest of bitter oranges
with the oiliest of rinds, also the sweetest of sweet oranges with
the finest flavoured fruit, but the soil cannot be persuaded or
forced into growing sour fruit. Lemons and limes planted there
turn sweet and die. Immense quantities of bitter orange-peel are
shipped from there every year, nearly all to Holland. In Amster-
dam there is a regular orange-peel market, where saucers full of
peels are set out as samples on long tables, and testers go among
them selecting for purchase. Such experience have these men
that they can tell by simply breaking and smelling a bit of peel
what part of the tropical world it comes from, and that from
Curaçao always commands a higher price than any other.

The bark and all the tender parts of the *Amyris acuminata*, on
being bruised or wounded, discharge a pale whey-coloured fluid,
which possesses a fragrance something like that of the orange-
leaf*.

Flowers of somewhat similar odour to orange-blossom are produced by the *Philadelphus coronarius*, known as the Syringa or
Mock orange, a shrub which is somewhat common in England,
and quite distinct from the genus *Citrus*. The perfume of the
Gardenia citriodora and of the *Cytisus laburnum* is also considered
of this type.

---

* Roxb. Fl. Ind. ii. p. 246.

Orange-blossoms (called by the Chinese "Chang-hwa") are used in the tea-factories of Canton for scenting the tea known as Orange Pekoe; they are gathered when fully expanded. The petals, when separated from the stamens, are mixed with the tea, which is apparently perfectly dry and finished, in the proportion of about 40 lbs. of flowers to 100 lbs. of tea. The dry tea and undried flowers are allowed to lie mixed together for a space of twenty-four hours; the flowers are then sifted and winnowed out, but sometimes a few stray leaves are left, and may be detected in the tea even after it arrives in England. The moisture which the tea has acquired from the flowers is expelled by placing the tea over slow charcoal fires, in baskets and sieves prepared for the purpose. The scent communicated by the flowers is very slight for some time, but, like the scent peculiar to the tea-leaf itself, comes out after being packed for a week or two. The peculiar volatile oil to which the fragrance of unscented tea is due does not appear to exist in the leaf in its green state, but to be formed by a slight fermentation which takes place in the leaf during the process of curing. Sometimes the scenting process is repeated when the bouquet is not considered sufficiently strong; indeed, it is sometimes scented twice with orange-flowers and once with *Jasminum Sambac* ("Mo-le-hwa"). Other flowers are similarly used by the Chinese, such as the *Aglaia odorata*, called "Lan-hwa" or "Yu-chu-lan," the proportion of the flowers to the tea being equal. The flowers of *Jasminum paniculatum*, "Sieu-hing," are frequently mixed with those of *Jasminum Sambac*, in the proportion of 10 lbs. of the former to 30 lbs. of the latter, and the 40 lbs. thus produced are sufficient for 100 lbs. of tea, but when *J. Sambac* is used alone, 50 lbs. of flowers are required for 100 lbs. of tea. The *Olea fragrans*, "Qui-hwa," is used chiefly in the northern districts as a scent for a rare and expensive kind of Hyson-pekoe, a tea which forms a most delicious beverage, the proportion of flower used being very large; but the tea scented with this flower will only keep well for one year, at the end of two years it has either become scentless or has a peculiar oily odour which is disagreeable. Teas scented with orange-blossom and with those of the *J. Sambac* will keep well for two or three years, and with the *J. paniculatum* for three or four years. The *Aglaia* retains the scent longer than any, and will preserve well for five or six years. The other flowers used by the Chinese for this purpose are the *Gardenia florida*, "Pak-sema-hwa," and (for their own use) the petals of a rose, "Tsing moi-qui-hwa." It has been frequently stated that the flowers of *Chloranthus inconspicuus* are used for scenting tea, but this has long been disproved by Fortune, a botanist who resided in China and studied the genus *Chloranthus*, one species of which, *C. Fortunei*, was specifically named after him.

Some flowers, such as those of the *Aglaia*, after being sifted out from the tea, are dried and used in the manufacture of the fragrant "Jos-stick," which is much used as an incense in the religious ceremonies of the country.

The fruit of the *Evodia fraxinifolia* yields by distillation about 4 per cent. of a thin fluid essential oil of a very pale yellow colour, and exceedingly agreeable and intense odour, similar to bergamot; so intense is the perfume that it is able to overcome the smell of iodoform, even when used in the very small proportion of two drops to the ounce. The sp. gr. of the oil is very low, not exceeding 0·840; it is soluble in alcohol and ether [*]. A description of *Evodia fraxinifolia* was first published under the name of *Rhus fraxinifolium* in Don's 'Prodromus Floræ Nepalensis,' 1825. The plant is described as a large tree, a native of Nepal. Sir William Hooker, however, subsequently pointed out that the floral characters did not agree with those of the genus *Rhus*. In the 'Icones Plantarum' 1848, plate 170, it is referred to Blume's *Philagonia*, and a good figure accompanies the letterpress. The plant is placed under the natural order *Terebinthaceæ*, and a reference is given to *Tetradium* (?) *fraxinifolium*, of Wallich, in Herb. Hook. 1821. In the recent work, Hooker and Bentham's 'Genera Plantarum,' this species is included in the genus *Evodia*, under the natural order *Rutaceæ*, and it is identified with the *Tetradium trichotomum* described in Loureiro's 'Flora Cochinchinensis,' p. 91, which is there mentioned as having trichotomous racemes of whitish flowers; the tree is of medium size, and inhabits the hills of Cochinchina [†].

---

[*] Helbing, in a paper read before the Pharmaceutical Conference, August 1887.

[†] Christy's 'New Commercial Plants,' no. 10.

CITRONELLA.

Oil of Citronella is derived from the *Andropogon Nardus* of Linnæus, which is figured in Bentley and Trimen's 'Medicinal Plants,' tab. 297.   Synonyms :—

    A. flexuosus and *A. coloratus,* Nees.

    A. Martini, Thwaites *.

    Cymbopogon Nardus, Linn. †

The plant is known under many common names, such as *Ganjní-ka-ghás,* Hind. ; *Kamá-Khér,* Beng., &c.

In Rimmel's Report of the Products exhibited at the 1862 Exhibition, he wrongly assigns Citronella to *A. citratus,* and he is wrong in his names of three out of four of the grasses.

This grass is very common in the plains of the Punjáb and North-west Provinces ; it is extensively cultivated in Ceylon and at Singapore for the manufacture of the oil from its leaves, and it is abundant at Travancore.   As cultivated in Ceylon on Winter's Estate near Galle, it often attains a height of 6 or 8 feet ; oil from this estate is considered as fine or finer than that from Singapore.   In Ceylon the Citronella is raised from seed, and planted like " Guinea-grass " ; it yields two or three crops a year ‡.

The principal plantations are in the Malara district, where about 16,000 acres are planted with Citronella grass, which flourishes upon the poor ground.   The total area devoted to its cultivation in Ceylon is estimated at from 25,000 to 30,000 acres ; principally in the southern provinces.   The 'Ceylon Mercantile Planting Directory' states the total number of distillatory apparatus to be 467.   The distillation seems generally to be effected by the farmers.   The total export of oil in 1890 was 14,559,075 ounces ; that of 1891 is estimated at 15,000,000 ounces.

Citronella grass is distinguished from other species by its peculiar reddish tint, short spikes, and narrow leaves.   The pure oil is thin, almost colourless or of a pale greenish-yellow tint, and strongly aromatic.   It is to this oil that the well-known odour of honey-soap is due.   Very interesting details of recent researches in the

---

    * Encycl. Ceylon Plants, p. 361.

    † Pharmacopœia of India.

    † 'Tropical Agriculturalist,' iii. p. 58.

chemistry of Citronella are described by Dodge *; mention being also made of Professor Flückiger's discovery of the peculiar property possessed by this oil (and that of *A. citratus*) of solidifying, with evolution of heat, when shaken for ten minutes with a saturated solution of sodium bisulphite.

The investigations of Kremers † show that citronelle-oil consists of an aldehyde, $C_7H_{14}O$, a terpene, $C_{10}H_{16}$, an isomer of borneol named *citronellol*, and acetic and valerianic acids. These two acids are said to be formed through the oxidation of the aldehyde, and to exist originally in combination with citronellol as a compound ether. The composition of the aldehyde, as ascertained by Kremers, does not agree with the results arrived at by Dodge ‡, who assigned to it the formula $C_{10}H_{18}O$, and found its boiling-point to be between 217° and 222°. This aldehyde has also been found in oil of *Eucalyptus maculata*, var. *citriodora*.

It is well known to the trade that Citronella is largely adulterated with kerosene, immense quantities of which are imported into Ceylon in great excess of the requirements for illuminating purposes.

The test recommended to discover this is that one part of the oil should give a clear solution with ten parts of 80-per-cent. alcohol when shaken vigorously. In the presence of 10 per cent. of petroleum or kerosene the mixture becomes milky. An addition of a fixed oil is still more perceptible, as little as 1 or 2 per cent. being recognizable, but the mixture becomes turbid rather than milky. The sp. gr. of the oil should not be below 0·895 at 15° C. (Schimmel).

LEMON-GRASS.

Oil of Lemon-grass.—This is derived from the *Andropogon citratus* of DeCandolle; syn. *A. Schœnanthus*, Wallich §. The vernacular names Gandha-bená (Bengal) and Malutrinukung-bhustrinung (Sanskrit) are by Roxburgh given to a plant he describes as *A. Schœnanthus*, Linn.; this description may be referable to *A. citratus*, De C., but it seems to agree equally well with

---

* 'American Journal of Chemistry,' xi. p. 456.

† Proc. Am. Pharm. Assoc. 1887, and Am. Chem. Journ. xiv. p. 204 (Mar. 1892).

‡ Am. Chem. Journ. xi. p. 458.

§ Plant. As. Rar. iii. tab. 280.

A. Laniger of Desfontaines. It is a large coarse, glaucous grass found under cultivation in various islands of the Eastern Archipelago, in the Island of Jamaica, and in gardens over an extensive tract of country in India. It very rarely flowers, but Dr. Dymock of Bombay states that he has seen it in flower more than once. It is largely cultivated in Ceylon and Singapore for the odoriferous oil distilled from the leaves, which is called Lemon-grass, Verbena-oil, or Indian Melissa-oil. The oil is employed in Europe as an ingredient in perfumes, very considerable quantities being used in the manufacture of Eau-de-Cologne. It is also used for adulterating the so-called " true verbena-oil " from the *Lippia citriodora*, a plant cultivated in Spain, also called *Aloysia citriodora*, but it certainly is not a verbena at all. The Ceylon oil of Lemon-grass is manufactured entirely by the natives. Oil of Lemon-grass is said to be called "Minjak sereh" in Java, but that word may be applied to the oil of *Tetranthera citrata*, a Javanese plant of similar odour. This "verbena" odour is also developed in *Eucalyptus Staigeriana*, *Eucalyptus maculata*, var. *citriodora*, and *Backhousia citriodora*, Australian plants from which oils are distilled. It is also noticeable in the leaves and green twigs of the Trinidad lime (*Citrus limetta*). Oil of *Eucalyptus Staigeriana* has a sp. gr. of 0·880, and boils from 170° to 230°. Oil of *Backhousia citriodora* has a sp. gr. of 0·900, and boils from 223° to 233°C. Both these oils are distinguished by an intense odour of lemon and verbena. The oil of the former contains a considerable quantity of a terpene, whilst that of the latter appears to consist principally of an aldehyde ($C_{10}H_{16}O$?). There is probably a great future for this oil (Schimmel). *Backhousia citriodora* is a shrubby Myrtaceous tree of 20 to 30 feet in height, very common in Queensland.

CHAPTER IV.

JASMINE, JONQUIL, AND HYACINTH.

JASMINE.

ABOUT 100 species of Jasmine are known, mostly natives of India, Arabia, and the tropical regions of the Old World, there being only one or two South-American species. Linnæus obtained a fancied etymology from ια, "a violet," and ὀσμή, "smell," but the odour of its flowers bears no resemblance to that of the violet; it is, in fact, so peculiar as to be incomparable, and is probably almost the only floral perfume which cannot be imitated by art, i. e. by carefully blending other perfumes. The species most commonly known in this country is *Jasminum officinale*, L., the Common Jasmine. Its native habitat is not known, although it has been found wild in the South of Europe and several parts of India ; neither is the date of its introduction on record, but it has long been inured to our climate so as to thrive and flower well. To produce a good crop, it should be pruned in the autumn, as the flowers only form on the young shoots. It may easily be propagated by cuttings or layers. There are golden- and silver-edged leaf varieties, and a double-flowered variety. The flowering-time is from June to October. The perfume is far less powerful than when grown in a hot climate; this, and nearly all varieties of Jasmine, love the sun and flourish in the hottest parts of the earth.

The *Jasminum Sambac*, Aiton, is a native of Arabia (where it is known as Ysmyn Zambak), but is found wild in many parts of India, birds eating the berries and dropping some of the seeds, which germinate. The seeds are only produced in hot countries. This plant is a climber, with white flowers, which, on account of their exquisite fragrance, are highly esteemed in the East. The perfume, although powerful, is said to refresh instead of oppress the head, as most strong perfumes do.

There are in all four varieties of this species—the single, double,
great double, and *trifoliatum*. The common double still retains
its twining habit, but the branches of the great double variety are
erect or spreading. Flowers of one of the double varieties, known
in India as " Moogree," are used as votive offerings in religious
ceremonies. The *J. Sambac*, var. *trifoliatum*, which in India is
called the Kudda-Mulla, differs from the other varieties of Sambac
in many points; its flowers are almost constantly solitary, the
calyx is divided into a greater number of segments, and the leaves,
instead of being regularly opposite, almost constantly grow three
together at the extremities of the flowering branches, and in other
parts indifferently—singly, opposite, or ternate. The perfume of
the flowers is very powerful; this is called the " Tuscan " Jasmine,
as it was first imported from India by the Grand Duke of Tuscany
about the year 1691. The bloom of this plant is much admired
by females in India, who, in the evening of the day, string it into
chaplets and necklaces. The flowers of both single and double
varieties bloom throughout the greater part of the year. The use
of the flowers for perfuming tea in China is described in the
chapter on Citrine odours.

J. odoratissimum is a native of Madeira; its flowers are yellow,
and have the advantage of retaining when dry their natural
perfume, which is suggestive of a mixture of jasmine, jonquil,
and orange-blossom. *J. azonicum* is also a native of the island
of Madeira, and has long been cultivated in greenhouses in
England, where, under favourable circumstances, it will continue
to produce its fragrant white flowers nearly through the whole
year.

The use of *J. paniculatum* for scenting tea is described above.
It is a native of China, with white flowers.

The *J. hirsutum*, a native of China and India, is a very beautiful
shrub. The large white flowers are very fragrant. The leaves
and stem are hairy, as its name indicates; the degree of pubescence
varies very much according to the age, the leaves especially being
much more hairy while young than in adult plants. It is a native
of China and Bengal. The dark green foliage, which is very abundant, covers whatever it grows against nearly as closely as ivy,
and forms a remarkable contrast to the snow-white blossom. The
flowers are sometimes nearly thirty in a bunch, showing in about
August and continuing to be produced for several months. The

flower is exceedingly fragrant, and does not turn purple in decay as does the Arabian Jasmine, nor is it so fugacious.

The *Jasminum revolutum* was imported from China in 1814; its flowers are yellow, prolific, and very fragrant.

J. grandiflorum is very like *J. officinale,* but the branches are shorter and stouter, the flowers very much larger and reddish underneath; they form at the extremities of the branches, and are abundant during summer and part of autumn; they are persistent, and sometimes, after drying on the plant, they will re-open. This is the Spanish or Catalonian Jasmine; it grows wild on the island of Tobago. By grafting it on to a two-year old plant of *J. officinale* an erect bush about 3 or 4 feet high can be cultivated, and so pruned as to require no supports. As generally grown in the South of France, the plant is reared from cuttings of the *J. officinale,* which are put into the earth in rows and trenched. Level ground is chosen; if hill-side only is available, this is formed into a series of terraces. When strong enough, the young stem is grafted with shoots of the *J. grandiflorum.* The first year it is allowed to run wild. The second year the long slender branches are trained along light poles supported horizontally and running the whole length of the rows, the branches being twined and interlaced between them. At the approach of winter the plants are banked up with earth to half their height. The exposed parts then die off. When the last frost of winter is passed the earth is removed, and what remains of the plant is trimmed up for the coming season. It forms rapid growth, and when necessary water is supplied to the roots by means of the trenches above mentioned. The blossoms, which are the size of a shilling and intensely fragrant, are produced from July to the middle of October, but those of August and September yield the greatest amount of odour. The flowers are gathered as soon as possible after they open; this occurs in the evening, and up to about August 15, early enough for the blossoms to be gathered the same day. They are delivered at the factories at once, where they are immediately put on to the glass "châssis," to be treated by the cold process of "enfleurage;" the work on them continuing very often till long after midnight. To obtain a good result, fifty successive enflowerings of the pommade are necessary. Later on in the year they are gathered in the early morning, directly the dew is off. The flower is gathered without any green part, as the corollas are monopetalous and the tube is but very lightly joined to the placenta. The picking is continued for perfumery purposes up to October 15; after that date the flowers are lacking in fragrance. It is said that an acre of land will yield about 500 lbs. of blossom during the season.

An essential oil is distilled from jasmine in Tunis and Algeria. Exhibition specimens of oils of Jasmine have been sent from Calcutta to Lucknow, but they were probably fixed oils, *i. e.* made by digesting the flowers in an oil expressed from seeds or nuts; the oils perfumed with *J. hirsutum* and *J. Sambac* were labelled "Motia-ka-utter" and "Bella-ka-utter;" that from *J. grandiflorum* was called "Chamelé-ka-utter;" all these were contaminated with the odour of santal.

Syrup of jasmine is made by placing in a jar alternate layers of the flowers and sugar, covering the whole with wet cloths and standing it in a cool place. The perfume is absorbed by the sugar, which is converted into a very palatable syrup.

The true perfume of jasmine, as represented by any of its varieties, is unique—not exhaled by any other flower or exactly reproducible by any combination of natural products or chemical compounds. There are a few flowers whose perfume bears a remote resemblance to it, also one chemically-prepared compound, hereafter described. An essential oil with a fragrance which has been compared to it was obtained by Gleim [*] from the fresh berries of the *Benzoin odoriferum*, Nees, syn. *Laurus Benzoin*, a deciduous shrub inhabiting damp shady woods in North America. It is a bush of 8 or 10 feet high, with oblong or elliptic wedge-shaped leaves, and bears clusters of small yellow flowers on naked umbels, appearing before the leaves. By distilling 8 troy ozs. of the fresh berries with water, 4 fluid drachms of a colourless volatile oil were obtained, having a sp. gr. of 0·87, of very fragrant odour resembling somewhat that of jasmine.

The compound known in chemistry as "secondary styrolyl acetate" is said to have a pleasant odour resembling that of jasmine; the method of its formation is as follows:—

Ethylbenzene, $C_6H_5 . C_2H_5$, is formed when ethylene is passed into a heated mixture of benzene and aluminium chloride [†], and in small quantities when benzene is heated to 100° C. with ethyl ether

---

[*] Amer. Journ. Pharm. [4] v. p. 246.
[†] Bull. Soc. Chim. xxxi. p. 540.

and zinc chloride *. It is also obtained by heating toluene with aluminium chloride in an open vessel †. Ethylbenzene is a liquid possessing a smell like that of toluene; it boils at $136°·5$ C., and has a sp. gr. of $0·8664$ at $22°·5$. It is oxidized to benzoic acid by dilute nitric acid or chromic acid.

Secondary styrolyl bromide, $C_6H_5 . CHBr . CH_3$, is obtained by the action of bromine on boiling ethylbenzene ‡. By subjecting this bromide to the action of silver acetate and glacial acetic acid, *secondary styrolylacetate* is formed, $C_6H_5 . CH(O . CO . CH_3)CH_3$. This liquid has a pleasant odour *resembling that of jasmine*; it boils at $217°–220°$, at which temperature it is partially resolved into styrolene and acetic acid.

JONQUIL.

Narcissus Jonquilla, L., a bulbous plant probably of oriental origin. The name of the species in Italian (*Giunchiglia*) is derived from the nearly cylindrical leaves, grooved on the upper side, recalling those of a species of *Juncus*.

At the plantations around Grasse the bulbs of this plant are set out in rows. The blooms, which are of a fine bright yellow, appear about the end of March, four or five on each stem. Each flower as it blooms is picked off at the calyx. The harvesting period is of very short duration, and it very often happens that it takes two seasons for the manufacturer to finish off his pommade of extra strength. The crop is also very uncertain, being abundant one year and scanty the next. The exquisite perfume is extracted by the processes of maceration and enfleurage—chiefly the latter. It seems that a comparatively small acreage is under cultivation, and that bulbous plants do not receive much attention in France, owing no doubt to the great care required for their successful cultivation; otherwise there are many bulbs producing flowers of exquisite fragrance which might be advantageously grown. In Holland, where flowers are more rare than in the Riviera, and where flori-culturists are more patient, such plants receive their due care, but not for the purpose of extracting their perfume.

* Bull. Soc. Chim. xxxii. p. 618.
† Compt. Rend. ci. p. 1218.
‡ Bull. Soc. Chim. x. p. 343.

JACINTHE OR HYACINTH.

This plant does not appear to be commercially grown for perfumery purposes, the extracts sold under its name being compounded from the extracts of other flowers or prepared artificially by the processes next described.

The delicate perfume of hyacinths is ethereal, suggestive of organic ethers. There are many ethers and synthetically prepared compounds resembling the odours of flowers, but some of them are also suggestive of "chemicals," the products being crudely manufactured or insufficiently purified from minute traces of chemicals employed in their fabrication. All synthetical products, especially those in which chlorine is employed in the process, should therefore be thoroughly purified from all remnants of adherent reagents employed.

It has been known to chemists for many years that an oil obtained by the oxidation of turpentine resembles the perfume of hyacinths, but it is only recently that this oil has been produced commercially for use in perfumery.

For the production of this oil it is necessary first to prepare *Terpine-hydrate*, $C_{10}H_{20}O_2 + H_2O$, known in modern chemical language as *Dipentenylene glycol*. Eight volumes of oil of turpentine are mixed with two volumes of nitric acid, sp. gr. $1\cdot25-1\cdot30$, and one to six volumes of alcohol. The mixture is shaken frequently during the first few days and then left to itself in shallow vessels for several weeks. Brown crystals are thereby formed, which must be pressed and then re-crystallized from boiling water with addition of animal charcoal.

Another process recommended by Hempel [*] says:—"Eight parts of oil of turpentine are mixed with two parts of alcohol and two parts of nitric acid, sp. gr. $1\cdot25-1\cdot30$, in flat basins. After a few days the mother-liquor is poured off from the crystals which have already separated, and is neutralized with an alkali, after which treatment another crop of crystals separates out [†]. The preparation only succeeds at the cool seasons of the year, as in summer a resinous mass is usually obtained [‡].

---

[*] Ann. Chem. Pharm. xxx. p. 71.
[†] Wallach, Ann. Chem. Pharm. ccxxvii. p. 284.
[‡] Ibid. ccxxx. p. 248.

According to Tilden, one volume of nitric acid, sp. gr. 1·4, is mixed with one volume of strong alcohol * and two and a half vols. of pure rectified oil of turpentine (French or American), is allowed to stand for two days until the smell of turpentine has disappeared, and then poured into flat dishes, small quantities of alcohol being gradually added to it at intervals of two days. About one third of the oil of turpentine is thus converted into the terpine hydrate, and a still larger yield may be obtained by continuing the operation.

Terpine hydrate crystallizes in large, transparent, monosymmetric prisms which dissolve in 200 parts of cold and 22 parts of boiling water, and are still more readily soluble in alcohol.

Wiggers, by the action of hydriodic acid on terpine hydrate (Dipentenylene glycol) obtained the compound $2C_{10}H_{16} + H_2O$, which was investigated by List and named *terpinol* †. It is also formed when terpine hydrate is boiled or distilled with very dilute hydrochloric or sulphuric acid, potassium sulphate, &c., also when hydrochlorate of terebenthene is boiled with water, alcohol, or alcoholic potash. It is a colourless, strongly refractive oil, which boils at 168°, slightly soluble in water, and optically inactive; its sp. gr. is 0·852. It has a pleasant *odour of hyacinths,* especially when diluted. By boiling, it suffers partial decomposition, in such a manner that the first part of the distillate contains less oxygen than the latter portion.

Tilden found that the above reaction yields a mixture of a terpene with a compound $C_{10}H_{18}O$, for which he retained the name terpinol ‡. It is shown, by the researches of Professor Wallach §, that terpinol is not a simple body, but a mixture in variable proportions of Terpinol, Terpinene, Terpinolene, and Dipentene; or, rather, that when terpine hydrate is boiled with dilute sulphuric acid, or phosphoric acid, the product varies according to the conditions of the experiment, but the principal constituent is *Terpineol,* which is an alcohol (Dipentenyl alcohol) of the composition $C_{10}H_{17}OH$, this name being substituted for " terpinol " in order to

---

* It is stated in Jahresb. Chem. 1878, p. 638, that Tilden used methyl alcohol, while he actually employed " methylated spirit."

† Ann. Chem. Pharm. lxvii. p. 367.

‡ Journ. Chem. Soc. xxxiii. p. 247; xxxv. p. 287.

§ Annalen der Chimie, ccxxx. p. 251.

H

correspond with "borneol" and "cineol." It is formed by simple elimination of water from the terpine hydrate.

In order to prepare *terpineol*, 25 grms. of terpine hydrate are boiled with 50 cubic centimetres of aqueous phosphoric acid. It is a colourless, very thick liquid, boiling between 215°–218°, and is optically inactive. Its sp. gr. at 15° C. is 0·940, and at 20° 0·935. Bouchardat and Voiry * produced it by heating terpine hydrate with very dilute sulphuric acid. It was also probably obtained by Deville as a by-product in the preparation of terpine hydrate †; it also appears to occur in oil of cardamom from Ceylon (*Elletaria major*).

The high boiling-point of Terpineol (Dipentenyl alcohol) speaks for its unusual resistance to heat, and permits, for perfumery purposes, of its being used in the manufacture of toilet soaps by the warm process. It is readily soluble in vaselin.

It is further adapted to this use by reason of the property it possesses of not being attacked or decomposed by alkalies, such as soda or potash. Observations extending over months have shown that the odour remains unaffected even in soaps made strongly alkaline. These experiments regarding its adaptability to soap were made by Schimmel and Co., who recommend, for the production of an extraordinarily fine Lilac perfume, 10 or 12 ozs. of Turpineol to 100 lbs. of soap, with the addition of Heliotropin (Piperonal), ylang-ylang oil, geranium oil, and East-Indian santal-wood oil. "The perfume of this soap very much improves after storing some time."

The odour of hyacinths is also produced artificially as follows :— The compound *α-Phenylchlorethylene* or *Phenylvinylchloride* was first obtained by Stenhouse in an impure condition by the distillation of cinnamic acid with bleaching-powder solution ‡. It is also formed by the action of sodium-carbonate solution on Phenyldichloropropionic acid, which is obtained by passing chlorine into a solution of cinnamic acid in carbon disulphide; it crystallizes in lustrous plates which melt at 162°–164°, and are insoluble in water, but gradually decompose into carbon dioxide and *α-phenyl-*

---

* Compt. Rend. civ. p. 996.
† Ann. Chem. Pharm. lxxi. p. 351.
‡ Ibid. lv. p. 3 and lvii. p. 79.

chlorethylene on boiling with water or on standing in the cold with sodium carbonate solution *.

Also : *α-Phenylbromethylene* or *phenylvinylbromide* is formed when phenyldibromopropionic acid is boiled with water. In order to prepare this acid, cinnamic acid is dissolved in carbon disulphide and a solution of bromine in carbon disulphide gradually added, the acid being thus precipitated. It forms small plates which readily dissolve in ether and melt at 195°. On boiling with water, phenylbromolactic acid, cinnamic acid, α-phenylbromethylene and carbon dioxide are formed. This decomposition is brought about by cold sodium-carbonate solution, but is delayed by an excess of this reagent. Phenylbromethylene is an oily liquid having a pleasant odour *resembling that of hyacinths,* as does also the phenylchlorethylene above described. The former boils at 219°–221° (a small quantity of hydrobromic acid being evolved) ; the latter compound boils at 199° †. The odour of hyacinths is also noticed intensely in a compound called *cinnyl alcohol,* which is prepared from styracin (see Storax). Cinnyl alcohol, $C_9H_{10}O$, was formerly called Styrone, Hydrate of cinnyl, Cinnamic alcohol, Styracone, and Storax alcohol.

As indicated by Toel ‡, cinnyl alcohol is obtained by cautiously distilling styracin with a strong solution of caustic potash or soda. A milky liquid then passes over, from which, when saturated with common salt, a creamy substance separates, gradually collecting on the surface in an oily layer and solidifying.

Wolff § dissolves styracin in boiling alcoholic potash, mixes water with the liquid, filters from potassium cinnamate, and separates the precipitated cinnamic alcohol from undecomposed styracin by distillation. Pure cinnyl alcohol forms beautiful soft, silky needles, having a sweet taste and an odour of hyacinths. It melts at 33° C., and volatilizes without alteration at 250° C. It is moderately soluble in water, very soluble in alcohol, in ether, in styrol, and in oils both fixed and volatile.

It has recently been indicated that the chief constituent of the non-aldehydes in cassia-oil is the acetic ether of cinnamyl. It can be separated by submitting the non-aldehyde portion of cassia-oil

* Ber. Deutsch. chem. Ges. xiv. p. 1867.
† Ann. Chem. Pharm. cxcv. p. 140, and ccvi. p. 33.
‡ Ibid. lxx. p. 1.
§ Ibid. lxxxv. p. 299.

to repeated fractional distillation *in vacuo*. The fraction boiling between 135° and 145° (at 11 mm. atmospheric pressure) has been found to consist entirely of this ether. The cinnamic alcohol obtained by saponification crystallizes from ether in white solid rystals.

When rectified oil of cajeput, the fraction boiling between 174° and 178° C., which forms three-fourths of the crude oil and is cajuputene dihydrate, $C_{10}H_{16} + 2H_2O$, is cohobated with phosphoric anhydride for half an hour and distilled, there passes over between 160° and 165° the hydrocarbon Cajuputene, $C_{10}H_{16}$, which is a colourless liquid possessing the odour of hyacinths; it is insoluble in alcohol but soluble in ether and in oil of turpentine. Its sp. gr. at 15° C. is 0·850. It is permanent in the air.

The perfume of hyacinth is also remarkable in Benzyl alcohol (Hydrate of Benzyl), C_7H_8O. Benzyl alcohol is formed by the action of sodium amalgam and water on benzaldehyde *, and on benzoic acid †. Benzoyl chloride is also reduced to benzyl alcohol by the action of sodium amalgam and hydrochloric acid ‡; it is also yielded in large quantity by adding sodium amalgam to an ethereal solution of benzamide which contains water and has been rendered faintly acid with hydrochloric acid §. It is most easily obtainable from benzyl chloride (which can readily be prepared from toluene) by boiling it for two hours with water and lead hydroxide ‖, or by simply boiling it for two days with 25 to 30 parts of water ¶. Mennier states ** it may be advantageously prepared by boiling equal parts of benzyl chloride and potassium carbonate with 10 parts of water for several hours. If 10 parts of benzaldehyde be shaken up with a solution of 9 parts of caustic potash in 6 parts of water until a permanent emulsion is obtained, and sufficient water to form a clear solution be then added to the semi-solid mass of crystals formed on standing by the separation of potassium benzoate, the benzyl alcohol can readily be extracted from the liquid with ether. The ether is then distilled off and the residue purified by rectification without being dried; 92 per cent. of the theoretical yield can thus be obtained. As only one half

* Ann. Chem. Pharm. cxxiv. p. 324.   † Ibid. cxxxii. p. 75.
‡ Ibid. cxxxvii. p. 252.
§ Ber. Deutsch. chem. Ges. vii. p. 1462.
‖ Ann. Chem. Pharm. cxliii. p. 80.   ¶ Ibid. cxcvi. p. 353.
** Bull. Soc. Chim. xxxviii. p. 159.

of the benzaldehyde is converted into the alcohol, benzyl chloride is a more economical source; it is, however, more difficult to obtain in a state of purity than benzaldehyde, and therefore does not yield a pure product so readily.

Benzyl alcohol can be prepared from Peru balsam as follows :— The balsam is agitated with 2 volumes of caustic potash of sp. gr. 1·2, the emulsion exhausted with ether, the extract separated and evaporated, and the residual oil heated with 4 volumes of caustic potash of sp. gr. 1·3 until a homogeneous liquid is obtained. The pulpy mass of crystals formed on cooling is pressed in linen, and the liquid diluted with water and distilled until the distillate ceases to appear milky. The alcohol is then separated from the aqueous distillate, and the portion which remains dissolved in the latter extracted by ether *.

Benzyl alcohol is a liquid which boils at 206°, and has a sp. gr. of 1·063 at 0°. It is slightly soluble in water, 100 parts of water at 17° dissolving 4 parts.

Benzyl alcohol, C_7H_8O, is also obtained when a mixture of pure Benzaldehyde, C_7H_6O, with its own volume of absolute alcohol is mixed with 3 to 4 vols. alcoholic potash of sp. gr. 1·02, heat is evolved, and the whole solidifies to a crystalline magma. The potassium benzoate is washed out with hot water, the alcohol distilled off, the residue mixed with water till it begins to be turbid, and then shaken up with ether. The brown oily residue obtained by evaporating the ethereal solution is dried over fused potash and repeatedly rectified. It is believed to be identical with Peruvin, which is obtained by the action of potash on cinnamein.

Benzyl acetate can be obtained by distilling benzyl alcohol with acetic and sulphuric acids, also by heating benzyl chloride with potassium acetate and alcohol. It is a liquid which possesses an aromatic odour, boils at 206° and has a sp. gr. of 1·057 at 16°·5 †. Benzyl propionate and benzyl butyrate can also be prepared.

THE MIGNONETTE,

Reseda odorata, as usually grown in English gardens, is poor both in aspect and in perfume compared with the state to which it can be brought by careful cultivation. In the South of France it

---

* Ber. Deutsch. chem. Ges. ii. p. 512.
† Liebig's Ann. cxciii. p. 298.

is largely grown for the perfume, which is extracted by the
" enfleurage " process. The plant is delicate, and the crops often
fail in consequence of late winds. The seeds are there sown in
December and commence flowering in March. The flowers
gathered in March and April yield the finest perfume.

The Mignonette is a native of Egypt, and has also been found
wild on the coast of Barbary. It does not appear to be longer
lived in its native climate than in England, where advancing
winter infallibly destroys it in the open air. It can, however, be
grown in pots under glass, and reared as a perennial shrub by
careful treatment and protection from frost; it is then called a
" tree-mignonette " and can be made to last for three years; a stick
of about two feet long is inserted in the pot to which the plant is
tied as it advances in height, the leaves being occasionally stripped
from the lower part so that a stem may be formed to the height
required. As soon as the seed-vessels begin to ripen they are cut
off, and a fresh crop of blossom soon makes its appearance. A
very proliferous monster variety of mignonette appeared accidentally a few years ago amongst some seedlings in a Nursery
at Hassock's Gate, Sussex, from which cuttings were rooted and
exhibited at the Royal Horticultural Society. At that time it
much resembled the ordinary kind, only the flowers were double,
forming little balls of minutely fringed petals. By careful propagation the strain greatly improved, the spikes developing into
panicles more than a foot in length, branching profusely to within
a few inches of the apex with elegantly depressed branches having
their apices ascending; the whole covered with double and richly
scented flowers. The proliferous character of this peculiar specimen consisted in the fact that every branch arose out of the centre
of an abortive flower and occupied the place of a pistil; occasionally two branches arising out of the same flower. In some cases
a whorl of open but coherent carpels appeared, the branch origin-
ating from the middle of that whorl. Each of the branches,
especially the lower, may have lateral ones; these also in the same
way rise out of the centres of similarly proliferous flowers. The
plant could not seed, but was readily propagated by cuttings.
This handsome variety was totally unlike any of the finest of the
ordinary kinds of mignonette and was very richly perfumed. It
was described by the Rev. G. Henslow at a meeting of the Linnean
Society in December 1881, and figured in that Society's Journal.

Perfumes somewhat analogous to Mignonette have been noticed in the flower of the vine and in that of " Henna."

TUBÉREUSE,

Polianthes tuberosa, flor pleno, sometimes called *Hyacinthus tuberosus* and *Hyacinthus Indicus,* is believed to be a native of the temperate regions of Mexico.

The tuberous bulbs are annually imported into England from Genoa and from North America in very large quantities and grown under glass for the sake of the fragrance of their white flowers (which is most powerful at night), but, even with the help of artificial heat, the flowers do not attain such perfection of perfume as in the South of France, where the plant is largely cultivated in the open air.

In the district of Grasse the roots are planted in April, being set 9 or 12 inches apart, in rows 2 feet apart. The land selected should be deep rich soil, as the roots penetrate downwards to a considerable depth in search of moisture. If cultivated on a dry soil the plants require well watering and manuring. The stems will bear 10 or 12 flowers, and under good cultivation even more. Each flower as it blooms is picked off. The harvesting for the factories takes place about the first week in July and lasts to the middle of October. There is a yield of flowers after this time, but it is only of service to the florist, the fragrance not being developed in sufficient strength to be of use to the manufacturer. In November the roots are taken out of the ground and packed away in dry sand to guard them from humidity and cold. Propagation is effected by offsets, which are produced in quantity.

The produce of flower in good years amounts to about 2500 kilos. per hectare. The perfume is extracted by the cold process of " enfleurage."

CHAPTER V.

THE ODOUR OF VIOLET.

VIOLET.

Viola odorata, Linn. A native in groves and hedges almost throughout the whole of Europe; it is also found in Siberia, China, and Japan.

There are nine distinct varieties of *V. odorata,* all finely scented :—

Var. *a. vulgaris* (DeCandolle); flowers deep-purple or purplish-blue, pale and streaked in the mouth. The flowers of this plant impart their colour and flavour to aqueous liquors; a syrup made from the infusion has long been used as an agreeable and useful laxative for children. The infusion is also valued as a delicate test for the presence of uncombined acids or alkalies, the former changing its blue to a red; the latter to a green.

Var. *b. cærulea* (Sweet); blue flowers.

Var. *c. purpureo-plena* (Sweet); flowers double, purple.

Var. *d. cæruleo-plena* (Sweet); flowers double, blue.

Var. *e. pallido-plena* (Sweet); flowers double, pale blue. This variety is commonly called *Neapolitan Violet.*

Var. *f. alba* (DeCandolle); flowers white. Very plentiful in Surrey.

Var. *g. albo-plena* (Sweet); flowers double, white.

Var. *h. variegata* (DeCandolle); flowers variegated.

Var. *i. cornuta;* all the petals horned.

The habit of growth of all these varieties is trailing; other species are found of the same habit of growth and sweet-scented, as the *V. suavis*, and amongst the 170 other species of *Viola* known some are scented; but those mainly cultivated for their perfume are the double purple and double blue varieties of *V. odorata*, the Neapolitan (*pallido-plena*), some hybrid varieties, "Russian," &c., being grown for bouquets.

On the shores of the Mediterranean the Violet is cultivated on a large scale, especially in the districts of Grasse and Cannes. They are planted out in October or April. October is preferred as it is the rainy season, and the young plants are not then exposed to the heat of the sun or to the drought, as they would be if starting life in April. The best place for them is in the olive-groves, where they are protected from the powerful rays of the sun in summer and from the extreme cold in winter. They are placed in long furrows and do not require watering except where the earth is extra dry. In September the ground must be broken up round them and manured. Specks of bloom appear among the plants during November; their number increases daily, until by December the green is quite overshadowed, and the whole plantation appears one glorious hue, for the leaves of the plant having developed themselves sufficiently for its maintenance, now rest in their growth and are completely overtopped by the young buds they have protected, and which now shoot past them and bloom in the open. The flowers are picked twice a week; they lose perfume if they are allowed to remain long on the plant, and are all gathered before the leaves start growing or would otherwise be completely covered in by the foliage. The flowers are gathered in the morning and delivered at the factories in the afternoon, where they are taken in hand at once, as they would lose considerably in perfume if held over till the next day. This explains why the Neapolitan violets or any other of the highly perfumed violets sent from the South of France to London do not seem more odoriferous than the same varieties grown in England. The variety known as "Double Parma" gives a good result.

The plant is somewhat delicate and the harvest is very subject to climatic influences, sometimes suffering to the extent of 75 per cent. The old plants are removed every five years and young roots substituted, planting them between the old rows, or spaces which were before vacant.

The perfume is extracted by the cold process of "enfleurage," and subsequent solution in alcohol as an "extrait." It is exceedingly fine, and rarely obtainable quite pure at the shops: the " extrait de violette," vended retail, being largely composed of tincture of orris-root, an odour approaching it, but almost as distantly as does that of the pelargonium to the rose.

ORRIS ROOT.

This is produced from three species of Iris :—

1. The *Iris germanica*, Linn. This plant has large dark blue flowers, and is common in the environs of Florence and of Lucca. It is found in various places in Central and Southern Europe, in the North of India, and in Morocco. This variety is the one mostly cultivated in gardens in temperate countries in Europe.

2. *Iris pallida*, Lamarck. This resembles the *germanica* except in the colour of its flowers, which are more pale. It is found wild in calcareous soils in Istria. It is abundant in the environs of Florence and of Lucca, in the olive districts.

3. *Iris Florentina,* Linn. This species bears large white flowers. It is a native of the coast of Macedonia and the south-west coasts of the Black Sea. It also grows in an indigenous state near the Gulf of Ismid and in the vicinity of Adalia in Asia Minor. It is also found in the environs of Florence and of Lucca, but in the opinion of Hanbury it is not indigenous but merely acclimatized to these two last situations. He is also of opinion that the three above-named species are clearly distinct, and gives botanical reasons for that opinion * .

As the bulk of the orris-root of commerce is derived from the *I. germanica* and *I. pallida,* and as the *Iris Florentina* is *not* a native of Florentine soil, the commercial term for distinguishing the best orris-root seems rather a mistake. The district may, and in fact does, produce finer roots than the district of Verona—hence the source of error in stating the root to be obtained from *Iris Florentina* and ignoring the two principal varieties. Formerly, commercial orris-root was supplied from the wild plant only, but during the last few years plantations have been established in the district of Verona ; and as the plants begin to yield after three

* Hist. des Drogues, ii. p. 472.

years' growth, the first crop will, it is expected, be available in 1892.

In Tuscany the cultivation, which until lately was restricted to a small district in the neighbourhood of Pontasieve, near Florence, and at Arizzo, has been considerably enlarged since 1888.

From Reggio, in Calabria, a further supply will be obtainable from new plantations laid down in 1890. Consequently the present high prices, caused by a bad crop and by the manipulations of speculators in Italy, may not maintain.

The three species of Iris above-named are all known to the Tuscan peasants by the name "Giaggiolo," and when gathered from uncultivated plants the roots are taken indiscriminately, the greater part evidently consisting of *I. germanica* and *I. pallida*, which are the most common. They are usually dug up in the spring; the flags are cut back to within a few inches of the root, and another cut is made across the first tuberous formation; this head part of the root with the clipped flag growing from it is then replaced in the ground, and soon starts a fresh growth, making offshoots and fresh roots. The remainder of the root is trimmed of its bark, cleaned, and dried in the sun.

The fresh orris-root has simply an earthy smell. The characteristic odour, which is somewhat similar to that of violets, is afterwards gradually developed during the process of drying, and does not attain its maximum for at least two years, and even intensifies after that time. Its essential oil may therefore be included in the class of so-called "ferment-oils."

Sometimes fresh roots are bleached by exposing them to the fumes of burning sulphur; this has a deleterious effect on the perfume.

The roots are sorted out into several qualities, sizes, and sorts. Some of it is used for the manufacture of beads, and there is a market for the chips and turnings from this manufacture. The culture of this plant is extremely simple—it seems to thrive both in calcareous soil and in damp soil—and could undoubtedly be grown to advantage in many other places than those which now supply the market.

Professor Flückiger's examination of the products of the root is as follows * :—When the dried root is submitted to distillation with water, there eventually comes over a crystalline odorous matter of yellowish-brown colour of the consistency of a firm

* Archives der Pharmacie, June 1876.

ointment, which is termed "butter of Iris," and possessing the characteristic odour of orris. This product only amounts to a little more than 1 per mille of the root used, and consists principally of *myristic acid*, which is perfectly odourless, and a minute quantity of a brown essential oil to which the entire fragrance of the root is due—the proportion of this oil in the root is estimated at being possibly not more than 1 in 10,000.

The myristic acid can be separated from the crude product (butter of Iris) by repeated crystallizations from alcohol and purification by animal charcoal. The obstinately adherent volatile oil is concentrated in the mother liquids, and the crystals become less and less odorous until finally they have lost all aroma. The pure crystals melt at 52° C., and their composition is represented by the formula $C_{14}H_{28}O_2$—which is that of myristic acid. The effect of the presence of the smallest quantity of the adhering volatile oil, or a trace of *lauric* acid melting at about 44° C., which may easily accompany the myristic acid, must be to lower the melting-point. The original details of the investigation contain the following annotation :—" The question arises how the myristic acid, which can only with difficulty be distilled without decomposition, passes over with the oil. The explanation of this is to be sought in the phenomenon of diffusion. (Rose-oil is similarly accompanied by a stereoptene which is difficult to volatilize by itself.) An alcoholic solution of the pure crystals possessed no rotary power, and energetically reddened litmus paper moistened with alcohol.

" The occurrence of myristic acid in the distillate is probably due to a fat which is present in the root, and is split up by the vapour of water. The quantity of this fat must be very small, since 300 grams of orris-root powder exhausted with carbon bisulphide gave a soft perfumed resin, but neither free myristic acid nor neutral fat could be detected. The carbon bisulphide extract was digested with sodium carbonate and alcohol, in order to obtain a solution of sodium resinate and myristate, from which the acid sought could be precipitated by acetic acid. If myristic acid were present, it would, on prolonged digestion of the turbid acid liquid, gradually rise to the top as an oily layer. This, however, did not take place even after several days ; the brown resinate slowly sank to the bottom as a pulverulent mass, and the liquid became clear without yielding an oily layer."

Carbon bisulphide appears to be unsuitable for the removal of the perfume from orris root, the quantity of essential oil being exceedingly small, and the solvent removing with it tannin, and a very soft resin.

An examination of a pure sample of the crude product was also made by Dr. Hager [*], according to which, this substance " at the ordinary temperature has the consistence and colour of the basilicon ointment of the German Pharmacopœia; it melts at 38° to 40° C., forming a clear brownish-yellow liquid, which begins to congeal at 28° C.; it is soluble in 5 to 6 parts of 90 °/₀ alcohol, forming a clear yellow solution at medium temperatures."

In the distillation of orris-root the practice of adding sulphuric acid to convert the starch into dextrine and glucose is frequently followed. In this way the yield of oil is increased—by reason of the starch granules which imprison the oil-cells being dissolved—but the delicacy of the odour of the oil is injured.

Iris germanica is cultivated in Kashmir. It is known in India by its common Persian name *Bikh-i-banafshah*, meaning " Violet-root." The correct Persian name is *Súsan-i-ásmánjuni.* Indian orris-root differs from the European inasmuch as the bark of the rhizome has not been removed. The rhizomes of different species of Iris hardly differ in structure. They consist of a brown epidermis, composed of compressed and nearly empty cells, covering a white cortical cellular tissue containing starch; this is separated by a layer of brownish compressed empty cells from the central, woody, yellowish tissue of the rhizome. The latter is built up of large, thick-walled, spherical porous cells, loaded with starch; here and there between the cells may be seen a prism of oxalate of lime. The vascular bundles are numerous, in each, irregular rings of spiral vessels surround a central bundle of jointed vessels. Bombay is supplied with orris-root from Persia and Kashmir.

COSTUS.

The perfume of this root very much resembles that of orris-root, and can, therefore, be classed in the Violet series.

This plant was well known to the ancients by the Greek name κόστος and the Syriac name Koshta. Botanically it is known as *Aplotaxis Lappa*, Decaisne (derived from ἁπλόος, *simple*, and τάξις, *series*; the word should therefore be Romanized as *Haplotaxis*).

[*] Pharm. Centralhalle, 1875, p. 153.

It is identical with *Aplotaxis auriculata*, DeCandolle, and *Auclandia Costus* of Falconer.

It is a composite plant inhabiting the North-western Himalayas. According to Stewart it grows at from 10,500 to 13,000 feet elevation, in parts of the basins of the Jhelam and the Chenab.

Under the name of *Saussurea hypoleuca*, Sprengel, it is included in Clarke's 'Compositæ Indicæ,' and is said to extend from Sikkim to Kashmir at an elevation of from 7000 to 12,000 feet. It is described by Dr. Falconer * as growing on the mountain slopes of the Kashmir valley at an elevation of 8000 to 9000 feet, and he was the first to trace the drug to its true source. The plant has no relation whatever to *Costus speciosus*, Sm., or *Costus Arabicus*, Linn.†, as some dealers in drugs imagine.

The native synonyms of the root are :—Arabic, *Qust* or *Koost*; Persian, *Kosht*, *Kost*; Sanskrit, *Koshtam*; Tamil, *Joshtam*; Telinga, *Goshtamu*; Bengalee, *Pachak*. In Bombay it is called *Ouplate*; in Kashmir *Koot*. In China it is known as *Muh-hiung* and *Kwang-muh-hiang*.

In 'Powell's Punjab Products,' the writer, quoting from the recorded observations of Dr. Johnstone, who collected the plant in Kashmir, says :—"As the snow melts at the end of March the rootstock appears, its caudal leaves develop in the beginning of June, and it comes to full fruition in September. It is a perennial, leaves and stem dying yearly to the rootstock ; the exstipulate caudal leaves rise in threes, the two lateral spathing the centre ; the centre sheathing the stem as it shoots above the ground. The stem, two or three of which may arise from the rootstock, stands in adult growth forty inches, is fluted, lined internally with pith, and sheathed with exstipulate tristichous leaves. The rootstock varies in size from nine to fifteen inches in length, and from three inches upwards in thickness. The caudal leaves spring straight from the rootstock and are supported on petioles eighteen inches long. The leaves are simple, obcordate, eight by five inches in adult growth and strongly veined." Guibourt says ‡ :—"The thickness of the root varies from that of the little finger to 54 millimetres." The roots are dug up in the months of September and October, when the plant begins to be torpid ; they are chopped

---

* Trans. Linn. Soc. xix. p. 23.

† Cooke, in Pharm. Jnl. [3] viii. p. 41.

‡ Hist. des Drogues, iii. p. 29.

up into pieces from two to six inches long, and exported without further preparation.

The quantity collected is very large, amounting, as far as Dr. Falconer could learn, to about two million lbs. per annum. Large quantities are imported into Bombay from Kashmir *viâ* Amritsir, and a portion is sent through to Calcutta.

The root, as met with in commerce, consists of irregular, crooked, twisted pieces, from two to three inches in length, and from half to one and a half inches in diameter, almost always split, with a rough, brown, and somewhat reticulated surface, marked by longitudinal ridges, and very compact and brittle. The pithy central portion is generally absent and appears to have decayed away before the root was collected. Internally it is dirty white, with radiating bundles, leaving numerous small cylindrical channels filled with a brownish resin.

When cut or rubbed it has a strong and definite odour resembling that of violets or orris-root. In taste it is at first camphoraceous, and then bitter with a slight pungency, but by no means unpleasant. The genuine root does not appear to be at all subject to the attack of insects, although fragments of foreign roots introduced as adulterations are nearly destroyed. Of the two varieties met with in the bazaars in India, that called *Kut tulkh* is probably the old, and *Kut shirin* the young root. Dr. Stewart hazarded the opinion that *Kut shirin* is the produce of a different and unknown plant.

A microscopic examination shows that Costus-root consists of two parts, viz. a thick cortical layer of close texture pervaded by a few lactiferous vessels, and an inner radiating portion, the parenchyma of which is not so dense. This portion is also provided with lactiferous ducts, and a very abundant scalariform vascular system which appears loaded with resinous matter. No trace of starch is to be seen, nor does the iodine test indicate its presence. The dried root yields 1 per cent. of a light yellow essential oil of sp. gr. 0·982; rotatory power $+15° 29'$ in 100 millim. tube. It begins to boil at 275° C., and about a half passes over below 315° when decomposition takes place (Schimmel).

Dr. Birdwood remarks that the root of a plant with the native name of *Poshkar*, believed to be a species of *Lingularia*, is used for adulterating *Koot*; the Kashmiris at Lahore make the same statement, and say that five or six kinds of roots are used as adulterants.

In 1859 a communication was made to the Agri-Horticultural Society of India, and published in their journal, concerning two roots, one called *Koot* and the other *Thooth*. They were from the hills of that part of the Kangra district which borders on Chumba. The *Koot* was identified as the Costus, the subject of the communication, and the other was believed by Dr. Thomson to be the root of *Salvia lancata,* which was said to be common also in Kashmir, where it is used to adulterate " *Kut.*" Subsequently Mr. Cope, of Umritsur, contributed some remarks to the same Society on the adulterations of this drug. He says :—" This adulteration is carried to such a pitch, with the assistance not only of the *Thooth* (which so closely resembles the genuine article in every respect but its qualities, that it is difficult to distinguish the one from the other after admixture, which imparts to the false the odour of the true drug) but with other foreign substances, of which cow-dung is one, that I have ascertained as a fact that the more unscrupulous dealers use some 20 seers of Koot to flavour 100 seers of trash. When *Thooth* was first found useful as an admixture it was sold at Rs. 1–8 per maund; being now the main ingredient of the Putchuk of commerce, it has risen to Rs. 4–8. I am told that two other substances resembling the genuine article in exterior appearance have been ascertained to serve as ingredients in the mixture sent to Calcutta and Bombay for exportation to China under the name of *Putchuk.* They are a root called *Chog* brought from the hills, which is generally reported to be a deleterious drug, and *Nirbisi,* the root of a species of *Aconitum,* probably a virulent poison " *.

Costus is universally employed by the shawl-merchants in Kashmir as a protector of Kashmir fabrics from the attacks of moth and insects. The dried root is an agreeable fumigatory and yields excellent pastilles which burn fairly. It is exported in enormous quantities to China, where it is used as an incense. Baden Powell in his 'Punjab Products' says :—" Lines of camels may often be met passing down to Multan, the ' Kut ' perfuming the air for a considerable distance. No mandarin will give an audience until the ' pachak ' incense smokes before him ; in every joss-house it smoulders before the Tri-Budh Deity; in every floating junk on the Chinese rivers (the only house of countless hordes), Budh's image is found, and the smoke of the ' pachak '

* Journ. Agri-Hort. Soc. India, xi. part i. p. 76, and xi. part iii. p. 3.

religiously wends its way heavenward; with the bulk of the Chinese this ceremony is regarded as sufficient to propitiate the gods. In India it is a crown monopoly; each village in the vicinity of the 'Kut' fields is assessed at a fixed amount yearly, which must be delivered in the capital. The Maharaja's agents buy up the surplus at one chilki rupee per maund, and retail it double rate." Besides the above-mentioned uses, it is employed as a medicament both internally and externally, and is smoked as a narcotic and stimulant.

The opinion was expressed by Guibourt * that the Costus plant is nearly allied to the *Carlines* and *Chamæleons*, the roots of which plants are split up on one side and have also the similar gnawed appearance. Attention is also drawn to the fact that the " White Chamæleon " found by Pierre Bélon on the island of Crete produced a root so powerfully odorous of violets that the room in which specimens were kept became perfumed to such a degree as to cause headache. This plant is the " White Chamæleon " of the ancient Greeks †. It has a root as thick as a man's thigh, and develops its powerful violet odour when dry. It is the *Carlina gummifera* of Lesson, and identical with the *Acarna gummifera*, Willdenow; the *Atractylis gummifera*, Linnæus, and the *Cnicus carlinæ folio, gummifer acauleatus* of Tournefort.

Several woods have a well-defined violet odour, as the *Acacia homalophylla*, generally known as "Myall wood." It gives off its fragrance as long as the wood remains unpolished. This tree is very common in the deserts of the interior of Eastern and South-eastern Australia, especially in the country along the River Murray and its tributaries. Samples exhibited in London from Queensland were said to be derived from *Acacia pendula*.

A substance having a similar odour has been extracted by benzene from the wood of the *Picrasma quassioides*.

Amongst flowers, the violet odour is very noticeable in the *Tritelia uniflora*, a lily from Buenos Ayres; *Dendrobium heterocarpum*, an orchid; also in the *Oncidium inosmum*, a beautiful orchid with yellow flowers spotted with brown.

* Journal de Chimie médicale, viii. p. 666.

† See Lefranc, "Sur les plantes connues des Grecs sous les noms de Chamæleon blanc et noir," Bull. de la Soc. Bot. de France, xiv. p. 48, and Journal de Pharmacie, 4e série, viii. p. 572.

A strongly perceptible odour of violets emanates from the young green parts of the *Geonoma Pamila* or " scented palm." It is a native of Brazil, and one of the 40 known species of *Geonoma*, which are reed-like palms whose natural habitat is confined to the tropics of the Western Hemisphere, where they form part of the underwood of dense forests. A specimen of this palm may be found in the Botanic Gardens, Regent's Park.

Cassie.

The well-known odour of Cassie is considered to approach that of the Violet. The trees cultivated at Grasse and Cannes for this perfume are the *Acacia Farnesiana*, Willd., a native of the West Indies and originally naturalized in Europe in the Farnesian Gardens at Rome ; the *Acacia Bertoloni*, a native of Chili ; and the Australian *Acacia lophantha*, Willd., producing flowers of similar odour.

The *Acacia dealbata* or " Silver Wattle," and the *A. pycnantha* of Tasmania, and *A. suaveolens* of New South Wales, do not appear to have been planted in the South of France, although they might be tried with advantage. There are other sweet-scented Acacias in the tropics, such as *A. odoratissima*, Willd., and *A. latronum*, Willd., both natives of Coromandel, and *A. lomatocarpa*, DeCand., native of Malacca, whose perfume might with advantage be utilized in our Colonies *.

The young plants of *Acacia Farnesiana* are raised from seeds, which, after being steeped in water, are sown in a warm and well-

---

* The recent Report of the Royal Commission of Enquiry into the Vegetable Products of Victoria states that a witness, resident in Melbourne, gave information as follows regarding the Australian method of manufacturing " essence of Wattle blossom ":—" The blossoms gathered after sundown, carefully separated from stalks and leaves, are macerated in olive-oil of very fine quality at ordinary temperature for twenty-four hours, after which they are strained out and pressed. fresh flowers being added. This operation is repeated for seven days. One part of this perfumed oil is macerated with one part of pure spirit, 60 o.p.. for seven days, being frequently agitated. Lastly, the oil is allowed to separate, and the spirit carefully poured off." (Elsewhere, in course of examination, this witness lays great stress on the superior value of *grape*-spirit for this use. For some perfumes, *grape spirit*, containing minute quantities of several organic ethers, is preferable to " silent " spirit.)

sheltered spot. Whilst quite young, the weak-looking plants are weeded out and the strongest ones left in the bed. In the third year they have generally reached a height of 2 or 3 feet ; they are then grafted and transplanted to the open ground in fields well exposed to the sun and sheltered from the cold winds, each tree requiring about 12 square feet of land. Before planting, the ground is deeply trenched and well dressed with manure. The tree flourishes best in the neighbourhood of Grasse and Cannes. The trees begin to flower after the third year from the time they were transplanted, and continue to grow till they reach a height of about 15 feet, with branches 6 feet long and a stem as thick as a man's wrist. The season of flowering is from October to January or February, according to the presence or absence of frost. The blossoms are successive, some being ready for plucking while others are scarcely formed, which is a great advantage to the cultivator, as he can thus manipulate his harvest by degrees. The flowers are gathered twice a week, in the daytime, and are brought to the factories in the evening. Each full-grown tree will produce two lbs. weight of flowers. The perfume is extracted by the processes of maceration and enfleurage.

It has been noticed that the seeds and root-bark of these sweet-scented Acacias, when chewed, taste and smell execrably of garlic. The association of the violet odour and the odour of garlic has been remarked in other plants, such as the *Tritelia uniflora*, and is very curious. It is probable that the peculiar odorous principle of many vegetable substances is newly formed during fermentation (instantly or after a time) of the saccharine juices of the plants, either by natural or artificial means. At all events it is a fact that very small quantities of the blossoms of the violet, elder, linden, or cowslip, added to a fermenting liquid, are sufficient to communicate a very strong taste and smell, which the addition of the water distilled from a quantity a hundred times greater would not effect. This fact is made use of in Bavaria in brewing various kinds of beer, distinguished by different flavours, which are given by allowing small quantities of the herbs and blossoms of particular plants to ferment along with the wort. Wines are also artificially flavoured in the same way, notably the German " May-wine " or " Maitrank," sprigs of *Asperula odorata* being put into the fermenting vat. The multiplication of odours by fermentation and their production from plants not originally containing them will be afterwards noticed, but before quitting the subject of *Violet* and the curious ways in which this odour is formed by Nature, mention may be made of a fact which was known to the Romans, viz. that when oil of turpentine is taken inwardly, it imparts the odour of violets to the urine, and the same effect is produced when the human body is exposed to its vapours for a considerable time.

CHAPTER VI.

YLANG-YLANG.—CHAMPA.—ARTABOTRYS.

YLANG-YLANG.

As erroneous ideas are current respecting the botanical source of Ylang-ylang oil, it is considered necessary to describe at some length the tree yielding it, also the trees yielding analogous perfumes. The facts hitherto published in foreign journals regarding the physical properties of the oil are also given *in extenso*, as the constitution of the oil indicates the possibility of its synthetic formation.

The tree whose flowers furnish the essential oil known by this name—sometimes spelt Ihlang-Ihlang and Alan-guilan (meaning " Flower-of-Flowers ")—is the *Cananga odorata* of Rumphius*, belonging to the natural order " Anonaceæ."

Rumphius gives a detailed description † of the *Bonga-Cananga*, as the Malays designate the tree, but his illustration is defective.

Lamarck gives short notices of this tree ‡ under the headings " Canang Odorant " and "*Uvaria odorata.*"

According to Roxburgh, who describes it under the name *Uvaria odorata* §, the plant was brought in 1797 from Sumatra to the Botanic Gardens of Calcutta, and he states that in 1809 it had a trunk 36 inches in circumference at 4 feet from the ground, and was tall in proportion.

Dunal gives a somewhat more detailed description of *Uvaria*

---

* Hooker and Thomson's ' Flora Indica,' i. p. 130, and Hooker's 'Flora of British India,' i. p. 56.

† Herb. Amb. ii. cap. xix. fol. 195, tab. 65.

‡ Ency. méthodique, " Botanique," i. p. 505.

§ Flor. Ind. ii. p. 661.

odorata, or, more properly, *Unona odorata*, as he himself corrected it, in his 'Monographie de la Famille des Anonacées' *, which chiefly repeats the statements of Rumphius.

A very fine coloured illustration of the *Cananga odorata* is given in the 'Flora Javæ' of Blume †, the accuracy of which illustration may be accepted from the fact of Professor Flückiger's acknowledgment ‡ of its perfect agreement with numerous specimens of Cananga which he had seen at DeCandolle's herbarium at Geneva, also at the herbarium of Delessert.

The unjustifiable appellation "*Unona odoratissima*," which has erroneously passed into many writings, originated with Blanco §, who, in his description of the intense perfume of the flowers, was induced to used the superlative " *odoratissima*."

Baillon ‖ defines " Canangium " to be a section of the genus *Uvaria*, from which he contends the Ylang-Ylang tree should not be separated.

The notice of Maximowicz, " Ueber den Ursprung des Parfüms Ylang Ylang " (" On the origin of the perfume Ylang-Ylang "), contains merely a confirmation of the derivation of the same from Cananga.

The *Cananga odorata* is a tree attaining a height of about 60 feet, the trunk being straight throughout, with a smooth ash-coloured bark, having few but profusely ramified diverging branches ; the young shoots are round and smooth. The leaves are short-petioled, drooping, ovate-lanceolate, acuminate, with the margins entire, but waved and slightly downy along the nerves of the underside ; they are arranged in two rows, and attain a length of 4 to 7 inches and a breadth of 2 to 3 inches. The handsome and conspicuous flowers are in fascicles of generally 3, sometimes 4, on short peduncles from the axils of the leaves or the shoots of the former year's growth and from the nodes of leafless branches. These large bell-shaped, gradually drooping flowers are of a pale yellow or greenish yellow, and possess a most exquisite perfume, which is frequently compared to a mixture of the hyacinth, narcissus, and clove, and sometimes to a mixture of

---

* Paris, 1817, pp. 108 and 114.
† Brussels, 1829, vol. i. fol. 29, tabb. ix. et xiv. B.
‡ Archiv. der Pharm. 1881, p. 218, and Pharm. Journ. [3] xi. p. 934.
§ 'Flora de Filipinas,' Manilla, 1845, p. 325.
‖ 'Dict. de Botanique.'

jasmine and lilac. The lobes of the tripartite leathery calyx are finally recurvate. The six lanceolate petals soon spread out flat and grow to a length of 2¾ inches and a breadth of about half an inch, and are longitudinally veined, The filaments are somewhat numerous. The elevated receptacle is slightly flattened or depressed at the summit. The green, berry-like fruit is formed of from 15 to 20 rather long-stalked separate carpels, which enclose from 3 to 8 seeds arranged in 2 rows.

According to Hooker and Thomson and to Bentham and Hooker* *Cananga odorata* is the only species of this genus; the plants formerly classed together with it under the names *Unona* or *Uvaria*, some of which likewise possess odorous flowers, are now distributed between these two genera, which are tolerably rich in species. The Cananga differs from the *Uvaria* by the valvate sepals and from *Unona* by the double-rowed arrangement of the seed.

Cananga odorata is a native of Ava and Tenasserim; it is distributed throughout all Southern Asia, and is very generally cultivated. In the wild state the tree grows to a much greater height, but the flowers are not so rich in perfume (according to Blume, almost odourless). In habit the Cananga resembles the *Michelia Champaca*, L. (a fine illustration of this is likewise found in Blume's ' Flora Javæ,' iii. Magnoliaceæ, tab. 1), a native of India, which is exceedingly prized on account of the very pleasant perfume of its yellow flowers, and which was very celebrated in ancient times by the Hindus. Among the admired fragrant flowers which are preferred to all others by the Javanese —who are in this respect very dainty—the most highly valued are those of the "Tjempaka" (*Michelia Champaca*) and the "Kenanggi wangi" (*Cananga odorata*) †.

The oil of Cananga seems first to have reached Europe in about 1864, and in Paris and London its choice fragrance found full recognition. At first only very small quantities were imported from the Indian Archipelago; but large consignments were soon received from Manila in the Philippine Islands (14° 53′ N. lat., 120° 52′ E. long.), where German pharmacists engaged in its distillation, also in that of the *Michelia Champaca*. These chemists were Oscar Reymann, Adolf Roensch, and Henri Julien, and the industry is now extensively carried on by Sartorius.

---

* 'Genera Plantarum,' 1864, i. p. 24.

† Junghuhn, ' Java,' Leipzig, 1852, p. 106.

The Cananga is also cultivated in the Malay Peninsula, Java, and the Moluccas, and probably its cultivation could be easily and profitably extended in many other localities of similar climate, so as to meet the continually increasing demand for the oil and enable consumers to obtain it at a moderate price. The Javanese oil is at present considered the finest and of superior perfume to the Indian.

According to Guibourt * the so-called "Macassar oil," which for many years has been celebrated as a "hair-oil," is cokernut oil perfumed by digestion with the flowers of *Cananga odorata* and *Michelia Champaca* and coloured by turmeric. Such unguents have been in use in India from ancient times, and are known as "Borbori" or "Borriborri," for anointing the hair and the whole body, particularly during the rainy season.

The yield of essential oil from Cananga is somewhat small, being, according to Reymann, of Manila, about 25 grams from 5 kilograms of flowers.

An examination of Ylang-ylang (Aylan-gilan) oil was made in 1873 by H. Gal, and the results of his investigations presented in a memoir to the French Academy of Sciences †. The specimen of Ylang-ylang oil examined by Gal is stated in his report to have been obtained "by distillation from the flowers of the *'Unona odoratissima,'* an Anonaceous tree growing in the Antilles and Jamaica." His interesting study may be abstracted as follows :—
"The oil has a density of 0·980 at a temperature of 15° C. It passes over entirely in distillation without leaving any carbonaceous residue, but within very extended limits of temperature, ebullition commencing at about 160° C. and the temperature continuing to rise till beyond 300° C. The oil is insoluble in water, only partially soluble in alcohol, but entirely soluble in ether. The portion insoluble in alcohol and which can be dissolved in ether appears after the evaporation of that solvent as a semi-fluid transparent mass ; this product amounts to about one fourth of the oil. Nitric acid acts upon the Ylang oil with great energy, intense vapours being disengaged in the cold, and by the addition of water a resin is obtained presenting a great analogy with that which is formed by oxidation of benzoin by means of the same reagent. Sodium

---

* Hist. des Drogues, 1850, iii. p. 675.
† 'Comptes Rendus,' June 16, 1873.

bisulphite is without action upon this oil. Potash, on the contrary, when sufficiently concentrated and used at a suitable temperature, gives rise to a kind of saponification. If the alkaline portion be removed and a fresh quantity of potash added, and the treatment be repeated until the oil is no longer attacked, a substance is left which is insoluble in water. The aqueous portion upon the addition of hydrochloric acid deposits a solid body having a crystalline aspect. This dissolves with facility in boiling water; the solution, being filtered to separate a small quantity of resinous matter, yields upon cooling white pearly plates. This body melts at about 120° C.; it volatilizes very readily, and is deposited upon the cool sides of the vessel in shining needles; it boils regularly at about 245° C. These are the physical characteristics of benzoic acid, and this substance has also its chemical properties. In fact, if a small quantity be heated in presence of an excess of lime, an oil is separated which is insoluble in water and possesses the odour and properties of benzene. Treated with perchloride of phosphorus an energetic reaction takes place, and the piquant and characteristic odour of chloride of benzene becomes manifest. A few drops of this latter body with alcohol yielded benzoic ether. For greater certainty Gal submitted some of this acid to analysis. 0·276 of matter, ignited with oxide of copper, gave 0·126 of water and 0·696 of carbonic acid.

	Found.	Calculated.
C	68·7	68·8
H	5·0	4·9

" It is therefore quite evident that the acid abstracted from the oil by saponification is none other than benzoic acid.

" The part insoluble in potash was distilled with water and then separated from the water in the distillate. After drying over calcium chloride this oil distilled at about 170° C. to 300° C., very nearly as the primitive oil. With so great a range of temperature it was useless to expect to separate from this matter definite products with a constant boiling-point; Gal therefore attempted to ascertain the nature of these bodies, which might be supposed to consist of hydrocarbons analogous to those so often met with in oils. The product was treated with phosphoric anhydride; a vigorous reaction took place and a liquid was collected which no longer possessed the odour of the essence. Iodide of phosphorus also reacted upon it with great energy, and a liquid was distilled more dense than water and possessing a piquant odour. These reactions showed that it was an oxygenated substance—or, rather, a mixture of oxygenated substances—resembling the alcohols in chemical properties." Gal considers it "probable that the acid referred to may be considered as forming in the oil benzoic ethers with these alcohols. On the one hand, the acid does not exist in the oil in a free state, and on the other hand he was unable to obtain any alcohol soluble in water by distillation of the oil in presence of potash."

An examination of the oil of Cananga was since made by Adolf Convert, of Frankfort, with the following result * :—" The oil did not affect litmus-paper moistened with alcohol ; at 170° C. a small portion distilled over, but the thermometer gradually rose to 290° C., and at a still higher temperature the oil was decomposed. As the portions passing over below 290° had a strongly acid reaction, the presence of compound ethers was very probable. Ten grams of the oil were then boiled with 20 grams of alcohol and 1 gram of potash during one day in a flask provided with an inverted condenser. The alcohol was then separated by distillation, the residue supersaturated with dilute sulphuric acid, and, together with much water, submitted to distillation until the distillate almost ceased to have an acid reaction. The distillate was then neutralized with barium carbonate and the filtrate concentrated, whereupon it yielded crystals which were recognized as nearly pure acetate. The acid residue containing the potassium sulphate was shaken with ether, and the ethereal solution, after being evaporated, left a crystalline mass of an acid reaction, which was coloured violet by ferric chloride. This reaction, probably ascribable to the presence of a phenol, failed to make its appearance after the mass had been recrystallized from boiling water. The aqueous solution of the purified crystalline scales then gave only a small flesh-coloured precipitate with ferric chloride. The crystals, forming small lamellæ, melted at 120°. In order to demonstrate the presence of benzoic acid, the crystals were boiled with water and oxide of silver, and the crystals that separated from the cooled filtrate were dried over sulphuric acid—0·0312 gram of them yielded, on combustion, 0·0147 gram, or 47·1 of silver. Benzoate

* Archiv der Pharm. 1881, p. 218.

of silver contains 46·6 of metal; hence the identity of the crystals with benzoate of silver was proved. For the separation of the alcoholic constituent, which is present in the form of an apparently inconsiderable quantity of benzoic ether, far more Ylang oil would have been required than was at the operator's disposal.

" Besides a benzoic ether, and probably a phenol, the oil gives indications of the presence of an aldehyde or ketone, since, on shaking it with acid sodium sulphite, the formation of a very small amount of crystals was observed. Like the benzoic acid, the acetic acid is no doubt present in Cananga oil in the form of a compound ether."

The name *Cananga* is frequently used in Amboyna to designate the *Unona tripetala*, DeC.* (*Unona tripetaloidea*, Dunal †; *Uvaria tripetala*, Lam.‡).

This is a tree of about 40 feet high, native of Amboyna in plains and on hills. In habit it resembles the *Michelia Champaca*. Its leaves are lanceolate, the upper surface wrinkled, the under surface tomentose; the flowers are on solitary peduncles, sweet-scented and greenish. The fruit is about the size of a plum. The carpels are stipulate, ovate, somewhat triquetrous, granulated, and 3-seeded. The name of this tree in the Malay language is *Cananga outàn*. The word *outàn* seems to mean *wild* or *uncultivated*, as it is also applied to an uncultivated sort of patchouli to distinguish it from the cultivated sort designated *wangi*; and the word *outàn* in ouràng-outàn (a monkey) means a " wild " man.

The *Unona hamata* of Roxburgh and *Unona uncinata*, Lamarck, are referred to *Artabotrys odoratissima*; and *Unona suaveolens*, Blume, to *Artabotrys suaveolens*.

The *Uvaria dulcis* is a large woody climber found in the Moluccas and Tenasserim. It is described by Dunal §. Its young branches are densely tomentose. Leaves oval or oblong, puberulous above, woolly beneath, 3 to 5 inches long, $1\frac{1}{2}$ to $2\frac{1}{2}$ inches broad, petiole $\frac{1}{6}$ to $\frac{1}{4}$ inch. Flowers 2 inches diameter, subumbellate, greenish, odorous; peduncle $\frac{1}{4}$ to $\frac{1}{2}$ inch. The petals are oblong, subacute, tomentose on both sides.

---

* Syst. i. p. 490.
† Mon. des Anon. p. 104; Rumph. Amb. ii. p. 197, tab. 66. f. 1.
‡ Dict. i. p. 597.
§ Mon. des Anon. p. 90, tab. 13; also DeC. Prodr. i. p. 88; Hooker and Thomson, Flor. Ind. p. 88; and Blume, Flor. Jav., Anon. tab. 3.

CHAMPAC.

This perfume of great delicacy is derived from the flowers of
Michelia Champaca, Linn. (Sp. 756), and under such name is
described by   ⟨
    DeCandolle, Prodr. i. p. 79.
    Roxb., Flor. Ind. ii. p. 656.
    Wight and Arnott, Prodr. i. p. 6.
    Wight, Illust. i. p. 13.
    Blume, Flor. Jav. ix. t. i.
    Tent., Flor. Nap. vii. t. 3.
The synonyma are :—
    Michelia aurantiaca, Wallich in Plant. As. Rar. ii. t. 157.
    M. pubinervia, Blume, Flor. Jav. Magn. xiv. t. 4.
    M. Rheedii, Wight, Illust. i. xix. t. 5. fig. 6.
        ,,   Rheede, Mal. i. t. 19.
        ,,   Rumph., Amb. ii. t. 67.
The name *Michelia* was given by Linnæus to this tree in honour
of Pietro Antonio Micheli, a celebrated Florentine botanist who
died in 1737, and the name *Champaca* is derived from Ciampa, an
island between Camboge and Cochin China, where the tree grows.
This island is also called Tsampa, hence also the appellation *M.
Tsiampàca*. The essential oil distilled from the flowers is known
in India as *Pand*, also as *Champa-Ka-utter* and *Keeula-Ka-utter*,
samples of which were so labelled at the London Exhibition, 1862.

The genus *Michelia* is entirely Indian. It is very nearly allied
to *Magnoliaceæ*, and although there is no broad line of distinction
between the two, the latter is distinguished from the former by
having terminal flowers, more densely spiked carpels, and definite
ovules.

In the shady forests of the Eastern Himalaya five species of
Michelia form a prominent feature in the vegetation of the temperate zone at elevations between 5000 and 6000 feet. They are,
however, impatient of drought, and one only (the *Champaca*)
extends as far west as Kumaon. In the Khasian hills and the
Malayan peninsula other species occur, and the latter locality,
when we become better acquainted with the vegetation of its
mountains, may be expected to yield many species.

The genus is common in Java and the islands of the Eastern Archipelago.

The *Michelia Champaca* is generally cultivated, but is found wild in the forests of the temperate Himalaya, from Nipal and Kumaon eastward in the forests of Pegu and Tenasserim; in the temperate regions of the Nilghiris, Courtalam, and Travancore, at altitudes of 3000 to 5000 feet, and has been distributed in the forests of Java, where it is cultivated; also the Philippine Islands and many localities throughout the tropics. By cultivation the perfume of its flowers is greatly developed.

It forms a large tree. The branchlets are pubescent. The leaves are 8 to 10 inches long by 2½ to 4 inches wide; the upper surface shining, the ribs of the lower surface rather pubescent or silky and paler in colour, the petiole being 1 to 1½ inch in length. The flowers are pale yellow to orange-coloured, on short, axillary, 1-flowered, silky peduncles. The buds are silky. The sepals are oblong and acute, the petals linear. The carpels congested and subsessile. The flowers are not unlike a double Narcissus.

This tree is celebrated for the exquisite perfume of its flowers, though some Europeans find it somewhat too powerful, and at night it becomes rather rank. The native women adorn their heads with the flowers for the sake of the perfume and for the elegant contrast of the rich orange colour with their black hair. The tree is highly venerated by the Hindus, who have given one of its names, "Tulasi," to a sacred grove of their Parnassus on the banks of the Yamuna. It is also dedicated by them to their god Vishnu.

The most fragrant of the other species of *Michelia* are:—

M. montana *.—A forest-tree of 60 feet in height; native of Java. Flowers pale yellow; buds, peduncles, and spathes almost naked; petals 9, lanceolate and acute.

M. pubinervia †.—A tree of 50 to 60 feet; native of Java. Flowers pale copper-coloured. The veins beneath the spathes are clothed with rufous villi; the outer petals spatulate and obtuse.

M. parviflora ‡.—A native of Java and Ternata. Flowers small, of a livid flesh-colour or cream-coloured. Leaves elliptical, a little

---

* Blume, Flor. Jav. fasc. 19, p. 15, t. 5.

† Blume, Flor. Jav. fasc. 19, t. 4.

‡ Rumph. Amb. ii. p. 109; DeC. Syst. i. p. 449; Blume, Flor. Jav. fasc. 19, p. 18.

acuminated, smooth; branchlets, buds, and spathes clothed with rufous tomentum. This is a shrub of 7 to 10 feet in height.

M. alba*.—This species is very little known. It is a native of Java and Baleya. Its flowers are white, and described as being smaller and more pleasant than those of *M. Champaca*.

M. Doltsopa, DeC.†—A native of Java and of Nipaul, about Harain-Hetty, where it is called *Doltsopa*. The leaves are oval-oblong, acuminated, smooth, but rather glaucous and rather puberulous on the under surface. The flowers are on long stalks and the flower-buds clothed with rusty tomentum. The flowers are yellow, the petals 6 to 9, are oval, the outer ones oblong, the inner ones narrower. Not only the flowers of this species are fragrant, but the wood is sweet-scented, and is described as the best in Nipaul for buildings.

The physical characters and chemical constitution of oil of *Michelia* do not appear to have been studied.

Bhúchampac, or " Ground Champac," is a common name incorrectly applied in India to the *Kœmpferia rotunda* belonging to a totally different order (Zingiberaceæ) to the true Champac. The vernacular name may have been derived from a fancied resemblance of its flowers to those of *Michelia Champaca*. Another denomination, *Malankuwa*, by which it is known in Malabar, according to Van Rheede, is by him explained as meaning " Mountain ginger " ‡.

The synonyms in Sanskrit are *Tamrapushpa, Sidd'ha*, and *Band'hu Drughana*. The first of these synonyms indicates that the flower is of the colour of copper, which may perhaps be reconciled to the purple hue within the blossom of this *Kœmpferia*.

The *Kœmpferia rotunda* is a species of Galangal (afterwards described). It is figured in Curtis's 'Botanical Magazine,' tab. 920, and is mentioned in the 'Asiatic Researches,' iv. p. 242, and xi. p. 328. The richly fragrant flowers open during April and May, one or two at a time on each plant, and wither entirely before sunset. In India it is frequently cultivated in gardens on account. of its beauty. The full botanical description is given by Roxburgh in his 'Flora Indica,' i. p. 15.

---

* Rumph. Amb. ii. p. 199.

† Syst. i. p. 448; Wall. Tent. Fl. Nep. t. 3.

‡ Hort. Mal. xi. p. 17, tab. 9.

KEORA.

The name *Keora* is applied in Hindustanee to a fragrant flowering species of Screw-pine, the *Pandanus odoratissimus*, Lin. fil., which is common on marshy land and in the vicinity of the sea in many of the warmest places in the tropics. It is described and figured by Roxburgh * ; as *Pandanus verus* by Rumphius †, and as *Kaida* by Van Rheede ‡ ; also as *Keyro* by Forskal §. The Sanskrit name is *Ketuka*, the Bengalee *Kea*. By Europeans on the Coromandel coast it is known as *Kaldera*.

The male and female flowers are always on separate trees. In the Telinga dialect the male tree is known as *Mugalik* and the female as *Gozdoogoo*. The Malay name of the tree (irrespective of gender) is *Pandang*. In Tahiti it is called Fara or Wharra, and in the Hawaiian Islands *Hala*.

In the East Indies the tree attains a height of 10 to 20 feet, in Mauritius 30 feet, and in some places, as in the Nicobar Islands, where it is known as *Mellore* or "Nicobar Bread-fruit" (which may be a variety of this species), it has been described as attaining a height of 35 to 40 feet ||.

Occasionally this tree may be found with a round branching head, but generally it is in the form of a large ramous spreading bush. From the stem or larger branches, long, fusiform, obtusely pointed roots ensue, descending till they come to the ground, which they enter and then divide. The substance of the most solid part of the wood is something like that of a cabbage-stem, which by age acquires a woody hardness, on the outside. The smooth glossy leaves are confluent, stem-clasping, and closely imbricated ; they form three spiral rows round the extremities of the branches, drooping, from 3 to 5 feet long, and tapering to a very long, fine, triangular point. The margins and back of the leaves are armed with very fine sharp spines, those on the margins pointing forwards, those on the back pointing sometimes one way sometimes the other. The leaves are composed of longitudinal, tough, useful fibres like those of the pineapple. The flowers of

* Corom. i. tabb. 94, 96.　　　† Amb. iv. p. 139, tab. 74.
‡ Mal. ii. tab. i. 5.　　　§ Ægypt. tab. 72.
|| As. Res. iii. p. 161.

the male tree consist of a large terminal, pendulous, compound
leafy panicle, 16 to 20 inches long; the leaves thereof are white,
linear-oblong, pointed and concave; in the axil of each there is a
single thyrse, composed of simple small racemes of long, pointed,
depending anthers, which are not sessile, but raised from the rachis
of these partial racemes by tapering filaments. The flowers of the
female tree are terminal and solitary, having no other calyx or
corolla than the termination of the three rows of leaves forming
three imbricated fascicles of white floral leaves or involucres like
those of the male racemes, only here they stand at equal distances
round the base of the young fruit. The fruit is oval, bright
orange-coloured, from 6 to 8 inches diameter and from 6 to 10
inches in length, weighing from 4 to 8 lbs. It is something in
appearance like a pineapple, and contains a rich-looking yellow
pulp intermixed with strong fibres. In Tahiti the natives prepare
a fermented drink from the juice of this fruit, called *Ava fara,*
Pandanus wine.

The flowers are produced in the rainy season, the tender white
floral leaves of the male tree being the most fragrant. Their
fragrance has been described as " the most delightful, the richest,
and the most powerful " of floral perfumes; it is also permanent,
being retained by the flower after complete desiccation.

This singular-looking tree sometimes covers large tracts of
country with an impenetrable mass of vegetation. It is particularly abundant in the Sunderbunds, growing on both sides of
creeks in such profusion as to render them impassable by its
thorny interlacing branches and aërial roots. It is very common
along the banks of canals and backwaters in Travancore, in which
places it is planted to bind the soil, the superfluous growth being
cut back. In Mauritius the trees are set to form fences or hedgerows around plantations or along the sides of the many roads which
intersect them. The tree is propagated readily from branches;
hence it is not unusual to find extensive hedges entirely composed
of male or of female trees, owing to there having been originally
a male or a female tree in the neighbourhood to propagate from.
At certain periods the trees are trimmed, the fibrous leaves being
useful for making mats and package-bags for the transport of
coffee, sugar, and grain. The fusiform roots are composed of
tough fibres, which basket-makers split and use to tie their work

with. The seeds are useful for food, also the tender white base of the leaves, either raw or boiled.

Samples of the essential oil distilled in India (there known as *Keora-ka-utter*), also of the fragrant aqueous distillate, have occasionally been exhibited in England, but they offered no fair criterion of the true perfume, being contaminated with oil of santalwood; indeed, it is almost impossible to obtain from India any essential oils which are not so contaminated, it being the custom there to mix santal with all odoriferous substances. The flowers are also used in the preparation of a kind of scented catechu paste, much esteemed by Hindu ladies for toilet purposes.

Considering the immense tracts of country in various parts of the tropics over which this valuable tree is distributed, also that it grows rapidly without any care, it is surprising that the essence of its flowers should not have been introduced into European commerce *.

NAG-KESUR.

Another essential oil which has been overlooked commercially, although a specimen of it was exhibited in London thirty years ago, is that known in India as Nag-Kesur-ka-utter, distilled from the flowers of the *Mesua ferrea* †, a handsome tree of the order *Guttiferæ* found in the southern Concan and Goa territory, common on the mountains of Eastern Bengal, the Eastern Himalaya, and the Andaman Islands. It is also under cultivation in India and Java for the beauty and fragrance of its white flowers, which are from ¾ to 3 inches in diameter, with a large globe of bright goldcoloured linear anthers in the centre. These flowers appear at the beginning of the warm season. The anthers of these flowers retain their fragrance in the dried state and are sold in the Indian bazaars under the name of *Nag-Kesur* for making satchets and for stuffing pillows, also for dyeing silk. Other local names for these dried anthers are *Nágacésura*, *Chámpéya*, and *Cánchana*, words which have a more or less direct reference to gold ‡. The tree is in some

---

* It is rather remarkable that the flowers, both male and female, of another species of *Pandanus*, the *P. fœtidus*, known in Hindustanee as *Keora Kanta*, possess a highly offensive odour, almost similar to that of the *Sterculia fœtida*.

† Roxb. Fl. Ind. ii. p. 605; Hooker and Thomson, Flor. Ind. ii. p. 277; Wight, 'Icones,' tab. 118.

‡ As. Res. iv. p. 295.

K

localities known as *Nag-champa*. The name "*ferrea*," or "iron," is in allusion to the hardness of the timber.

ARTABOTRYS ODORATISSIMUS.

The odour of the strongly-scented flowers of this plant is closely allied to that of Ylang-Ylang, the *Cananga odorata*.

The plant is figured and described by Brown in the 'Botanical Register,' v. p. 423.

The synonymy given by Brown is a valuable contribution to the 'Botanical Register,' and presents a critical view of the scientific history of the species. The synonyms are as follows :—

> *Uvaria odoratissima*, Roxb., Flor. Ind. MS. ined.
> *Unona uncinata*, Dunal, Anonacées, p. 105, tt. 12 & 12*a*.
>   „   „   DeCandolle, Syst. Nat. i. p. 491.
>   „   „   Dunal, Anonacées, p. 107.
> *Uvaria esculenta*, Rottler, in Nov. Act. Soc. Nat. Cur.
>   Berol. iv. p. 201.
>   „   *uncata*, Loureiro, Cochin, p. 349.
> *Anona unicata*, Lamarck, Ency. ii. p. 127.
> *Annona hexapetala*, Linn. Supp. p. 270; Hort. Kew. ii.
>   p. 253 ; Ed. 2, iii. p. 335.
>   „   „   Willdenow, Sp. Pl. ii. p. 1266.

The shrub is a native of China and the East Indies, where it is cultivated as an ornamental covering for walls; it is also distributed in Java and Ceylon. It attains a height of about 6 feet. Its leaves are oblong or lanceolate and glabrous, from 2 to 8 inches in length by 1 to 2 inches in breadth. The flowers are yellow, solitary or in pairs, and extremely fragrant. The petals are 1 to $1\frac{3}{4}$ inch in length, when young pubescent, especially at the base, glabrous when expanded. The carpels are obovate-oblong and glabrous.

When cultivated under glass in England the flowers of this shrub do not attain the yellow colour natural to them, but remain of a pale sickly green.

The following distinctive features of *Artabotrys* are mentioned by Brown to separate this genus from the plants formerly known as *Unona* or *Uvaria* :—" In *Artabotrys* the petals are of equal depth, the germen 2-seeded, growing up to a 2-seeded fruit (or sometimes accidentally to a solitary-seeded fruit) ; the seeds are without an arillus, placed side by side, not one above the other, and the peduncles furnished with a grapnel or crooked tendril for its peculiar support, not, as in most of the other tendril-bearing plants, for the assistance of the branches in their ascent."

The vernacular names in India for this plant are :—

Bombay.—*Vilayaté-Champa.*

Tamil.—*Manoranjitam.*

Telinga.—*Phala - sampenga ; Sakala - phala - sampenga ; Manoranjitam.*

Malay.—*Madura-Kaméshvari ; Manuranjitam.*

Madras.—*Manoranjatam.*

The *Artabotrys suaveolens,* Blume (Flor. Jav., Anon. p. 62, tt. 30, 31), is a large woody climber met with in the forests of the eastern peninsula from Sylhet to Malacca, and distributed eastwards to the Philippines. Its leaves are oblong-lanceolate, acuminate, shining above, glabrous on both surfaces, or sparsely hairy on the midrib beneath; peduncles woody, recurved, many-flowered; flowers fascicled, bracteolate; limb of petals cylindric or subclavate; carpels sessile, smooth.

CHAPTER VII.

ODOUR OF THE HAYFIELDS.

The Odour of Coumarin.

The fragrant bean called Tonka, or Tonquin bean, is the seed of the *Dipterix odorata*, a large leguminous forest tree, native of Cayenne. In British Guiana this tree is called "*Kumara*," from which name the word "coumarin" is probably derived. It grows plentifully in some localities, especially above and on the islands in the rapids of the Esquibo River. The seeds contain a fine fixed oil obtainable by expression, which is used by the natives of the colony as a perfumed hair-oil. This oil may have been the basis of a once well-known hair-restorer called "Balm of Columbia." The tree is also an inhabitant of Martinique. The average height of it is 70 feet, but in Guiana it is said to attain 90 feet. The pods contain but one seed, which is shaped like an almond, but much larger, and covered with a shining black skin. The fruit-pod does not open naturally at maturity, as does that of most leguminous plants. The seed when ripe and detached from the stalk rapidly acquires a powerful and aromatic odour, suggestive of new hay, although while actually growing it is nearly odourless; the hardening of the mature seed developing *coumarin*, its odorous principle, the small white crystals of which are visible beneath the covering of the kernel and between the lobes. The evaporation of this substance is prevented by the fixed oil in the kernel. The perfume is extracted in the form of a tincture, by cutting the beans very small and macerating them in spirit of wine.

Coumarin was discovered in the Tonka bean in 1825 by Boullay and Boutron-Charland *, and was studied and analysed

* Journ. de Pharm. xi. p. 480.

by Delalande *. Coumarin is also abundantly developed on the dried leaves of the *Liatris odoratissima*, commonly called " Deer's tongue " or " Hound's tongue," an herbaceous plant found plentifully in North Carolina, in portions of Lower Georgia, and throughout the east and south of Florida, principally along the St. John's River and its tributaries. It is mostly found on the edges of what are called " bays," *i. e.* low-lying places in the pine woods, which are partially covered with water, and overgrown with " bays," a species of magnolia †. Its root-leaves are from 8 to 12 inches long, by 2 or 3 broad ; those on the stem being very much smaller. The stem divides above into a broad branching panicle of purple flowers, which make the plant an attractive one.

The leaves of the *Liatris* when green are very slightly odorous, due to the presence of a small quantity of essential oil contained in glands sunk in funnel-shaped depressions on both the upper and lower surfaces of the leaf. Dr. Paschkis found that the proportion of oil yielded by the leaves was only 10 drops to 250 grams. By drying in the shade the leaves rapidly develop a pleasant odour very similar to that of the Tonka bean, and mainly due to the same principle—coumarin, crystals of which are abundantly deposited on the matured leaves when dried. In the early season the leaves are said to be very succulent, losing 60 or 70 per cent. of moisture in drying, but in September the loss hardly exceeds 20 per cent. It has been observed by Dr. Wood that they retain their fragrance for many years after they are gathered. The odour is given off more intensely on a damp day than on a dry one. A damp atmosphere will develop the perfume months after all sign of its activity has disappeared.

The odour is very volatile and diffusive, being most perceptible in the upper stories of warehouses where it is kept. A single leaf of Deer's tongue will for many years preserve and manifest its perfume, and yet the particles of coumarin volatilized would hardly be appreciated in the most delicate balance.

According to the experiments of Dr. Wood, the yield of coumarin from Deer's tongue is about 2 drachms to 2½ drachms from 1 lb. of leaves.

---

* Ann. Ch. & de Phys. [3] vi. p. 345. † *Magnolia glaucus.*

Coumarin also occurs in the leaves of the *Angræcum fragrans* *. This is an orchid somewhat resembling Vanilla. It is synonymous with *Acrobion* or *Anethum fragrans* (Sprengel) and *Æranthus fragrans* (Reichenbach). It is found in Madagascar, Bourbon, and Mauritius. In the former island it is known as " Fanave," and in both the latter as "Faham" and "Bourbon tea." The leaves have an agreeable odour when green, and both leaves and fruit when they turn yellow on the plant or have been artificially dried acquire a much stronger odour. The long thin fruit-pods possess a stronger and more delicate perfume than the leaves, and when cured with boiling water in the same way as Vanilla, they blacken, remain entire without splitting, and retain their perfume for a great length of time. The odour resembles a mixture of Vanilla, Tonquin beans, and Melilot. Commercially we know only the leaves; they contain an aromatic principle soluble in alcohol, ether, and boiling water. It has been isolated by Gobley †, in the form of small white silky needles, which, on being pressed between the fingers or slightly warmed, develop the characteristic odour of Faham and bitter almonds. The fruit is supposed to contain a larger proportion of this principle than the leaves. The plant is propagated by seed.

Coumarin is contained in considerable quantity in the pods of the Balsam Peru tree, *Myroxylon Pereiræ.*

The *Ceratopetalum apetalum*, an Australian Saxifragaceous tree growing to a height of 50 feet, contains coumarin in large quantities, both in its bark and fruit. It is also found in the Australian *Alyxia buxifolia*, a straggling sea-side shrub, which, however, does not produce wood of any size.

The occurrence of coumarin in the hay of *Anthoxanthum odoratum*, or "sweet vernal grass," is well known, and was first studied in 1846 by Dr. Bleibtren ‡. It has also been found in six other grasses :—

Ataxia Horsfeldii.	}	From the leaves.
Cinna arundinacea.		do.
Hierochloa Alpina.		do.
do. Australis.	Grasses.	do.
do. Borealis.		Root.
Milium effusum.	}	Leaves.

* Dupetit and Thouars, " Orchidées d'Afrique."
† Journ. de Pharm. [3] vii. p. 348.
‡ Ann. der Chim. und Pharm. lix. p. 177.

Besides the above-named plants, its presence has been discovered in :—

Adiantum pedatum.	⎫	Leaves.
do. Peruvianum.	Ferns.	do.
do. trapeziforme.	⎬	do.
*Drynaria Wildenovi.	⎭	
Phœnix dactylifera.	A palm.	The date.
Aceras anthrophora.	⎰ Orchids.	Leaves.
Nigritella angustifolia.	⎬	do.
Orchis fusca.	⎱	do.
Herniaria glabra.	Leguminous tree.	do.
Ruta graveolens.	Herbaceous plant.	do.
Dipterix oppositifolia.	⎱ Leguminous trees.	The ripe seed.
do. pteropus.	⎰	do.
Alyxia stellata.	Evergreen tree.	Bark.
Asperula odorata.	⎱	Leaves.
Galium trifolium.	⎬ Herbaceous plants.	do.
Liatris spicata.	⎰	do.
Prunus Mahaleb.	Evergreen tree.	Bark.
Melilotus officinalis.	⎱	Flowers and leaves.
do. leucanthus.		do.
do. hamatus.	⎬ Herbaceous plants.	do.
do. altissimus.		do.
do. albus.	⎰	do.
Ageratum Mexicanum.	Annual.	Leaves.

It was at first thought that the sole odorous principle existing in *Melilotus officinalis* was coumarin, but the investigations of Zwenger and Bodenbender showed that the odorous principle is coumarin-melilotic acid, $C_{18}H_{16}O_5$, a compound of melilotic acid (hydrocoumaric acid) with coumarin[†]. It has since been shown by Phipson that the odorous principle of the plant *at the time of its maturity* is principally melilotol and melilotic acid.

A note by T. L. Phipson on the discovery of melilotol appeared in the ' Chemical News,' 1875[‡] ; his further observations on this subject appeared in the ' Journ. de Pharmacie,' 1878[§], to the effect that in endeavouring to ascertain how much coumarin could be extracted from the *Melilotus officinalis*, which grows abundantly

---

* A specimen of this fern is in the Royal Botanic Gardens, London.

† Ann. der Chem. und Pharm. cxxvi. p. 257, and Bull. de la Soc. Chim. 1864, p. 145.

‡ xxxii. p. 125. § xxviii. p. 300.

in sheltered places in the neighbourhood of Weymouth, he gathered
a quantity of the plant while in full bloom in August, and having
dried it at the ordinary temperature in the air, distilled it entire,
leaves, flowers, and stalks, with water; then by washing the distillate with ether, and evaporating the ether, obtained about 0·2
per cent. of pure melilotol in the form of a brownish oil giving an
acid reaction, slightly soluble in water and very soluble in alcohol
and ether.

Melilotol is of an extremely agreeable odour, not precisely that
of coumarin, but exactly resembling that of recently cut hay or
dried *Anthoxanthum odoratum.* He therefore infers that the odour
of dried melilotus and that of the hayfield is principally due to
melilotol and not to coumarin. By boiling melilotol with a concentrated solution of caustic potash melilotic acid is formed, and
a slight odour of bitter almonds given off. Phipson's analysis of
melilotol resulted in figures agreeing exactly with the formula
$C_9H_8O_2$; that of coumarin being $C_9H_6O_2$. He does not deny
the existence of coumarin ready formed in the plant, but affirms
that the sweet odour is due to melilotol, and that melilotic acid,
$C_9H_{10}O_3$, also exists in the plant, especially in the month of
August ; the theory being that coumarin is first formed earlier in
the year, and then by the action of nascent hydrogen is converted
into melilotol, which, in turn, taking up two equivalents of water
is converted into melilotic acid. In confirmation of this theory :—
If coumarin be subjected to the action of nascent hydrogen (by
sodium amalgam) melilotic acid is formed, the reaction not
stopping at melilotol, but taking up the 2 equivalents of water as
fast as they are liberated.

To prepare coumarin from substances which contain it in an
uncombined state,—exhaust with alcohol, distil off the alcohol
from the filtered tinctures, and let the remnant crystallize and
re-crystallize in water or alcohol with addition of animal charcoal.
It forms brilliant, colourless, silky needles and leaflets of very
aromatic odour, which fuse at 67° to 67°·5, boil at 270°, and sublime unaltered. It is of neutral reaction, dissolves scarcely in
cold, readily in boiling water, most readily in alcohol, volatile,
and fixed oils. It crystallizes from boiling water in small rhombic
plates. By fusion with caustic potash it yields salicylic acid.

The solubility of coumarin in alcohol of various strengths and at various temperatures has been found by Schimmel & Co. to be as follows :—

100 parts of alcohol	dissolve		
	at 0° C.	at 16° to 17° C.	at 29° to 30° C.
of 90 vol. per cent.	7·1 parts	13·7 parts	42·5 parts
80 ,,	6·0 ,,	12·3 ,,	38·3 ,,
70 ,,	4·4 ,,	9·1 ,,	26·0 ,,
60 ,,	3·2 ,,	6·0 ,,	16·0 ,,
50 ,,	1·7 ,,	3·4 ,,	8·9 ,,
40 ,,	0·7 ,,	1·5 ,,	3·9 ,,
30 ,,	0·3 ,,	0·6 ,·	1·7 ,,
20 ,,	0·2 ,,	0·4 ,,	0·8 ,,
10 ,,	0·15 ,,	0·25 ,,	0·5 ,,
100 parts of water	0·12 ,,	0·18 ,,	0·27 ,,

Coumarin is readily soluble in vaselin.

The announcement of the artificial preparation of coumarin was made by Perkin in a paper read before the Chemical Society 16th May, 1867. He succeeded in forming it by causing sodium-salicylol to react on acetic anhydride, in which it dissolves with considerable evolution of heat[*]. When the violence of the re-action moderates, the mixture is boiled for a few moments, then, on the addition of water, an oil separates and floats on the surface. On distilling this oil, there first passes over a little acetic anhy-dride, then a little salicylol, and finally coumarin, which crystal-lizes on cooling in the receiver. He obtained it even more simply by gently heating for some hours a mixture of 3 parts salicylic aldehyde, 5 parts acetic anhydride, and 4 parts of sodium acetate; the whole solidifies on cooling to a crystalline mass, from which, on treating it with water, there separates an oil smelling of acetic acid and coumarin. An ethereal solution of this oil, when shaken with a solution of sodium carbonate, gives up thereto a crystal-lizable acid, whilst coumarin remains dissolved in the ether. The acid dissolves easily in hot water, alcohol, and ether, and crystal-lizes from water in white needles which melt at 146°, and have the composition of acetylcoumaric acid.

When acetylcoumaric acid is gently heated above its melting-

---

[*] Journ. Ch. Soc. xxi. pp. 53, 181.

point, it evolves pungent fumes of acetic acid, and ultimately leaves a thick oily body, an ethereal solution of which deposits crystals of coumarin on evaporation, which can be purified by crystallization from alcohol. It agrees in all its properties with the coumarin of the Tonka bean, $C_9H_6O_2$. Its vapour has a powerful action on the brain[*].

The formation of coumarin by Perkin's method, as above described, depends on the simultaneous formation of sodium-acetate, for when a mixture of acetyl-salicylol, acetic anhydride, and sodium acetate is heated to the boiling-point, a considerable quantity of coumarin is produced; acetylcoumarin being first formed, and then decomposed on heating into acetic acid and coumarin.

Zwenger says[†] that when sodium amalgam is added in small quantities to a solution of coumarin in water containing a little alcohol at $40°$-$60°$, the coumarin is first converted by assumption of water into coumaric acid $C_9H_8O_3$, and this, by further assimilation of nascent hydrogen, is transformed into melilotic (hydrocoumaric) acid, $C_9H_{10}O_3$.

Coumarin is also formed in small quantities when malic acid is heated with phenol and sulphuric acid[‡].

Coumaric acid occurs, associated with coumarin, which is its anhydride, in common melilot and in Faham leaves[§]. It can be prepared by dissolving 3·5 grm. of sodium in $60-70$ cb. cm. of alcohol, adding 10 grm. of coumarin, and heating for one or two hours. Water is then added, the alcohol boiled off, and hydrochloric acid added. The precipitate is dissolved in a cold solution of sodium carbonate, the small quantity of coumarin which remains undissolved by ether, and the coumaric acid again precipitated with hydrochloric acid, and finally purified by recrystallization from hot water[‖]. It crystallizes in long needles which melt at $207°$-$208°$, are slightly soluble in cold, more readily in boiling water, and readily in alcohol. It may be sublimed if carefully heated, but decomposes on distillation, with the forma-

---

* Ber. Deutsch. chem. Ges. x. p. 284.
† Jahresb. 1865, p. 343, and 1867, p. 448.
‡ Ber. Deutsch. chem. Ges. xvii. p. 929.
§ Ann. Chem. Pharm. Suppl. viii. p. 30.
‖ Ber. Deutsch. chem. Ges. x. p. 284.

tion of phenol and other substances. On fusion with potash it is decomposed into salicylic and acetic acids.

Ethylcoumarin was obtained by Perkin, who named it Butyric coumarin. It smells like ordinary coumarin, and at the same time like fresh honey *. It is produced by heating sodium salicylol with butyric anhydride, boiling the product for a few minutes, and then pouring it into water, distilling the oil which separates, and collecting the portion which passes above 290°. It melts at 70°–71°, solidifies to a crystalline mass on cooling, and distils with slight decomposition at 296°–297°. It dissolves sparingly in boiling water, easily in alcohol and ether. From alcohol the compound crystallizes in large translucent prisms. Ethylcoumarin is converted by boiling with caustic potash into *Butyrocoumaric acid*, which crystallizes from dilute alcohol in flat prisms and melts at 174° with decomposition †.

β-Methylcoumarin, $C_{10}H_8O_2$, is formed by the action of concentrated sulphuric acid on a mixture of phenol and aceto-acetic ether; this substance is very similar to coumarin, and crystallizes from benzene in needles which melt at 125°–126° ‡.

Dimethylcoumarin is formed by the action of sulphuric acid on a mixture of paracresol and aceto-acetic ether, and crystallizes from dilute alcohol in long, strongly refractive needles, melting at 148° §.

Paracoumarhydrin, $C_9H_8O_3$.—This substance, which is metameric with coumaric acid, is formed when paracoteïn, $C_{19}H_{12}O_6$, is heated with caustic potash, and crystallizes in plates which smell like coumarin and melt at 82°–83°. An acid very similar to piperonylic acid is also formed in the reaction; it is very similar to the piperonylic acid which occurs with paracoteïn in Paracoto bark ‖. Paracoumarhydrin appears, therefore, to be homologous with *piperonal*.

There are other artificially prepared compounds possessing this perfume, amongst which are Thallin and its salts. These salts are manufactured for medicinal purposes as antipyretics and anti-

---

* Journ. Chem. Soc. (2) vi. pp. 53, 472.

† Ibid. 1881, i. p. 439.

‡ Ber. Deutsch. chem. Ges. xvi. p. 2127.

§ Ibid. xvi. p. 2119, and xvii. p. 2187.

‖ Ann. Chem. Pharm. cxix. p. 30.

septics. Thallin is a liquid base, first prepared in 1885 by Prof. Skraup, its chemical formula is $C_9H_{10}N(OCH_3)$, and its systematic name Tetrahydroparachinanisol. It has a strong odour resembling coumarin, and forms well-defined salts with acids. These salts, which are yellowish-white crystalline powders, are all freely soluble in water, and have the property of forming intensely emerald-green solutions; hence, for brevity, the adopted name Thallin, from θαλλὸς, a green twig. Ferric chloride produces the colour. Sodium thiosulphate changes the green tint into violet, and then into wine-red; oxalic acid, at ordinary temperatures, into pale yellow, deepening into saffron on heating.

According to the Patent Specification, parachinanisol (from which Thallin is formed by hydrogenation) is obtained by heating together paramidoanisol, paranitroanisol, and acrolein *.

By reducing agents Parachinanisol takes up four atoms of hydrogen, forming Thallin †. At ordinary temperatures thallin is an oily liquid, solidifying when cooled to yellowish crystals.

Thallin sulphate is a yellowish-white crystalline powder, with a bitter and intensely aromatic taste, and a peculiarly persistent odour similar to coumarin. Like the base itself, this salt in a one per cent. solution is coloured emerald-green.

Thallin tartrate occurs as a yellowish-white crystalline powder with an odour reminding one of anise and coumarin.

The odour of coumarin is observed in anisic aldehyde, a heavy oil (sp. gr. 1·09) of an amber colour, produced by the action of weak nitric acid on oil of anise. The crude product contains anisic acid, which may be removed by washing the oil, carefully distilling it, and agitating the distillate with weak potash lye.

---

* Acrolein is prepared by distilling in a capacious retort a mixture of glycerine with phosphoric anhydride. The vapours must be condensed in a properly cooled receiver, luted on to the retort, and provided with a tube opening into a chimney having a good draught. The distilled liquid separates into two layers, the upper one consisting of acrolein, and the lower one of an aqueous solution of the same mixed with a quantity of acrylic acid. The distillate, after digestion with finely powdered litharge to neutralize the acid, must be rectified by the heat of a water-bath, and submitted to a second rectification from calcic chloride. All these operations must be conducted in vessels filled with carbonic anhydride, because acrolein becomes rapidly oxidized when exposed to the air. It is a clear colourless liquid, lighter than water, and when pure is neutral to test-paper. It boils at about 125° Fahr.

† Archives der Pharmacie, xxii. p. 840, and Pharm. Journ. [3] xv. p. 575.

This odour has been compared to *Hawthorn*. More complete details of the method of manufacture are given under that head.

The odour of coumarin is observed on heating Umbelliferon, Daphnetin, and Methylumbelliferon.

Umbelliferon, $C_9H_6O_3$, is a neutral glucoside, obtained by the dry distillation of various resins, chiefly of those derived from umbelliferous plants. Crude galbanum yields 0·83 per cent., sagepanum 0·32, and asafœtida 0·28 per cent. It is likewise obtained from the resins of Sumbul root and Angelica root. It is also obtainable from the alcoholic extract of the bark of *Daphne Mezereum* (the spurge-laurel), which also contains *daphnetin*.

Umbelliferon may be easily prepared from galbanum by boiling the crude gum with water, dissolving the resinous residue in milk of lime, and precipitating the filtered solution with hydrochloric acid. By distilling this purified resin with water, an oily distillate is obtained which, on standing, deposits crystals of umbelliferon, to be purified by recrystallization. Umbelliferon is also formed when a concentrated alcoholic solution of pure galbanum resin saturated with hydrochloric-acid gas is heated for some time to 100°. Umbelliferon may be synthetically prepared by heating resorcinol and malic acid with sulphuric acid, the reaction corresponding to the formation of coumarin from malic acid and phenol.

Umbelliferon forms colourless rhombic prisms having a faint silky lustre; it is tasteless, inodorous when cold, dissolves very slightly in cold water, and very abundantly in boiling water. It also dissolves in alcohol and chloroform. The aqueous solution is colourless by transmitted light, but exhibits by reflected light a splendid blue fluorescence. When umbelliferon is warmed it emits an odour like coumarin. Its aqueous solution when boiled emits the same odour. It melts at 224°, sublimes below its melting-point, and volatilizes without residue. It dissolves in concentrated sulphuric acid without decomposition, the solution showing a dark blue fluorescence.

Water and sodium amalgam convert it into hydro-umbellic acid. By boiling with dilute acids or by the action of emulsion it is converted into glucose and *daphnetin*, $C_9H_6O_4$*.

Daphnetin has been obtained synthetically by heating pyro-

---

* Ann. Chem. Pharm. cvi. p. 1.

gallol and malic acid with sulphuric acid. It crystallizes from hot water in yellowish needles or prisms which melt at 255°–256° and smell like coumarin on heating.

Methylumbelliferon is formed when umbelliferon is heated with methyl iodide, caustic potash, and wood-spirit. It crystallizes in plates which have a strong odour of coumarin when heated, and melts at 114°*. Its boiling aqueous solution emits the same odour.

* Tiemann and Reimer.

CHAPTER VIII.

VANILLA.

THERE are several species and varieties of the Orchids producing this valuable bean. Delteuil * gives a list of the plants cultivated in various countries, from which it appears that in Mexico are found the *V. planifolia* (yielding the finest fruit), *V. sativa*, *V. sylvestris*, and *V. pompona* (with short thick fruit called " vanillon "). Guiana and Surinam produce the *V. Guyanensis*; Bahia, *V. palmarum*; Brazil and Peru, *V. aromatica* (which is the least aromatic of all); Réunion, two sorts, which appear to be varieties of *V. planifolia*, the one usually called the small, the other the large vanilla, characterized by the stalk being thicker, the leaves much larger, the flowers larger and of a deeper yellow tint, the fruit thicker, shorter, and triangular; but this being of inferior aroma its cultivation has been almost generally abandoned.

In a paper communicated to the " Société d'Emulation " by Jaillet † on the culture and preparation of Vanilla, the author infers, from the writings of various botanists on this subject, also from personal observation, that the *V. planifolia*, *V. sativa*, and *V. sylvestris* are identical, the distinctive specific characters not being clear and decided, but depending upon the effects of age, climate, and vigorous growth. He concludes that although there really exist several species of Vanilla, and that there may be many varieties of the same species, all, or nearly all, the commercial vanilla is furnished by the *V. planifolia*. This would appear to be the opinion of the authors of ' Pharmacographia,' as *V. planifolia* alone is mentioned as the commercial source, but perhaps referring only to the products of Mexico and Réunion. According to

---

* ' Etude sur la Vanille,' Paris, 1874.
† Répertoire de Pharmacie, viii. 357.

Bentley and Trimen ('Medicinal Plants') there are several varieties of Vanilla found in commerce, as Mexican or Vera Cruz, Bourbon, Mauritius, Java, La Guayra, Honduras, Brazilian, &c., the finest being the Mexican, of which there are different qualities. They further state that these varieties of vanilla are doubtless derived from different species of the plant. The finest, as the Mexican, is commonly said to be the product of *V. planifolia*, Andrew *, synonymous with *V. claviculata*, Swartz; *V. sylvestris* and *V. sativa*, Scheide; *V. viridifolia*, Blume; and *Myrobroma fragrans*, Salisbury †. Scheide, a writer on the botany of Mexico ‡, considered the *V. sylvestris* as probably a synonym of *V. planifolia*; but Morren states § that the exactness of that opinion is not clearly demonstrated, and declares that the writings of Blume ‖, Swartz ¶, Plumier **, and Desvaux †† merely render more difficult the question as to which species, varieties, and sorts are to be attributed the vanilla of commerce; also that this question could only be solved by an experienced naturalist who should examine the plants in the localities where they actually grow, compare the different length, thickness, shape, colour, flavour, and value of the fruit yielded by each species and variety, and accompany the diagnosis with drawings made on the spot. This argument does not authorize Jaillet to quote Morren as an authority for stating that the *whole* of the commercial vanilla is derived from *V. planifolia*; Morren simply says that the finest vanilla closely resembles the fruit of that plant.

Vanilla is also found on the west slope of the Cordilleras, in Java, Mauritius, Seychelles islands, Tahiti, Cochin-China, Jamaica, Guadeloupe, St. Marie, Mayotte, and Madagascar. It is cultivated at Mysore, in India. The principal centre of the production is the littoral of the State of Vera Cruz, in the hot low-lying ground called the "terra caliente," also in the Mexican

* 'Botanist's Repository,' ii. p. 538.
† 'Paradisus Londinensis,' t. 82.
‡ 'Botanische Berichte aus Mexico,' 1820.
§ Bulletins de l'Académie Royale de Belgique, xvii. 1ᵉ partie, p. 130.
‖ 'Flor. Javæ,' Bijdragen, p. 422, and 'Rumphia,' i. pp. 197–8.
¶ Nova Acta Upsal. vi. p. 66; Fl. Ind. Occid., and Schrader's Journ. Bot. ii. fig. 1.
** 'Plantarum Americanum,' and Catesby's Hist. Nat. de la Caroline, iii. tab. 7.
†† Annales des Sciences Naturelles, 3ᵉ série, 1846, p. 117.

province of Oaxaca. It flourishes especially at Papantla, and such warm humid districts as Misantla, Jicaltepec near Nantla, Calipa, and Tuntla. Also in the States of Tabasco, Chiapas, and Yucatan. The Mexicans call this plant *Telxochitl*, and the Spaniards *Baynilla*.

The writings of Kunth describe the vanilla plant as being found attached to trees and in the crevices of rocks in the hottest regions of Central America, in sheltered situations adjoining springs of water, on the banks of the Orinoco, near to Carichana, at the cataracts of Maypur and Atur, at Javita, and at Esmeralda; in New Andalusia near the Convent of Caripa, at San Fernando, Bardones and Carupano; in Venezuela, between Porto Cabello, Guayguaza, Aroa and Nueva Valencia; in the Valley of Capaya and near the promontory of Codera; in the Andes of New Grenada, Quito and Peru, near Turbaco, d'Almagner and Popayan; on the eastern slope of Mount Pichincha; in the valleys of Loxo and near the River Amazon, between Tomependa and Jaen de Bracamoros; in the island of Cuba, near Elmariel, Bahia, and in Honduras.

A fine collection of many of the commercial varieties of vanilla has been lately completed at Kew, including samples from Réunion, Mauritius, Seychelles, Bahia, Mexico, and Java. A species of Vanilla from Tatia, New Granada, which has until recently remained unnamed in the Daniel Hanbury Herbarium at the Pharmaceutical Society, has been described by Rolfe from this specimen and from one in the Kew Herbarium as a new species under the name of *Vanilla ensifolia*, Rolfe. It is probably the source of some of the vanilla occasionally imported from South America.

From a recent Consular Report (1892) by G. H. Portal, of Zanzibar, we learn that Vanilla has grown well and freely on the island wherever it has been planted. The priests of the French mission at Bagamayo, on the opposite coast, cultivate it extensively and reap a handsome profit.

The culture and preparation of vanilla, requiring great care and management, are described by Jaillet in his paper above referred to as follows (some slight remarks from other authors being added, and one or two words corrected) :—

In Mexico, plantations are established either in virgin forests or in open fields. In the former case it is necessary to cut down

all shrubs, climbers, and such large trees as would cause an excess
of shade, leaving only young trees suitable to serve as supports to
the plants. Preference is given to those containing a milky sap,
as the orchid attaches itself to the bark by means of aërial roots
(produced from the nodes), which are its veritable organs of nu-
trition; the subterranean roots being very insignificant in
comparison to the size of the plant. It is not uncommon to
observe the gradual decay of the stalk near the root which is in
the ground, and at the same time a remarkable development of
the same stalk as it increases in length. Close to each tree, two
cuttings are planted side by side in the following manner :—in a
shallow trench about an inch and a half deep and fifteen or twenty
inches long, is imbedded a cutting as far as three joints or eyes,
the three leaves being first stripped off ; the trench is then
covered up with dried leaves, leaf-mould, coarse sand, brush-wood,
&c. The bed should be slightly raised above the level of the soil
in order to prevent a collection of stagnant water which might rot
the plants. The remainder of the shoot, 3 or 4 feet long, is
tied against the tree. The supporting trees should be quite 12 or
15 feet apart to allow sufficient room for the development of the
plant, the growth being very rapid. After a month the cuttings
will have taken root, and must be carefully kept free from weeds
and underwood of all kinds. In the third year these plants will
commence to bear fruit. Planting takes place in the rainy
season ; in default of sufficient rain, the cuttings must be frequently watered. In the case when it is desirable to plant a field,
plain, or low-lying ground, the method in Mexico is first to
thoroughly plough up the land and sow it with maize. While
this is growing, a quantity of young lactescent trees, of the fig
tribe, make their appearance over the field ; these, after a year or
eighteen months, are large enough to support the vanilla plants,
which are set in the manner above described, and from them the
finest product is obtained.

There are five sorts of vanilla plants locally known in Mexico
by the following names :—

1. *Vanilla coriente.* This is held in greatest esteem for the
quality of its pods, which are classed commercially into five
classes, the first being long fleshy pods full of pulp and seed, and
with a very fine skin ; the second, called *V. chicafina,* about half
the length of the preceding and with a thick skin, not so fine in

flavour, but still sometimes sold with it, also bearing the same vernacular name " lec," " leq," or " leg," abbreviated from " legitimate." The third sort is the *V. sacata*, having a finer cuticle than the first. The fourth, *V. resecata*, is small, dry, and only a fourth the length of the preceding. The fifth quality is called *basura*, and is a very inferior product.

2. *Vanilla sylvestris* or *simarona*, a wild species, with smaller fruit than the *coriente*. This plant appears to be botanically identical with the *coriente*, but as it is found growing in dense woods, whose foliage deprives it of sunlight, its pods cannot develop.

3. *Vanilla mestiza* (*mestiza*, meaning in English *middle, medium*, or *average*). This bears a rounder pod than other varieties. The green unripe pods are spotted with brown, and the ripe pods are very apt to split open.

4. *Vanilla puerca* (La Porcine, Vanille cochon, Swine vanilla). This variety bears much smaller pods than those of *V. coriente*; they are also rounder in form, of a darker green colour when unripe, and exhale an unpleasant odour during the process of curing —hence the name.

5. *Vanilla pompona*. This plant bears a very thick short fruit covered with a very thin skin. When this fruit begins to dry it acquires a very fine perfume (recalling that of heliotrope). The perfume is, however, considered less sweet than that of the " coriente," and it is apt to go off, or disperse by evaporation if tied in bundles alone, so it is sometimes packed up with the No. 1 variety. It is comparatively low-priced, and is commercially designated " Vanillon " in French.

The method of cultivation adopted in the island of Réunion is different; the plant being so trained that all the flowers may be within easy reach of the hand of the cultivator, not so much for facility of gathering the fruit as for the purpose of artificially inoculating the flowers. The plantation may be started in the forest or in an open field. In the first case, the cuttings are set at the foot of trees, and the trunks are connected together transversely by sticks of wood or bamboo attached horizontally, so as to form a sort of trellis on which the plant can spread freely. In no case are the trees lopped to allow too much sun, for the plant loves a humid soil and is injured by the direct burning rays. It is under large trees that the vanilla plant is seen in its typical form, vigorous and richly productive. When an open field is selected as the site of a plantation, the necessary supports for the plant must first be grown. For this purpose mangoes and fig-trees are preferred, also the *Jatropha Curcas*, the tree producing the " physic-nut," which strikes readily from cuttings, is of rapid growth, and furnishes an abundance of lactescent juice, well suited to supply the necessary nourishment to the vanilla plant. (There is, however, a possible danger of the acrid matter contained in the *Jatropha Curcas* being absorbed by the parasite plant *.)

When these young trees have attained a size sufficient to afford the necessary shade, cuttings of the orchid are set in the following manner :—Between the trees and following the lines in which they are planted, a trench 8 inches deep is dug, the cuttings are placed in it and covered with a little leaf-mould, dry leaves, and straw. The rainy season is preferred for this operation, as success in striking the cuttings depends essentially on moisture and shade. When the young shoots begin to grow, they only need to be guided and spread along the trellises previously arranged to receive them, and to allow the adventive roots to connect with the trench between the supporting trees. In two years the plantation is in full bearing.

The following cultural instructions were contributed by David de Floris, of Réunion, to the ' Journal of the Agricultural Society of India' † :—The cuttings must have at least three knots, but may have more according to the disposition of the protecting trees, or the shade which they can give. All trees are good as protectors with the exception of those which change their bark ; the best are the Mangoe-tree, the blackwood (*Acacia Lebbeck*), the Dragon-tree (*Dracœna Draco, or Pterocarpus Draco*), the Jack-tree‡, the Ouatier (*Bombax Malabaricum*), and the Pignon d'Inde (*Jatropha Curcas*) ; but this last should not be planted alone, on account of its shedding its leaves when the vanilla plants are in bearing, the sun then striking upon the vanillas and on their pods, being very injurious to both. It is necessary to plant the " Pignon d'Inde " between the Dragon-trees and the Ouatier or other trees, the leaves of which may serve to shade it as well as the vanilla plant, to which it only serves for a protection

---

* *Vide* Pharm. Journ. [3] xi. p. 430.   † Vol xi. part iv.

‡ *Artocarpus integrifolia*, L.

during a certain period of the year. The protecting trees ought to be planted six feet apart, in rows from east to west. They should be occasionally pruned, so as to produce a *half-shade* or chequered shade, and they should be sufficiently grown to produce this *demi-jour* before the vanillas are planted. In case, however, one should require to plant before the necessary shade exists, the plants should be surrounded with palm leaves in preference, and watered much more often than if they had their natural shade; the cuttings should be planted at the side of the supporting tree opposed to the sun. The longer the cutting, the more knots must be put into the ground; one knot when the cutting has three, two when it has four, and four or five knots when long creepers are planted. These cuttings should be laid in the ground the tendrils towards the tree, and well fixed with one, two, or several flat ties according to their length. They should not be tied with round string, which would eventually strangle the plants, but with a sort of bast or fibre from the leaf of the *Pandanus vacoa*. Manure to the cuttings would be hurtful, but rooted plants may be manured with rotted dung if the soil be poor. Vegetable manure composed of rotted leaves is preferable to dung, being less heating; but the stuff must be well rotted, as the young roots are very tender and delicate. Watering in the first few days after planting is always an absolute necessity, particularly in a dry locality. Plants put in in the middle of the cold season languish, lose their buds, and often perish. The earth should be trodden down on each plant after having been watered. The plantation should not be made very near the sea-shore unless protected by trees from the direct action of the salt air blowing over the plants, as such would render them poor and sickly. A ground sloping to the west is preferable, as permitting more warmth to the plants and less exposure to the wind. It is advised to manure the plants once a year, a little before the flowering-season, and to cover the manure with stones to prevent its evaporation; the stones also serving to keep the roots cool, and prevent the rains washing the earth away. Too much shade, or shade badly applied, seem almost as prejudicial to a good crop as the other extreme of exposure. Pods which have been too much shaded are long, soft, thin, and difficult to ripen; whereas, on the contrary, when they are sufficiently exposed to the sun they are fat, round, firm, and contain much more flavour.

No plant should be allowed to bear too freely, the quality and size of the pods suffer thereby. The pruning out of pods should be performed after the fruit is fairly set, and should be proportioned to the age and health of the creeper ; not more than five or six pods being allowed on a single cluster. A plant of three or four years' growth has hundreds of blossoms thereon, but the quantity of pods taken from the same should not be more than will yield half a pound of dried produce.

Of course in its native place of growth, the method of propagating by striking young shoots of three feet or so in length is the most rapid method ; but stock could probably be reared from seed taken from pods which have matured naturally by being left on the plant ; such pods split open and drop some of their seed.

Fecundation of the Flower.

In the flower of the vanilla the male organ is separated from the female organ by the light membranous skin of the labellum (the upper lip of the stigmatic orifice), this totally covering the female organ, and as the anther rests on that valve of the stigma, it is evident that notwithstanding the dehiscence of the anther, the orifice which allows passage of the pollen is closed by the labellum, thus rendering spontaneous fecundation comparatively rare. It does, however, sometimes occur, and may be attributed to the passage of a winged insect in search of food, or to the action of the wind detaching the pollen from the anther ; but it seems more rational to suppose that the brush-shaped appendage on the labellum is solely intended for the purpose of collecting the pollen and then depositing it on the stigma at the moment when the flower begins to droop and fade. Still, the natural fecundation 'is a rare occurrence, for in Guiana, Mexico, and all other countries where the plant is left to itself, it has been observed that a length of 12 to 26 inches of vine will only produce one pod, the number of flowers growing on such length of stalk being about forty, all of which can be artificially fecundated. The flowers are produced in clusters in the axils of the leaves. A plant in full health and strength may produce as many as two hundred clusters at a time, each cluster consisting of from fifteen to twenty flowers. A single plant, therefore, may bear three or

Fig. 7.

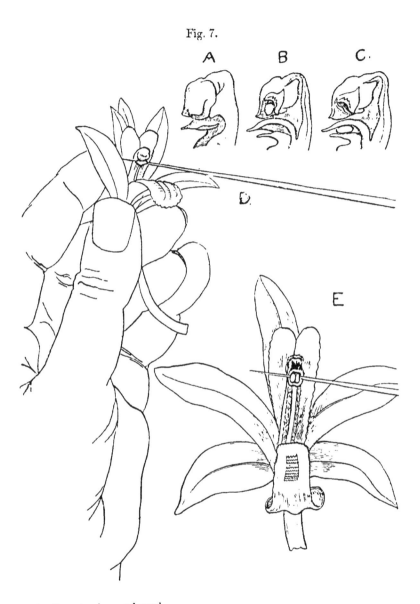

A. Gynostemium, enlarged.
B. Section of ditto before the operation, showing the pollen in the anther.
C. Ditto ditto after the operation, showing the pollen in the stigma.
D. First part of the operation.
E. Second part of the operation.

four thousand flowers. The flowers in a cluster expand one after the other, and only last a day. By some cultivators it is thought desirable not to fertilize more than two or three flowers in each cluster, and to select for the operation the largest and finest flowers; these are generally to be found amongst those which are the first to open. (Other cultivators fertilize five or six flowers.) If this rule be observed, it will be found that the quality of the pods will largely compensate for the quantity which might be obtained by fertilizing a larger number of flowers.

The old process for performing the operation of artificial fecundation consisted in cutting the labellum which is the obstacle to the natural process, but this plan was not always successful; and it was improved upon by a Creole slave, who discovered that a more rapid and sure way was to lift up or tear away the labellum from beneath the anther, and so bring that organ in direct contact with the stigma. In this way it is possible to obtain more than 3500 pods on a single plant, but such a demand on the plant would cause it to perish before the fruit could mature; therefore, as before observed, it is advisable to fecundate only the finest flowers on each bunch, selecting those which present a large fleshy peduncle. It is an ascertained fact that the handsomest fruit are produced from the first flowers, but the best fruit from the last flowers which open on each bunch. Fecundation is assured when the flower is persistent, and dries at the extremity of the fruit. This result obtained, the remainder of the bunch with all its buds should be cut off.

The flowers of the vanilla begin to appear in June, and are fecundated up to September. (In India from February to April.) The fecundation should be made from 8 to 9 o'clock in the morning till 3 in the afternoon, and the earlier the better. The operation should be done with great care, using as little force as possible. The instrument employed is simply either a small bamboo about 3 inches long, cut very thin and rounded off at one end, or the dorsal rib of the leaves of palms, cocoanuts, or latanias (these are Réunion palms). With anything like the point of a penknife the danger of wounding the delicate organs of the flower is incurred. The operation is quite simple, and may be executed with great rapidity by a light and practised hand. An expert will fertilize as many as a thousand flowers and upwards in the course of a forenoon. The rule is as follows :—Seize the base

of the flower between the thumb and middle finger of the left
hand, placing the forefinger on the back of the gynostemium to
support it. Or, between the fore and middle fingers of the left
hand, held horizontally, place the three upper petals of the flower,
raising the thumb and keeping it close to the anther. Now, with
the little instrument held in the right hand, tear the piece of the
corolla resembling a hood, in order to expose the organs of fecundation. The end of the instrument is then introduced under the
upper valve or operculum of the female organ. When this operculum is completely raised, straight up, the stamen, which at first
rises with it, tends to return to its original position, bending
towards the female organ; this inclination must now be assisted
with the thumb of the left hand, lightly pressing the stamen
against the stigma, to which it will adhere. Nothing now
remains to be done but gently to withdraw the instrument, and
the flower is fertilized. If, at the end of the third day, the flower,
which begins to wither immediately after the operation, maintains
its position on the summit of the ovary, the operation has been
successful.

Harvesting the Fruit.

The fecundated flower decays and dries at the extremity of the
ovary, and after a few days falls off, leaving the persistent
gynostem attached to the fruit, which continues to grow for a
month, but must be left on the stem for six months longer to
allow it to ripen completely. The first pods to ripen are generally
inferior to those which mature later. It is most important that
pods should not be gathered before they arrive at the proper stage
of ripeness, otherwise they ferment and rot in a few months after
preparation. The end of the pod begins to turn yellow when it
is approaching ripeness, but the only certain indication of maturity
is the crackling sound produced when the pod is pinched between
the fingers. It is quite as important to avoid gathering the fruit
too late as too early; if over-ripe it is apt to split on the stalk; and
if not so found, it will split in the curing. It is advisable to visit
the plantation frequently when the time for ripening approaches
and pluck the pods from day to day as they reach the required
degree of maturity, and not detach the entire bunch as is done in
some countries.

Some precaution is required in separating the pods from the

stem. The fruit should be grasped with the right hand towards the but-end, and removed from the stem by a gentle twist from right to left. Some persons take the pod by the middle or by the end and draw it roughly towards them ; when so treated it often breaks or the entire bunch is detached from the tree with the pods still unripe. Other persons gather it by pinching it off with the nails, but then the but-end no longer existing prevents uniformity in the packets and raises difficulties for the sale.

Curing the Fruit.

The odour of vanilla does not exist in the fruit as it is gathered, but is developed by a process of fermentation in the curing. When a pod is allowed to fully ripen on the plant it splits into two unequal parts, becoming first yellow, then brown, and finally black. While it is drying it exudes an unctuous liquid of a dark red colour, called *balsam of vanilla,* and when quite dry the pod becomes brittle and devoid of all perfume.

The following are the various processes for curing vanilla and preparing it for the market :—

Guiana process.—The beans are placed in ashes, and there left till they begin to shrivel ; they are then wiped, rubbed over with olive oil, and their lower end having been tied they are left to dry in the open air.

Peruvian process.—The beans are dipped into boiling water, tied at the end, and hung in the open air. After drying twenty days they are lightly smeared over with castor oil, and a few days afterwards tied up into bundles.

Mexican process.—As soon as gathered the beans are placed in heaps under a shed protecting them from sun and rain, and in a few days, when they begin to shrivel, are submitted to the " sweating" process ; this is carried on in two different ways according to the state of the weather. If it happens to be warm and fine the beans are spread out in the early morning on a woollen blanket and exposed to the direct rays of the sun. At about midday, or one in the afternoon, the blanket is folded around them, and the bundle is left in the sun for the remainder of the day. In the evening all the vanilla is enclosed in air-tight boxes so that it may sweat the whole night. The next day the beans are again exposed to the direct action of the sun; they then acquire a dark coffee-colour, the shade being a deeper brown in proportion to the success of the sweating operation. Should the weather be cloudy, the vanilla is made into bundles, and a number of these are packed together in a small bale, which is first wrapped in a woollen cloth, then in a coating of banana leaves, and finally the whole is enclosed in a thick matting and sprinkled with water. The bales containing the largest beans are now placed in an oven heated to 60° C. (140° F.). When the temperature of the oven has fallen to 45° C. (113° F.) the smaller beans are introduced, and the oven closed tightly. Twenty-four hours afterwards the smaller beans are taken out, and twelve hours later the larger ones. During this process the vanilla has *sweated* and acquired a fine chestnut colour. The delicate operation of drying has now to be commenced : the beans are spread on matting and exposed every day to the sun during nearly two months; when the drying is nearly complete sun-heat is no longer needed, and the beans are spread out in a dry place until the necessary degree of desiccation is arrived at. Finally they are tied in small bundles for the market.

Réunion process.—The beans are sorted according to length before being subjected to the treatment. The long ones are steeped in water at 90° C. (194° F.) during ten seconds, the medium size during fifteen seconds, and the short ones fully a minute. They are then exposed to the sun between woollen blankets daily until two or three o'clock in the afternoon until they acquire the characteristic chestnut colour. After this exposure, which may last from six to eight days, the beans are spread out under sheds to dry gradually. The sheds in this colony being roofed with zinc, they really constitute drying-closets, through which a current of hot air continually circulates. This desiccation takes about a month, during which time the only care necessary is to turn the beans frequently, so that they dry evenly. At the moment when it is found that the beans may be twisted easily round the finger without cracking—that is to say, when they have acquired a degree of dryness which can be judged only by experience, a fresh operation is commenced which requires the most minute and vigilant care; this is termed the *smoothing* process. The operator must pass every bean between his fingers repeatedly, for, on drying, the beans exude from their entire surface a natural fatty oil. It is to this oil, which exudes as the fermentation proceeds, that the lustre and suppleness of the bean is due. When the beans are sufficiently dry they are tied into bundles, each of them being composed of fifty pods of uniform length. In this manner three commercial sorts are obtained, and termed as follows :—1. " Fine vanilla," 8 to 11 inches long, very dark brown or nearly black, unctuous, glossy and clean-looking, and finely furrowed in a longitudinal direction. These soon become covered with an abundance of the frost-like efflorescent crystals technically called *"givre."* 2. " Woody vanilla," 6 to 8 inches long, lighter in colour, more or less spotted with grey, not glossy. These are generally the produce of pods gathered in an unripe state. They frost or " givre " very little, if at all. 3. " Vanillons," of which there are two sorts, those obtained from short but ripe fruit, which are excellent and frost well, and those from abortive and unripe fruit, whose perfume is simply the result of absorption from the fine beans with which they have so long been in contact.

There are modifications of these processes, but they do not materially differ. Of course under different climatic conditions different modes of curing are adopted, but the sweating or fermentation must be effected by one means or another.

The finished product being sorted and tied up into bundles according to the length of the pods, is finally packed into tin boxes of different dimensions according to the length of the bundles ; each box containing 10 to 12 kilogrammes ; no paper or wrapper whatever being enclosed, as such might be injurious to the preservation of the pods. The boxes are soldered up and labled according to the quality of the contents.

Adulteration.—Vanilla is subject to frequent falsification. It is sometimes found that the greater part of the odoriferous principle has been abstracted by alcohol, and an inferior odour substituted by rubbing the pods with Balsam of Peru. The pods are sometimes even filled with sand and other matters to give them weight, and it is not an uncommon fraud to dust them over with benzoic acid to imitate the fine qualities which are naturally frosted with the crystals of vanillin which form on them.

Aromatic Principles.

The fruit of the vanilla, whether matured by nature on the plant or finished by artificial process, exhales one of the finest odours produced in the vegetable kingdom. Some odours may be cited which somewhat recall it, as those of the *Pothos odoratissima, Heliotropium Peruvianum, Eryobotrya Japonica,* Tonka bean, *An-*

græcum fragrans, Capparis spinosa, Cereus grandiflora, and *Cestrum vespertinum,* but in an inferior degree, also strongly in the flowers of *Azara microphylla* *.

In the green state the pod consists of an acid pulp containing raphides in needles, and crystals of oxalate of lime (these bodies being also found in the stalk and leaves); the pods also contain a *citron-coloured oil* surrounding the seeds. When this oil is isolated by ether, it is found to possess an odour somewhat similar to that which the bean will furnish at maturity. As maturity progresses, the lower end of the pod begins to turn yellow and disengage a characteristic penetrating odour recalling that of bitter almonds; the valves of the pod crack open and permit the escape of a small quantity of a balsamic oil. By degrees the colour darkens, the epidermis softens, and the true odour of vanilla begins to develop. This natural fermentation gradually progressing up the pod, the proportion of balsamic oil increases and exudes in thick reddish drops, especially if the split pod has been tied together at the end. This oil is known by the name of *Balsam of Vanilla,* and in Peru it is very carefully collected by the planters, but not sent to Europe. The ripening process, thus slowly progressing upwards, does not reach the top or stalk part until about a month. The chemical changes which have meanwhile taken place inside the pod, through the combined action of the air and the sun, remains a secret process guarded by Nature.

Some chemists are of opinion that the odorous principle is localized in the centre of the fruit, in proximity to the seeds and the placenta, but others think that the entire fruit is concerned in its formation, as it undergoes such great changes.

The artificial methods of hastening the maturity, either by hot water or by the exposure to sun-heat or stove-heat, have for object the production of a *uniform* ripening of the pod over its entire length all at once, and not by degrees up the pod as happens in nature. The artificial processes also prevent the splitting of the pod and consequent loss of perfume which occurs when the pod is left on the vine.

The *givre* or crystals which form on good vanilla were formerly thought to be benzoic acid, but the researches of Gobley disclosed the fact that they are quite a different body, which he termed Vanillin. The conditions favourable for its formation are : pre-

---

* Described by Dean Hole, in his 'A Book about the Garden,' p. 234, as a hardy shrub.

servation of the beans in a dry place and in a box which is not hermetically sealed; also the removal from a warm locality to a cold one, which occurs when the produce is sent from the West Indies to Europe. The Réunion vanilla becomes frosted with crystals in about two months from the time of its preparation; being soldered up in tin boxes in the colony, it is found frosted when the boxes are opened on their arrival in France. These crystals assume two distinct forms :—they appear either in very thin laminæ or (which is generally the case) in needles so extremely fine and so close together as to appear like a minute hoary mouldiness or tufts of cotton—an aspect which, until the cause was understood, depreciated the value, but which is known to indicate the finest quality, or the ripest and most perfumed beans. Vanillin exists in the ripe fruit, but only appears on the surface under conditions favourable to its efflorescence.

The reactions distinguishing vanillin from benzoic acid are :— Sesquichloride of iron gives to vanillin a dark violet coloration ; cold sulphuric acid gives to it a green coloration, and the hot acid colours it red.

Quantitative Estimation of Vanillin in Vanilla-pods.

The following method was devised by Tiemann and Haarman, and is abstracted from a paper read by them before the Berlin Chemical Society *. It is based upon the fact that vanillin, in common with other aldehydes, combines with acid sulphites of the alkalies to form compounds which are readily decomposable by acids. Vanillin being the only aldehyde present in vanilla, it is isolated without difficulty :—

" 30 to 50 grams of vanilla cut small are placed with 1 to $1\frac{1}{2}$ litre of ether in a large stoppered bottle and left in contact during from 6 to 8 hours, being frequently shaken. The clear liquid is then decanted off and filtered through a plaited filter into a large flask. A fresh quantity of 800 to 1000 c. c. of ether is introduced into the stoppered bottle, shaken very frequently, and, after from 1 to 2 hours, filtered. The operation is repeated for a third time with 500 to 600 c. c. of ether. The now exhausted fragments of vanilla are thrown upon the filter with the very last portion of the third quantity of ether, and then washed with a small quantity of fresh ether. After this treatment the fragments of vanilla are absolutely taste-

* Ber. der Deutsch. chem. Ges. viii. p. 1115, and Pharm. Journ. [3] vi. p. 603.

less and odourless, so that it may be inferred that the whole of the vanillin has passed into solution. The united ethereal extracts are now distilled from the flask upon a water-bath to 150 or 200 c. c. The residue is placed in a tall narrow stoppered glass, then 200 c. c. of a mixture of equal parts of water and a nearly saturated solution of acid sodium sulphite added, and the closely-stoppered vessel shaken during from 10 to 20 minutes. It is necessary to open the flask from time to time, especially at the commencement of the shaking, and during the operation to hold the stopper firmly to prevent loss of the liquid.

"After the yellow-coloured ethereal layer and the almost colour-less solution have become sharply defined, they are separated from one another by means of a stoppered separating-funnel. The ether is returned to the stoppered flask and again well shaken for five or ten minutes with 50 c. c. of a concentrated solution of acid sodium sulphite and 50 c. c. of water. The aqueous solution is separated as before, and placed with the result of the former operation in a clean stoppered vessel. This saline solution, which contains all the vanillin, is now shaken for a short time with 180 to 200 c. c. of pure ether, in order perfectly to remove a small quantity of impurities derived from the original ethereal extract. After the ether has been again separated, the saline solution is poured into a large flask with a long, but not too wide, neck. The neck is closed with a cork pierced in three places. Through the middle hole passes a funnel-tube, reaching nearly to the bottom and dipping into the solution. A second tube, equally long, connects the flask with the steam apparatus or a vessel of boiling water. The third tube, for carrying off the sulphurous acid evolved, passes from the underside of the cork to a vessel containing soda-crystals and water. An empty wash-bottle should be introduced between this vessel and the flask to retain a small quantity of the solution that may be carried over from the flask. By using such an apparatus, the decomposition of the alkaline acid sulphite can be effected without inconvenience from the sulphurous acid evolved. Dilute sulphuric acid, in the proportion of 150 c. c. of a mixture of three volumes of concentrated sulphuric acid with five volumes of water to each 100 c. c. of acid sulphite solution, is poured gradually through the funnel-tube, and when the consequent evolution of sulphurous acid becomes less, steam is introduced to remove it as completely as possible.

" As soon as the wash-bottle becomes much moistened on the

inner side, the operation is stopped. When cooled, the contents of the flask are removed into a well-stoppered bottle, and shaken three or four times with not too small quantities of ether (from 400 to 600 c. c.) ; the ether takes up all the vanillin present. The ethereal extracts are separated from the aqueous solution and distilled together from a large flask, down to 15 or 20 c. c. The last part of the operation is conducted very cautiously, the temperature of the contents of the flask being raised with a little steam to a point not beyond 50° or 60° C. The residue, which is of a faint yellow colour, is placed in a weighed watch-glass, washed carefully with pure ether, and the ether allowed to completely evaporate at the ordinary temperature. If the operation has been properly conducted, and especially if a too strong heating of the concentrated vanillin solution has been avoided, pure crystals of vanillin are formed, melting at 81° C. These are dried over sulphuric acid until they no longer lose weight. The solution of acid sulphite of sodium can be recovered.

" The above process may also be used for the detection of adulteration of vanilla, by perfuming it with such substances as benzoin &c. These remain behind in the ethereal extract after the removal of the vanillin, and may then be easily recognized."

Vanillin.—The results of the analyses made by Tiemann and Haarman * show the proportion of vanillin in various pods to be as follows :—

MEXICAN.

1. First quality.	Harvest 1873.........	1·69 per cent.
2. „ „	„ 1874.........	1 86 „
3. Medium „	„ „	1·32 „

BOURBON.

1. First quality.	Harvest 1874.........	2·48 per cent.
2. „ „	„ „	1·91 „
3. „ „	„ „	2·90 „
4. „ „	„ 1875.........	1·97 „
5. „ „	„ „	2·43 „
6. Medium „	„ „	1·19 „
7. Inferior „	„ 1874.........	1·55 „
8. „ „	„ 1875.........	0·75 „

JAVA.

1. First quality.	Harvest 1873.........	2·75 per cent.
2. Medium „	„ 1874.........	1·56 „

* Ber. Deutsch. chem. Ges. viii. p. 1118, & ix. p. 1287.

The appearance, more or less fine, by which various qualities are commercially distinguished is far from being a sure indication of richness in vanillin. The "best" qualities contain on the average 1·5 to 2·5 per cent. According to these eminent authorities, vanilla contains no other aromatic principle but vanillin and a little vanillic acid, but after what has been remarked about " balsam of vanilla" their statement is open to considerable doubt; and it is doubtful whether vanillin, which is now manufactured artificially on a large scale, will ever beat vanilla-pods out of the market *.

The practice of estimating the value of vanilla by the amount of vanillin actually existing in it has been criticised by a German chemist †, by reason that "good fresh vanilla may be devoid throughout of crystalline matter, though it contain another body from which vanillin is ultimately formed" (a glucoside),—thus confirming the opinion of other observers ‡ that the chemical changes which develop the full odour of the fruit take place not only *during* its preparation for the market, but continue in action for some time after the pods are tied up in bundles.

Vanillin also occurs in Siam benzoin §, in asafœtida ‖, and frequently in small quantities in beet-sugar ¶, as the beet-sugar contains *coniferin*, which has been found, together with vanillin, in asparagus **.

The West-Indian Vanilla, called " Vanillon," exhibits different characters, its odour strongly resembling that of a dilute solution of *piperonal*. It contains from 0·4 to 0·7 per cent. of vanillin to which there obstinately adheres an oily substance, probably another aldehyde; the oil absorbed from it by bibulous paper smells not like piperonal, but like bitter-almond oil. The vanillic acid prepared from " vanillon " was contaminated with another substance, probably benzoic acid. The strong heliotrope-like odour of vanillon is perhaps due to vanillin mixed with a small quantity of benzal-dehyde; but yet it is surprising that when a small quantity of oil of bitter almonds is added to a solution of pure vanillin, the individual odours of the two oils are obstinately and persistently manifest for

* The Mexican crop of 1890–91 was the largest grown. The Réunion crop, from 3000 acres of land, was 506,462 lbs., as against 462,660 lbs. in 1888, and 417,230 lbs. in 1887. (Am. Journ. Pharm. June 1892.)

† C. Rump, ' Studien über Benzoë.' Hanover, 1878.

‡ Bentley and Trimen's Med. Plants.

§ Ber. Deutsch. chem. Ges. xi. p. 1635.   ‖ Ibid. xix. p. 705.

¶ Ibid. xiii. pp. 335, 662.   ** Ibid. xvi. p. 44, xviii. p. 3335.

a long time, each being clearly distinguishable, and it is only after several months of contact that they unite or blend into a single odour, which is that of heliotrope.

Vanillin forms white needles, generally occurring in stellate aggregates, which possess a very strong taste and smell of vanilla. It melts at 80°–81°, sublimes readily, boils at 285° C. without decomposition when heated in an atmosphere of carbon dioxide, and dissolves in 90 to 100 parts of water at 14° and in 20 parts at 75°–80°. It is scarcely soluble in cold, more readily in hot petroleum spirit. Its aqueous solution is coloured *bluish violet* by ferric chloride; if this solution be heated, white needles of *dihydrovanillin* separate out.

Glucovanillin is formed by the oxidation of coniferin with a dilute solution of chromic acid. It is readily decomposed by emulsin into grape-sugar and vanillin.

Vanillin is frequently adulterated with benzoic acid to a very large extent. Upon treating such a mixture with dilute solution of sodium carbonate, the benzoic acid is dissolved and can be precipitated from the solution by adding excess of water; or the filtrate, after neutralization with hydrochloric acid, will give with ferric chloride a *red-brown* precipitate of ferric benzoate. The acid, or the benzoate, can be reduced by means of sulphuric acid and magnesium ribbon to benzaldehyde, which is recognizable by its characteristic odour of bitter almonds *.

Vanillin should be stored in well-stoppered bottles, as by exposure to a damp atmosphere it is converted into vanillic acid, which, when pure, is odourless. In the natural state as it exists in the pods the aromatic resinous substances with which it is in contact help to prevent this change; but it is always advisable to keep the pods in well-stoppered glass jars.

Artificial Vanillin.

The crystalline coating of vanilla pods (*givre de vanille*) was first prepared artificially by Tiemann and Haarmann from Coniferin ($C_{16}H_{12}O_8$), which occurs in the cambium sap of the fir-tree and is decomposed by emulsin in the presence of water into grape-sugar and the compound $C_{10}H_{12}O_3$, forming odourless crystals which, after standing in the air for some time, have a faint smell of vanilla. The investigators therefore oxidized coniferin with chromic acid,

---

* Deutsch. Amerik. Apotheker-Zeitung, July 1888, p. 103.

and thus obtained vanillin. The elaborate and important original paper * was abstracted in French in the 'Comptes Rendus' of the French Academy, lxxviii. p. 1365, and in English in the Pharmaceutical Journal, [3] iv. p. 996.

Among the numerous glucosides of vegetable origin there is one, Coniferin, which for a long time escaped the notice of chemists and physiologists, although it is found in considerable quantity in the members of one of the most widely distributed orders of plants. This body was first found in 1861 in the juice of the cambium of the *Larix Europæa* by Hartig, who for that reason called it "Laricin." Afterwards, its presence being recognized in the cambium of all the pines, the name of "Abietin" was conferred upon it. Finally, as it appeared to exist in all the Coniferæ, Kubel, who was the first to study it chemically, with the consent of Hartig again changed the name to Coniferin.

Coniferin may be prepared in the following manner. At the time of the formation of the wood, in the spring or early summer, such conifers as *Abies excelsa* and *A. pectinata, Pinus strobus* and *P. cembra, Larix Europæa,* etc., are felled, and the trunks are sawn into several pieces and afterwards barked. The juice of the cambium is collected by scraping the wood with a sharp instrument, such as a piece of glass. This juice is boiled and filtered to eliminate albuminous matters, then evaporated to about one fifth its original volume. After a time it deposits brown-coloured crystals, which are pressed, purified, and decolorized by repeated recrystallizations and treatment with animal charcoal. The greater part of the impurities may be removed by treating the brown solution of coniferin, whilst still warm, with small quantities of acetate of lead and ammonia; the resinous and colouring matters being precipitated whilst the coniferin remains in solution. Excess of acetate of lead may be easily removed by means of a current of carbonic-acid gas. Coniferin is slightly soluble in cold water, more soluble in hot water and in alcohol, but is not soluble in ether. It crystallizes upon cooling in white, transparent, brilliant, sharp-pointed crystals. These crystals become opaque and dull in contact with air, losing part of their water of crystallization, which is driven off completely at 100° C. The aqueous solution of coniferin is bitter, lævogyrous, and does not reduce Fehling's solution, even after prolonged boiling. Moistened with carbolic acid or concen-

---

* Ber. Deutsch. chem. Ges. 1874, p. 608.

trated hydrochloric acid, after some time coniferin acquires an intense blue colour; in the sunlight this coloration is almost instantaneous. It is upon this reaction that the use of pine-wood as a test for carbolic acid is based.

In order to determine the chemical constitution of coniferin, the investigators sought first to determine the nature of the products resulting from its decomposition with elimination of glucose. Dilute hydrochloric or sulphuric acid, aided by heat, splits it up into a resinous matter and glucose; but the properties of the resinous matter so obtained not appearing sufficiently definite, it was determined to effect the decomposition by fermentation by means of emulsin. For this purpose 50 grammes of pure coniferin were placed in 500 grammes of water, 0·2 to 0·3 gramme of dry emulsin added and the mixture kept at a temperature between 25° and 26° C. The action commenced immediately, and in a few hours the presence of glucose in the liquor could be detected. The undissolved crystals of coniferin gradually disappeared and in their place were deposited at the bottom of the vessel white flocks, which were distinguishable from coniferin by their solubility in ether. After six or eight days the process of fermentation terminated, and by that time the bottom of the vessel was covered by a thick layer of this crystalline matter, the supernatant liquor being clear and slightly coloured. The liquor containing the precipitate was shaken with ether, which removed the flocks, and upon evaporation left a residue of well-formed white prismatic crystals. Sometimes, however, it left an oily residue, from which crystals were obtained upon cooling by a freezing-mixture. The crystals were pressed between filtering-paper and purified by recrystallization from ether. The aqueous solution, having had any remaining emulsin removed by coagulation by heat and filtering, was found to contain in solution only glucose and possible slight traces of undecomposed coniferin. The pure crystalline product, when recently prepared, is quite inodorous, but by exposure to the air gradually acquires a feeble odour of vanilla. The pure product was reduced to fine powder and triturated with water, sulphuric acid and potassium dichromate in solution added, and the mixture distilled. At first a liquid smelling strongly of ethylic aldehyde was obtained; the next portions did not present this character, but were strongly acid and diffused a well-characterized odour of vanilla. From these portions of the distillate, ether removed a body which crystallized in stellate groups of crystals, possessing in a high degree the odour and taste of vanilla.

Operating in this way the yield was very small, in consequence of the rapid resinification of the decomposition-product under the influence of the sulphuric acid, and in this state it was only slowly and partially attacked by the oxidizing mixture. It was found more easy and advantageous to operate directly upon coniferin. This was done by pouring an aqueous solution of coniferin into a warm mixture of potassium dichromate and sulphuric acid, and heating the whole together for several hours in a flask connected with an inverted condenser. After cooling, the liquid was filtered to remove a little resinous matter which was deposited, and then agitated with ether. Upon evaporating off this solvent a yellowish oil was obtained, which, after some days, formed a crystalline mass. By re-crystallizing this from hot water, and decolorizing with animal charcoal, fine crystals were obtained, identical in all respects with vanillin produced by nature *.

In the course of the investigations made by Tiemann and Haarmann in the laboratory of Dr. Hofmann at Berlin, it was found that on fusion with caustic potash, vanillin is converted into protocatechuic acid, and on heating to 200° C. with hydrochloric acid, under pressure, it is decomposed into methyl chloride and protocatechuic aldehyde, thus proving that vanillin is the methyl ether of protocatechuic acid or, to use the inconveniently long name applied to it by Dr. Tiemann, "monomethylprotocatechuic aldehyde" †.

Protocatechuic acid is obtainable from various carbon compounds by fusion with caustic potash ‡, and its synthetical formation in various ways is of theoretical interest §. It is best prepared from East Indian Kino, as per note at foot ‖.

* Ber. Deutsch. chem. Ges. vii. pp. 609, 614.

† Ibid. vii. p. 608.

‡ Ann. Chem. Pharm. cxxiv. p. 118, cxxvii. p. 357, cxxviii. p. 285, cxxx. p. 346, cxxxiv. p. 277, cxxxix. p. 78.

§ Ibid. clii. p. 109, cliv. p. 364, clix. p. 232.

‖ *Protocatechuic acid* is most readily prepared from East-Indian Kino, which is obtained by making incisions in the bark of *Pterocarpus Marsupium*; the sap flows out and dries to a dark red transparent mass. "One part of the finely powdered kino is gradually added to three parts of fused sodic hydrate, stirring the mixture constantly during the introduction of the kino and keeping the

The synthetic formation of the odorous principle of vanilla can be effected from the starting-point of *coal,* the stages of the process being as follows:—

Tar.

Carbolic acid.

Phenol.

Potassium phenate.

Paraoxybenzoic acid,—produced by the reaction of carbonic acid on potassium phenate (the analogous reaction with sodium phenate giving salicylic acid *).

---

mixture at a low temperature. The most convenient method being to sift the fine powder over the surface of the hydrate, which is kept fused at a gentle heat. When the mixture has acquired a bright orange-brown colour, it is poured on to a stone or an iron plate, and allowed to cool. The cake is then broken up and dissolved in about 20 parts of hot water, to which dilute sulphuric acid is added from time to time, so as to render the solution slightly acid. The dark brown solution is then allowed to stand for 24 hours, when it deposits a large quantity of sodic sulphate in crystals; the mother liquors which contain the protocatechuic acid, after being filtered to remove a small quantity of tarry matter, are agitated with ether, the ethereal solution is then separated, and the ether recovered by distillation in the water-bath: the dark syrupy residue on standing deposits the protocatechuic acid in crystals. Or,—the mother liquors may be readily separated from the crystals by means of the vacuum-pump, and the acid is then easily purified by one or two crystallizations from a small quantity of hot water, with the addition of some freshly precipitated lead sulphide, which removes the brown colouring matter." It crystallizes in monoclinic needles, containing one molecule of water which is lost at 100°, and melts at 194°; it dissolves in 53–55 parts of water at 14° and is very soluble in alcohol.

* *Salicylic acid* is manufactured according to Kolbe's process as follows:— The calculated quantity of pure phenol is dissolved in strong caustic soda solution, the whole evaporated to dryness and the residue rubbed into a dry powder; this is then gradually heated up to 180° C. in a metal retort, in a current of carbon dioxide which has been previously warmed. After some time phenol commences to distil over and is subsequently given off in larger quantity; the temperature is then raised to 200° and the operation continued until no more phenol comes over. The residue is dissolved in water and fractionally precipitated with hydrochloric acid; resinous and colouring matters are first thrown down, followed by tolerably pure acid, which is re-crystallized from water and purified by distillation with superheated steam (Compt. Rend. viii. p. 537). According to another patented process, carbonyl chloride, which is now manufactured on a large scale, is passed into a mixture of sodium carbonate and phenate heated to 140°, the temperature being finally raised to 200°.

Protocatechuic acid, produced from the above.

Dimethylprotocatechuic acid [*], prepared by heating the above with caustic potash, methyliodide, and wood spirit [†].

When this last compound (which is identical with veratric acid) is heated with dilute hydrochloric acid in a closed vessel to 130°–140° C., among the products of decomposition is found mono-methylprotocatechuic or vanillic acid, of which vanillin is the corresponding aldehyde.

The retrograde conversion of the acid into the aldehyde can be effected by the dry distillation of an intimate mixture of vanillate and formate of calcium. An oily distillate is obtained consisting of guaiacol and a small quantity of vanillin. If an ethereal solution of this distillate be agitated with a concentrated aqueous solution of sodium bisulphite, the latter, after decomposition with sulphuric acid, will yield the vanillin to ether. By crystallization from hot water it is obtained in a perfectly pure state. The other products of the decomposition of vanillic acid being carbonic acid and guaiacol, this latter corresponding in every respect with that obtained from beech-wood tar [‡].

The specification of Dr. Haarmann's English Patent, dated Feb. 25, 1874 (No. 709), claims the following processes :—" In order to obtain the artificial production of vanillin, by means of coniferin or the sap of plants belonging to the species of conifera as an extract of all those parts containing coniferin, take either, first, coniferin; or, secondly, the sap of the plants which has been purified from albumina and other impurities; or thirdly, an extract of the parts; or fourthly, the products obtained from

* *Dimethylprotocatechuic acid*, identical with veratric acid, exists in the seeds of *Veratrum Sabadilla*. It can also be prepared by shaking up 1 part of methyl-eugenol with 10 to 15 parts of water, and a solution of 3½ parts of potassium permanganate in 20 to 30 parts of water heated to 80°-90° C. gradually added. The filtrate is concentrated by evaporation and precipitated by hydrochloric acid (Ber. Deutsch. chem. Ges. ix. p. 937). Veratric acid dissolves in 2100 parts of water at 14°, and in 160 parts at 100°, and crystallizes from a concentrated solution at a temperature above 50° in anhydrous needles, while crystals containing a molecule of water are obtained from very dilute solutions at any temperature below this. It melts at 174°-175° and can be sublimed. It dissolves readily in alcohol and ether : ferric chloride produces no coloration.

† Ber. Deutsch. chem. Ges. viii. p. 514.

‡ Ibid. viii. p. 1124.

coniferin by means of fermentation, and treat one or other with oxidizing agents. For the first production I use the following method :—Ten parts of coniferin are dissolved in hot water and the concentrated solution is allowed to flow slowly into a gently heated mixture of ten parts of bichromate of potassium, fifteen parts of strong sulphuric acid, and eighty parts of water; the whole is heated for about three hours almost up to the boiling-point of the liquid. The vanillin produced in this way may be isolated either :—1. By shaking the solution, when it has been filtered and become cold, with ether, separating the etherate from the aqueous solution, evaporating the ether, and purifying the dark coloured residue by repeated crystallization. Or, 2. By combining with a boiler in which coniferin has been treated for about three hours with the oxidizing mixture, a distillatory apparatus, and then by distilling the contents till only one fifth of their original volume remains, and by treating the distillate in the before described manner with ether to obtain a pure product, and by treating the residue in the same way to obtain the last traces of an impure vanillin. The artificial vanillin can also be obtained from an aqueous extract of all those parts of conifera in which coniferin is present, the impure solutions being boiled for some time to separate the dissolved albumina, filtered, and strongly evaporated.

" For a quick, but sufficiently exact determination of both the coniferin and sugar which are contained in the concentrated solutions, I evaporate them perfectly in a water-bath, and the residue is dried at 100° C. On every ten parts of what remains in the concentrated solutions, fifteen parts of bichromate of potassium and twenty parts of concentrated sulphuric acid must be taken, and the quantity of water must be administered according to the concentration of the solution. The further process is as above.

" The coniferin decomposes under the influence of ferments into sugar and a crystallized substance which, in contact with damp air, also changes slowly and incompletely into vanillin. To obtain vanillin from this substance quicker, I treat it with an oxidizing mixture of bichromate of potassium and sulphuric acid. The vanillin obtained and purified in either way is proved in all cases completely identical to the natural vanillin."

Shortly after the above discovery, it was found that vanillin could be artificially produced from oil of cloves. On the 18th of March, 1876, a patent was taken by de Laire in France * for this purpose. According to his specification the process is thus described :—" The starting-point of the manufacture is the preparation of *Eugenol*; I obtained this from oil of cloves, which contains it in large proportion. I dissolve one kilo of this oil in four or five kilos of ether, and agitate the mixture with a weak aqueous solution of sodium hydrate, which absorbs all the eugenol ; the hydrocarbon which constitutes the rest of the oil remaining in solution in the ether. I separate the sodium solution and saturate it with sulphuric acid to set free the eugenol, which I collect by agitation with ether. On evaporation of the ether, the eugenol is left pure. This body is then heated for two or three hours with an equivalent proportion of acetic anhydride in an apparatus connected with an inverted condenser, being thereby converted into acetyleugenol. The mass is then allowed to cool, and when cold is agitated with several times its weight of water. The liquid is then gently warmed, and to it is gradually added a weak solution of potassium permanganate (1500 grammes of permanganate altogether).

" The hydrate of manganese is separated out by filtration, the remaining solution gradually saturated with soda and its volume reduced by evaporation. When the liquid is cold it is acidified with sulphuric acid and agitated with ether, which takes up the vanillin. On evaporation of the ether the vanillin remains in crystals."

A patent was taken out in England for the production of vanillin from oil of cloves by Dr. Tiemann on the 20th of April, 1876. The process, which is very similar to de Laire's, taken out a month previously in France, is described as follows in the specification † :—" To separate the two constituents (oil of cloves and the hydrocarbon) one from another, the oil is diluted with three times its volume of ether. The etheric solution is then agitated with a weak solution of hydrate of sodium or potassium, which takes up all the eugenol and leaves the indifferent hydrocarbon in the ether. After having acidulated the alkaline

---

* Brevet 111930.   † No. 1661.

solution of eugenol by means of sulphuric acid, it is again agitated with ether to take up the eugenol, which can then be obtained in a pure state by distilling. Pure eugenol is heated for about two hours with acetic anhydride and thereby changed into acetoeugenol, and this liquid when cooled is to be carefully diluted with about twenty litres of warm water, and then into this mixture is allowed to flow gradually a solution of one and a half kilos of permanganate of potassium in about 200 litres of warm water, during which time the liquid is kept constantly stirred. By means of permanganate of potassium the aceto-eugenol is oxidized and the hydrate of the manganese dioxide is separated out. To the liquid filtered from the latter, hydrate of sodium is added in a small excess to give it a weak alkaline reaction, when the liquid is to be evaporated to the volume of about 25 litres. The liquid, concentrated in this way, is acidu-lated with sulphuric acid and agitated with ether, which last then takes up the vanillin formed in the described manner. It is then purified by any of the known methods."

Vanillin from Bran.—A patent dated Dec. 27, 1876 *, was granted in France to Eugène Sérullas, of Paris, for an artificial product having the odour of vanilla obtained from the husk of oats. The patentee claims to have discovered in the pericarp of oats an inodorous principle which he termed " *aveneine,*" a sub-stance very soluble in boiling water and in alcohol.

Bran, which is a commercial residue in the preparation of oatmeal, is exhausted by any of the known methods which are employed for the preparation of *populin* and other glucosides. The resulting aveneine is to be purified and oxidized by any of the usual processes, but to produce the complete transformation the mixture is to be boiled for two hours and a half. After cooling it is to be agitated with ether to extract the product of oxidation, and on evaporation of the ether the product can be collected and purified in the ordinary way.

Vanillin from Siam benzoin.—Two parts of Siam benzoin and one part of slaked lime are boiled with water in an iron kettle, stirring constantly ; the resulting solution of benzoate of calcium is filtered, the filtrate acidulated with sulphuric acid and the

* Brevet 116200.

precipitated benzoic acid separated by filtration, the acid filtrate shaken with ether and the ether evaporated spontaneously. The impure vanillin thus obtained is purified by re-crystallization from water, from which it separates in long thin white needles, turning yellow on exposure to the air *.

Vanillin from Asafœtida.—The actual preparation of vanillin from asafœtida is described by Schmidt as follows † :—An ethereal extract of the gum-resin is shaken with a strong solution of sodium bisulphite, adding to the solution excess of sulphuric acid, and after driving off sulphurous acid, extracting the liquid with ether. Upon evaporation of the ether, crude vanillin is left as a residue, which after purification corresponds in every respect with vanillin from other sources. The possibility of deriving vanillin from asafœtida was suggested fifteen years ago by Tiemann, who showed the connection between that compound and *ferulic acid.* Vanillin has an acid reaction and forms salts, which have been investigated by Carles, and by Tiemann and Haarmann. By the action of an ethereal solution of acetic anhydride upon the sodium salt of vanillin, a coumarin-like body is produced which has been namcd " vanillin-coumarin." This body when boiled with caustic potash is converted into an acid identical with the ferulic acid obtained from asafœtida ‡.

Vanillin has been obtained by heating guaiacol § with caustic

* Ber. Deutsch. chem. Ges. 1878, p. 1634.
† Archives der pharmacie, June 1886, p. 434.
‡ Pharm. Journ. [3] vi. p. 813.
§ *Guaiacol.*—Produced from the products of the dry distillation of guaiac resin, which is a natural exudation or the result of incisions made in the bark of the Hujacum tree, *Guajacum officinale,* a tree growing in Jamaica, St. Domingo, and other West Indian Islands. In the Island of Gonave, near Port-au-Prince, another method is used : a log is supported in a horizontal position above the ground by two bars, and each end of the log set on fire, the melted resin running from a large incision which has previously been made in the middle. Guaiacum resin is a brittle dark green to brownish-black mass, which readily dissolves in alcohol. When submitted to dry distillation it yields guaiol or tiglic aldehyde, guaiacol, creosol, and pyroguaiacin.

When the crude oil obtained by distillation is washed with water and rectified at a moderate heat, guaiacene passes over first, and afterwards, when the heat is increased, the guaiacol distils. It may be purified by repeated rectification or by dissolving it in potash and boiling it with water as long as any light oil passes over, then mixing it with sulphuric acid in quantity not quite sufficient to combine with the whole of the potash, redissolving the

soda and chloroform * ; and from olivil † the crystalline constituent of the Lecca gum, or resin of the wild olive ‡, which is
used in Italy as Incense; also by the oxidation of ferulic acid,
which can be prepared on a large scale without difficulty §.
Vanillin has been obtained from opium. Narcotin is first
extracted and converted into opianic acid by heating 100 grammes
with 1500 grammes of water and 150 grammes of sulphuric acid
until the mixture boils; 150 grammes of finely powdered pyrolu-
site (corresponding to 90 grammes of manganese dioxide) are
then added somewhat rapidly and the hot solution filtered.
Opianic acid separates out on cooling and is purified by crystallization. It is slightly soluble in cold, readily in hot water,
alcohol, and ether, and crystallizes in thin narrow prisms, or

---

separated oil in potash and boiling the solution in a retort till the milky oil
which passes over becomes perfectly clear on the addition of a small quantity
of potash, again separating the oil by sulphuric acid and drying it *in vacuo*
over sulphuric acid. It is a colourless oil, having a sp. gr. of 1·119 at 22°
(Sobrero), 1·125 at 16° (Volckel); it boils at 210°.

Guaiacol can also be obtained from beech-wood tar. The crude compound
obtained from beech-wood tar creosote is repeatedly shaken with moderately
strong ammonia, washed, and rectified. The oil is then dissolved in an equal
volume of ether, and a small excess of alcoholic potash (concentrated) added to
it. The potassium salt separates out and is then re-crystallized from alcohol
and decomposed by dilute sulphuric acid.

   * Bull. Soc. Chim. ix. p. 424.

   † Scheidel, Ber. Deutsch. chem. Ges. xviii. p. 685.

   ‡ *Olivil*, $C_{14}H_{18}O_5$, is the crystalline constituent of the resin of the wild olive.
It forms a thick vapour on heating and has a pleasant odour resembling those
of benzoïn and cloves. It is employed in Italy for the fumigation of sick
rooms. In order to obtain the olivil, the resin is extracted with ether and the
residue treated with boiling alcohol. The olivil, which separates on cooling, is
washed with cold dilute alcohol and re-crystallizes from alcohol. It forms
white needles which are odourless, and have a sweetish, bitter taste, melt at
120° and solidify to an amorphous mass, which forms a strongly electrified
powder on trituration. This mass melts at 70°, but after re-crystallization from
alcohol, regains the original melting-point. It crystallizes from hot water in
stellate groups of prisms, which contain one molecule of water. It is readily
soluble in alkalies and precipitates the metals from solutions of gold chloride
and silver nitrate. On dry distillation it yields an oily liquid, to which Sobrero
has given the name of pyro-olivilic acid (Ann. Chem. Pharm. liv. p. 67). It
has the composition and properties of eugenol and is either identical with this
or iso-eugenol.

   § Ulrich, Ber. Deutsch. chem. Ges. xviii. ref. 682.

silky needles, which melt at 150°, and decompose on further
heating, giving off a vapour which smells like vanilla (Wohler).
It has a faint acid reaction and slightly bitter taste. When its
sodium salt is heated with soda lime, *methylvanillin* is formed;
and *isovanillin* when it is heated with dilute sulphuric acid to
160°–170° *. The methylvanillin is slightly soluble in hot water,
readily in alcohol, and crystallizes in needles which smell like
vanilla. The isovanillin, which can also be formed by heating
opianic acid to 160°–170° with dilute hydrochloric acid, crystallizes from hot water in monosymmetric prisms of vitreous lustre,
which sublime when heated, undergoing slight decomposition;
its vapour has a pleasant smell, resembling that of vanilla and
anise †.

Processes for the production of vanillin from pyrocatechin and
from guiacol have been patented in France by Alfraise, dated 23
Dec. 1891 ‡. The process he describes for converting guiacol
into aceto-ferulic acid and thence into vanillin does not appear to
be particularly novel, or to offer any solid advantage over already
known methods : any economy in working expenses would be best
known to manufacturers.

It may here be remarked that whereas the abstracts of specifications of French patents were formerly rather costly to obtain,
abstracts *in extenso* are now supplied at moderate price by the
Proprietors of the ' Revue de Chimie Industrielle,' 53 bis, Rue des
Grands Augustins, Paris. An abstract of the last-named specification is published in that Journal of May 1892, p. 154.

This description is but a brief *résumé* of the very extensive
literature of this valuable product. For an earnest study of the
subject the reader is referred to the following works, but to an
intending planter the pamphlet by Delteuil is a very fair guide :—

DELTEUIL., ' Etude sur la Vanille.' Paris, 1874.
JAILLET, "Culture et preparation de la Vanille." 'Repertoire de Phar-
    macie,' viii. pp. 357 & 411 (Aug. & Sept. 1880).
' Vanilla, its cultivation in India.' J. E. O'CONNOR. Calcutta, 1875.
' Annals of Natural History,' iii. p. 1. "On the production of Vanilla in
    Europe," by Professor MORREN.

---

* Journ. Chem. Soc. 1876, i. p. 287.
† Tiemann, Ber. Deutsch. chem. Ges. viii. p. 1135.
‡ Brevet 218232.

'Annales des Sciences Naturelles,' 3ᵉ série, v. p. 117 (DESVAUX).

' Revue Coloniale,' 2ᵉ série, ii. p. 383.   This article is not signed, being probably editorial.

'Bulletins de l'Académie Royale de Belgique,' xvii. 1ᵉ partie, p. 103 (MORREN).   (Excellent coloured plate of the pods.)

Consular Reports, from Consul Seagrave, Réunion.

" Botanische Berichte aus Mexico."   Linnæa, iv. pp. 514–583 (SCHIEDE).

' Bijdragen tot de Flora van Nederlandsch Indie,' p. 422 (BLUME).

' Rumphia,' BLUME, p. 196, tab. 67.

CATESBY, ' Hist. Nat. de la Caroline,' iii. p. 7, tab. 7.

ANDREWS, ' Botanical Repository,' tab. 538, vol. viii.

Mauritius ' Blue Book,' 1872.

VRIESE, ' De Vanielje.'   Leyden, 1856, p. 22.

LODDIGES, Bot. Cab. 733.   (Finely coloured plate of the flowers.)

PLUMIER, ' Plantæ Americanum,' p. 183, tab. 188.

PLUMIER, ' Flora Javæ,' edit. Burmanni, 25.

SWARTZ, ' Nova Acta Upsala,' vi. 66, tab. 5. fig. 1.

R. BROWN, ' Hortus Kewensis,' v. p. 220.

LINNÆUS, ' Spec. Plantarum,' 1347.

GARDNER, ' Travels in the Interior of Brazil,' p. 296.

SWARTZ, in Schrader's Journ. Bot. ii. fig. i (1799).

KUNTH, ' Synopsis Plantarum,' i. p. 359.

F. BAUER, ' Illustrations of Genera and Species of Orchidaceous Plants,' tabb. 10 & 11.

SALISBURY, Paradisus Londinensis, 82.

SWARTZ, Prodr. 120.

MORREN, ' Bulletin de l'Académie Royale des Sciences,' tab. iv. no. 5, p. 225.

MORREN, ' Ann. de la Soc. Roy. d'Horticulture de Paris,' tab. 20, 1837, p. 331.

LINDLEY's ' Flora Medica,' p. 579.

*LINDLEY's ' Genera and Species of Orchidaceous Plants,' part vi. p. 435.

LINDLEY, in Botanical Register, 1838, and Bot. Misc. p. 58.

' Pharmacographia,' 2nd ed. p. 657.

BENTLEY and TRIMEN, Med. Plants, tab. 272.

' Comptes Rendus de l'Académie,' May 11, 1874.   " Récherches sur la Coniferine " (TIEMANN and HAARMANN).

' Berichte der Deutsch. Chem. Ges.' ix. pp. 409–423, contains a summary of various papers bearing on the subject of Vanilla compounds published by Tiemann and other chemists.

Report of U.S. Consul KNOWLES of Bordeaux, Sept. 1891, partly abstracted in American Journ. Pharm. June 1892.

---

* The plates in Lindley's ' Orchidaceous Plants ' are indifferently reproduced from the drawings of Francis Bauer, 1807, which are now in the British Museum.

CHAPTER IX.

ODOUR OF BITTER ALMOND.—CHERRY LAUREL.— HELIOTROPE.

THE ODOUR OF BITTER ALMOND.

THE tree producing bitter almonds (*Amygdalus communis*, L., var. *amara*, DC.) is not distinguished by any botanical character or habit of growth from the tree which produces the sweet almond. Both the bitter and sweet almond form trees 20 to 30 feet in height. The leaves resemble those of the peach, but the lower serratures are glandular, which has given rise to the conjecture that glandular-leaved peaches have sprung more immediately from the almond than such as are without glands, as is generally the case with nectarines. Their flowers vary in colour from a fine blush to snow-white. The chief distinction is in the fruit, which is flatter, with a coriaceous dry covering instead of the rich pulp of the peach and nectarine, opening spontaneously when the kernel is ripe. The peach and the almond have been crossed by dusting the stigma of an almond with the pollen of a peach*.

The almond is propagated like the peach, by seed for varieties or stocks, and by budding on its own or on a plum stock for continuing varieties. Plum stocks are preferred for strong moist soils, and peach and almond stocks for dry situations. Light, sandy soil seems most suitable to the tree.

The almond-tree bears chiefly on the young wood of the previous year, like the apricot and peach, and in part upon small spurs on the two-year old, and three-year old and older branches; it should therefore be pruned like these trees.

There are two varieties of bitter almond, one with a hard brittle shell to the nuts, and one with a tender shell.

* Hort. Trans. iii. p. 41, tab. i.

The bitter almonds resemble sweet almonds in shape; but are generally somewhat smaller. In the order of their relative value they are known as French, Sicilian, and Barbary almonds. The fixed oil contained in the bitter almond is identical with that contained in the sweet variety, but in a slightly smaller proportion; the average yield being 43 to 44 per cent., obtained by hydraulic pressure. After the almonds have been freed from fixed oil by pressure, an odourless cake is left which yields the characteristic smell of oil of bitter almonds on the addition of water. The chemists who made this discovery thereby concluded that oil of bitter almonds must be a compound of water with a peculiar principle which they endeavoured to isolate. The use of water being impossible, they extracted the pressed almonds with boiling alcohol, and obtained, together with resin and a liquid sugar, a crystalline compound containing nitrogen, to which they gave the name of *amygdalin*. This compound, to which the taste of bitter almonds is due, gave no smell of bitter almonds when treated with water, nor did either of the two other compounds, nor the residue, nor either a mixture of them all*. The prussic acid and the oil of bitter almonds had vanished from their hands (Robiquet and Boutron-Charlard). The problem was solved by Liebig and Wöhler†, who also accurately determined the composition of amygdalin. They showed that both sweet and bitter almonds contained a peculiar nitrogenous substance, *emulsin* ‡, which converts amygdalin, in presence of water, into oil of bitter almonds, prussic acid, and grape sugar. To the oil of bitter almonds they gave the name of benzoyl-hydride, which was later

* Annales de Chim. et de Phys. xliv. p. 352.

† Ann. Chem. Pharm. xxii. p. 1; xxv. p. 175; xxv. p. 190.

‡ To prepare emulsin:—Sweet almond paste, well freed from fixed oil, is macerated in three times its weight of pure water; the mass is pressed, and the emulsin thus obtained is left to itself at 20° to 25° C. After the lapse of a day, the emulsin is found to have separated into two layers, the upper of which is coagulated, and looks like cream, while the lower is watery and transparent. After two or three days this watery liquid no longer gives a precipitate with acetic acid, but it forms with alcohol a precipitate perfectly soluble in water. This last precipitate consists of emulsin. After being washed with absolute alcohol and dried *in vacuo* over sulphuric acid it forms a white, opaque, friable mass, soluble in water.

Emulsin completely loses its power of transforming amygdalin into benzaldehyde when its aqueous solution is boiled, but it retains its power when heated to 100° C. in the dry state, even for several hours.

changed to benzoic aldehyde, and is now known by the name of *benzaldehyde* (C_6H_5CHO).

The action of the ferment is destroyed by boiling water and by heating with alcohol, so that when dried and powdered bitter almonds are shaken up with boiling water and distilled, none of the essential oil is obtained, and the same result occurs when, as in Robiquet and Boutron-Charlard's process, they are treated with boiling alcohol.

Amygdalin, which is the first example of a glucoside (a large number of which bodies are now known), occurs in many plants, chiefly the Amygdalaceæ, Drupaceæ, and Pomaceæ, which all yield benzaldehyde and prussic acid when distilled with water[*]. The kernel of the peach also yields an oil resembling the oil of bitter almonds in every respect[†]; while that obtained from the leaves, flowers, seeds, and bark of the cherry contains both oil of bitter almonds and another oil which has a penetrating, repulsive odour[‡]. According to Winkler, the fresh leaves of the cherry-laurel (*Prunus Laurocerasus*), the cherry (*Prunus Padus*), and the peach contain a small quantity of free oil of bitter almonds, varying in amount with the water present[§], which can be extracted by ether. Of Syrian peach and apricot kernels considerably over a million pounds weight are annually exported from Damascus, which is the principal locality of the production. The cracking of the shells and removal of the kernels is done partly by hand and partly by machinery.

In the distillation of oil of bitter almonds a difficulty is experienced by the formation of large quantities of froth, by reason of

[*] To prepare amygdalin, bitter almonds are freed as far as possible from fatty oil by pressing, and the mass repeatedly extracted with boiling alcohol ; the alcohol is then distilled off, and the residue recrystallized from boiling alcohol. It crystallizes in lustrous plates or scales, or from water in transparent, rhombic prisms, which become anhydrous at 120°, solidifying to an amorphous mass on cooling. At 8°–12° it dissolves in 12 parts of water, while it is soluble in every proportion in boiling water. It is slightly soluble in cold, more readily in boiling alcohol, and insoluble in ether. On the addition of emulsin to its aqueous solution, it decomposes into grape-sugar and phenyl-hydroxyacetonitril, which is partially decomposed by distillation into benzaldehyde and hydrocyanic acid. The same decomposition takes place on boiling with dilute hydrochloric acid. (*Roscoe and Schorlemmer.*)

[†] Righini, Ann. Chem. Pharm. x. p. 359; Geissler, ibid. xxxvi. p. 331.

[‡] Winkler, Repert. Pharm. lxvii. i. p. 56.

[§] Jahresb. Chem. iv. p. 519.

the albuminoids present. In order to remedy this difficulty, and to get all the amygdalin into solution, 12 parts of the coarsely powdered cake are immersed in 100 to 120 parts of boiling water, and kept at the boiling-point for 15 to 30 minutes; this coagulates the albuminous matters and dissolves the amygdalin, then, after cooling the mixture, an emulsion of one part of powdered almonds (either sweet or bitter, or a portion of the first-mentioned cake) stirred up in six or seven parts of cold water is added to the mass. This one part contains sufficient emulsin to produce the desired decomposition, at a temperature not exceeding 40° C. The entire bulk is then rapidly distilled. By this process the oil is prepared on a large scale, the yield varying from 0·74 to 1·67 per cent. of oil from the cake. Therefore, if 100 lbs. of almonds yield 57 lbs. of cake, the yield of oil is 0·42 to 0·95 per cent. on the weight of the almonds. The aqueous distillate contains some oil in solution which is removed by a subsequent distillation. The great variation in the figures of the yield is partly accounted for by the variability of quality of the almond used, and the consequent varying amount of amygdalin present, and it is partly due to admixture of *sweet almonds*; *an adulteration which frequently causes much loss to the manufacturer*, whose profit greatly depends on the percentage of essential oil which he can distil from the residuary cake.

Natural Oil of Bitter Almonds.

This is chemically known as *Benzaldehyde*. It is a colourless, strongly refractive liquid, having the well-known characteristic smell and a burning aromatic taste. It dissolves in more than 300 parts of water, boils at 179°, and has a sp. gr. of 1·0636 at 0°, and of 1·0504 at 15°.

This oil is generally adulterated; in many cases with alcohol. When pure, sulphuric acid produces a clear crimson-red colour without visible decomposition. Mixed with an alcoholic solution of potash, crystals are eliminated. Iodine dissolves only partially and slowly in it. Nitric acid, sp. gr. 1·42, causes no immediate reaction, and in the course of three or four days crystals of benzoic acid begin to appear, but if only 8 per cent. or 10 per cent. of alcohol or rectified spirit is present a violent effervescence speedily commences, and nitrous fumes are evolved. By using nitric acid sp. gr. 1·5 the smallest quantity of alcohol may be detected. Chromate of potash does not affect it.

For the detection of the adulteration of nitrobenzene in oil of bitter almonds, it has been recommended to shake the suspected sample in a test-tube, with one-half its weight of solid caustic potash. The yellow colour of the oil is not changed if it is pure; but if nitrobenzene be present, the colour will soon change to a characteristic red. If a considerable quantity of nitrobenzene be present, the red colour is changed to a more or less fine green colour, which, on the following day, again becomes red. The quantity of nitrobenzene is determined by agitating the adulterated oil violently and repeatedly with four volumes of a concentrated solution of bisulphite of sodium; after some time rectified ether is added, which dissolves the nitrobenzene, and by evaporation permits its estimation. To prove the residue of the evaporation to be nitrobenzene, it can be converted into aniline.

To distinguish the natural oil from the artificial oil which is frequently substituted for it, and to detect the presence of the latter, a simple process has been recommended, based on a reaction produced by the organic chlorinated compounds always contained as impurities in the artificial oil. It consists in saturating a piece of folded filter-paper with the oil to be examined, and after placing it in a porcelain dish standing in a larger one, igniting it, and covering it over with a large inverted beaker, the sides of which have been wetted with water. The combustion-gases become absorbed on the moist sides of the beaker, from which they are washed on to a filter with a little distilled water. The filtrate, when treated with solution of silver nitrate, should give no turbidity, much less a precipitate of silver chloride. Genuine essential oil of bitter almonds, distilled in the ordinary way from almonds or peach-kernels, never gives a chlorine reaction. (*Schimmel.*)

Some manufacturers free the oil from hydrocyanic acid; the purified oil, however, oxidizes much more readily than when in the crude state, so that others add hydrocyanic acid and warm gently in order to make it keep better, the nitril being formed[*]. When, however, it is desirable to store the oil freed from hydrocyanic acid, the oil should be very carefully dried from all traces of water by agitation with fused calcium chloride.

[*] Bull. Soc. Chem. [2] viii. p. 459.

To free the crude oil of bitter almonds from hydrocyanic acid, the following methods have been adopted :—

1. (*Liebig.*) Agitate the crude distilled oil with red oxide of mercury in slight excess, and after a few days' contact rectify the oil from a little fresh oxide of mercury. The product is quite pure when the process is properly managed. The cyanide of mercury thus formed may be either employed as such or re-converted into mercury.

2. (*Mackay.*) Crude oil of almonds 1 lb. ; fresh slaked lime *q. s.* to form a milk-like liquid ; afterwards add $1\frac{1}{2}$ lb. solution of potash and 3 pints of water ; agitate occasionally for 48 hours, then distil over the oil and rectify it from a fresh mixture of lime and potash.

3. (*Redwood.*) The oil is mixed with an equal quantity of water, and the mixture is digested in a water-bath with red oxide of mercury and small quantities of fresh slaked lime and proto-chloride of iron, with as little access of air as possible ; as soon as the decomposition of the acid has taken place the whole is introduced into a copper retort and submitted to distillation. The product is perfectly free from hydrocyanic acid.

The first process is considered the simplest, cheapest, and best. The usual method of testing Bitter-almond oil for prussic acid is given in the description of Artificial Benzaldehyde, further on.

Artificial Oil of Bitter Almonds.

Benzaldehyde is manufactured artificially on the large scale by boiling 2 parts of Benzyl chloride with 3 parts of lead nitrate (or, preferably, copper nitrate) and 10 parts of water for several hours in an apparatus connected with an inverted condenser, the operation being conducted in a current of carbon dioxide; half the liquid is then distilled off, and the oil separated from the water.

It is obtained from benzidene chloride by heating it under pressure in an iron vessel with caustic soda. It is said to be possible to heat without pressure in an apparatus connected with an inverted condenser if milk of lime be used, or if whiting or some other finely divided insoluble substance be added and the whole stirred into an emulsion, which boils at a higher temperature and thus facilitates the decomposition of the chloride.

Jacobsen * recommends a process in which benzidene chloride is heated with glacial acetic acid and zinc chloride, benzaldehyde and acetyl chloride being formed ; the necessary amount of water is then allowed to flow in, and the acetic acid which is formed recovered.

The artificial benzaldehyde of commerce, which is used in the colour industry and prepared from benzyl chloride, is always more or less impure, retaining traces of chlorinated compounds of pungent repulsive odour which render it unfit for perfumery purposes †.

The artificially prepared benzaldehyde comes into the market nearly always free from prussic acid ; but yet, it is occasionally met with containing that acid—which has of course been added by the manufacturer to supply an order. The usual method of testing bitter-almond oil for prussic acid is as follows :—From 10 to 15 drops of the oil are shaken up with 2 or 3 drops of 30-per-cent. soda solution. To this is added a few drops of a solution of slightly oxidized sulphate of iron. After another vigorous shaking, the liquid is slightly acidified with dilute hydrochloric acid. When the precipitate is dissolved, the presence of prussic acid is manifest by the appearance of the characteristic blue deposit. The least trace of prussic acid may be detected in this way. Or, the oil may be dissolved in alcohol and a solution of potash and sulphate of iron added ; then, on the addition of a dilute acid, Prussian blue is formed.

A sure and delicate test-paper for indicating the presence of

---

* Ber. Deutsch. chem. Ges. xiii. p. 2013 ; xiv. p. 1425.

† *Benzyl chloride*, $C_6H_5CH_2Cl$, is prepared on the large scale by the action of chlorine on boiling toluene (Ann. Chem. Pharm. ccxxi. p. 365), the toluene being contained in large glass balloons heated by a bath of calcium chloride, and the chlorine passed through in such a manner that it chiefly comes in contact with the vapour of the toluene. This is effected by only allowing the leaden conducting tube, which terminates in a short piece of glass tubing, to dip a small distance below the surface of the boiling liquid. The vapours of toluene are condensed by a cooling arrangement and the hydrochloric acid evolved is led into water. The product is washed with water containing a little caustic soda, and the benzyl chloride freed from unaltered toluene and higher substitution-products by distillation. Benzyl chloride is a colourless liquid, the vapour of which has a penetrating aromatic smell, rapidly produces a flow of tears, and attacks the mucous membrane most violently. It boils at 176° and has a sp. gr. of 1·107 at 14°.

prussic acid can be made by dipping bibulous paper in fresh tincture of guaiacum, then drying and dipping in an aqueous solution of cupric sulphate (1 in 2000) and once more drying. The paper will become intensely blue if moistened and afterwards dipped into oil containing prussic acid.

There are other chemically prepared compounds having the odour of bitter almonds, such as Nitrobenzene, Nitrotoluene, Meta-nitrotoluene, &c.

MIRBANE.

Nitrobenzene, $C_6H_5NO_2$, is used in perfumery as a *substitute* for oil of bitter almonds.

Nitrobenzene can be readily distinguished from oil of bitter almonds by the following test :—

Pour a few drops of each on a plate and add a drop of strong sulphuric acid. The oil of almonds acquires a rich crimson colour with a yellow border,—the nitrobenzene produces no colour. Also it yields no Prussian blue when mixed with sulphate of iron, alcohol, and potash.

It was first introduced in commerce by Collas under the name of " Essence of Mirban," and incorrectly called " artificial oil of bitter almonds." In 1874 a process was patented by Mansfeld for its preparation from coal-tar. It is now prepared on a very large scale and employed for a variety of purposes.

To prepare it in small quantities, equal parts of fuming nitric acid and benzene are gradually mixed. The apparatus consists of a large glass worm, the upper end of which is divided into two branches gradually dilating so as to form two funnel-shaped tubes : into one of these the nitric acid is poured, and into the other the benzene. These bodies meet at the point of junction of the two tubes, and the rate of their flow is regulated by an appropriate means. Chemical reaction instantly takes place and the new compound is cooled in its passage through the worm, which is refrigerated for the purpose. It has then only to be washed with water or a very weak solution of carbonate of soda or caustic soda for the process to be complete.

It is a light yellow, strongly refractive liquid, having at 0° a sp. gr. of 1·200. It has a peculiar smell, similar to that of bitter almonds, at the same time reminding one of oil of cinnamon, and possesses a sweet and burning taste. It boils at 210°, and at a low temperature solidifies in large needles, melting at 3°. In water it is scarcely soluble, but it dissolves readily in alcohol, ether, benzene, and concentrated nitric acid, and is itself an excellent solvent for many organic substances which are sparingly or not at all soluble in the ordinary solvents.

Nitrobenzene is poisonous, especially when the vapour is inhaled.

It is manufactured on a large scale by allowing a well-cooled mixture of fuming nitric acid, free from chlorine, and concentrated sulphuric acid to flow into benzene, contained in cast-iron vessels provided with agitators. The mixture must be kept very cool. Towards the end of the reaction, however, the temperature may rise to from 80° to 90°. When the reaction is over the product is run into tanks; the acid mixture separates as a layer at the bottom, whilst nitrobenzene, being insoluble in the acid, goes to the top. The uncombined acid layer is drawn off and the nitric acid recovered.

Crude nitrobenzene contains more or less benzene which has escaped the reaction. To remove the latter, the crude product is treated with steam, while the benzene distils over with a small quantity of nitrobenzene. The residual nitrobenzene is washed with caustic soda and water, and if necessary purified by distillation in high pressure steam.

Messrs. Schimmel & Co. report that "the competition of the cheap English nitro-toluol, the Pseudo-Mirbane of commerce, has made the laborious and dangerous manufacture of this poisonous body (nitrobenzene) quite unremunerative. The mania for cheapness has brought about a condition which deserves to be more closely considered. It should be premised that the odour of bitter almond is peculiar to nitrobenzene, so that those persons who buy an impure preparation mixed with nitrobenzene deceive themselves if they expect to obtain with it the same effect as with the pure compound. Of this they may convince themselves by making a comparative examination of the strength of the two qualities." This Report, dated April 1891, further states that the opinion of an expert employed to estimate a commercial sample of this kind in connexion with the prosecution of the consigner, was given as follows :—"The so-called Mirbane oil contains no noteworthy proportion of nitrobenzene, of which it

should alone consist. It is made up of 40 to 50 per cent. of the
higher homologues of this series, the nitrotoluenes, particularly
the metanitrotoluene and nitroxylols, of about 8 per. cent of
benzene hydrocarbons which have escaped nitration, and 40 to
45 per cent. of a residue boiling above 360° C. The indifference
of this last constituent to powerful reagents, as well as the fact
that a paraffinoid body was isolated from it, make it probable that
we have here a mineral oil. The specimen is an instance of bare-
faced and clumsy adulteration, and its value does not in the least
correspond to the price at which it was bought." It is added
that " Mirbane oils exist and circulate in commerce which practically
contain no nitrobenzene at all." A remark worthy of note.

Pseudo-Mirbane.

Nitrotoluene, $C_6H_4(NO_2)CH_3$, is obtained by dissolving toluene
in fuming nitric acid and precipitating with water. It is a
colourless liquid, boiling at 225°, possessing a smell of bitter
almonds, and a very sweet, somewhat biting taste. At the same
time, a small quantity of metanitrotoluene and the ortho- and
para-compounds are formed *. The relative quantities of the
two chief products depend upon the concentration of the acid and
the temperature at which the nitration is effected. When a very
concentrated acid is employed, and the temperature allowed to
rise, paranitrotoluene is chiefly obtained, while the yield of the
ortho-compound is greatly increased by employing a weaker acid
and cooling the mixture well.

The nitrotoluenes are manufactured on the large scale by
mixing 10 parts of toluene with 11 parts of nitric acid sp. gr. 1·22
and 1 part of sulphuric acid sp. gr. 1·33 with continual agitation,
in the apparatus used for the manufacture of nitrobenzene; this
mixture is then either cooled or kept warm, according to the product
desired. The crude product is washed with water and caustic
soda solution, freed from unattacked toluene by distillation with
steam, and then distilled with superheated steam. The distillate
is then repeatedly fractionated ; the larger portion of the fraction,
distilling above 230°, solidifies on cooling, and the crystals, after
purification by draining and pressing, yield pure paranitrotoluene
on distillation; the fraction boiling between 222°–223° consists

---

* Ber. Deutsch. chem. Ges. xii. p. 445.

chiefly of orthonitrotoluene, while the intermediate fractions contain some of the meta-compound *.

THE CHERRY-LAUREL.

The Cherry-Laurel, *Cerasus Laurocerasus*, syn. *Prunus Lauro-cerasus*, Linn., is a native of the Levant, Caucasus, the mountains of Persia, and the Crimea. The varieties of this evergreen shrub form bushes 6 to 10 feet high. Several of them have been naturalized in England as ornamental shrubs and for the preparation of the aqueous distillate. The Colchican laurel (*P. L. Colchica*) is a hardy dwarf spreading bush, with narrow, sharply serrated, pale-green leaves; the Versailles laurel (*P. L. latifolia*), with large leaves; the *P. L. rotundifolia*, with short broad leaves; the Grecian, with very narrow leaves; the Alexandrian, with very small leaves; and the Caucasian (*P. L. Caucasica*), which is superior to them all, being not only the most robust and hardy, but yielding the largest quantity of volatile oil of any of these varieties, 75 grammes per 50 kilos; this is a vigorous grower, and is easily propagated from cuttings planted in September in a sheltered situation.

Cherry-laurel water is used medicinally; but it is a dangerous medicament, owing to the uncertain quantity of hydrocyanic acid it may contain. The officinal preparation is directed to be made by distilling 1 lb. of the leaves with $2\frac{1}{2}$ lbs. of water and drawing over 1 pint of distillate. Perinelle† has pointed out the importance of always ascertaining the strength of *aq. laurocerasi* before it is placed in stock. His experiments have led him to the conclusion that the strength of the liquid in hydrocyanic acid will *vary considerably according to the time of year at which it is prepared and to the variety of the cherry-laurel employed.* The two periods of the year when the leaves are most readily obtained in quantity in England are May and November, when the shrubberies are clipped. The water distilled in May yielded only 39 milligrammes of hydro-

---

* *Toluene*, $C_6H_5CH_3$, which on the Continent is called Toluol, is a strongly refractive liquid possessing a smell similar to that of benzene (which in French is called benzol). It is obtained on the large scale from light coal-tar oil, and is chiefly employed in the colour industry.

† Rép. de Pharmacie, Aug. 1887, p. 331.

cyanic acid per 100 grammes of water, but that prepared in November yielded 134 milligrammes per 100 grammes *.

The oil of cherry-laurel is distilled in the south of Switzerland and in Italy. In England it is only obtained as a bye-product in the distillation of cherry-laurel water; in this way a quantity of about 2½ ozs., obtained during the distillation of 300 lbs. of leaves, was examined by Tilden in 1875, and the results of the investigation read before the Bristol Pharmacy Assoc. in that year. According to that report, the oil was found to be of a pale yellow colour and held in suspension a few crystals, probably consisting of benzoic acid. The sp. gr. was found to be 1·0615. It contained rather under 2 per cent. of prussic acid, and had an odour much resembling that of oil of bitter almonds. When shaken with excess of strong solution of acid sulphite of sodium, all the aldehydic constituents of the oil were dissolved, leaving only 1 or 2 per cent. of an oily substance containing a brown resin. The sulphite, when crystallized out from the solution and distilled with sodium carbonate, yielded an essential oil which, after drying by calcic chloride, presented all the characteristics of pure benzoic aldehyde. It distilled without residue between 174° and 178° C. and its sp. gr. at 17° C. was found to be 1·0492. The small amount of viscid oil left by the bisulphite when distilled with potassium bichromate and sulphuric acid, yielded no volatile product, and the crystalline deposit obtained after this treatment proved, on examination, to be benzoic acid. These experiments therefore indicate that the essential oil of cherry-laurel leaves consists mainly of benzoic aldehyde accompanied by hydrocyanic acid and about 1 per cent. of a volatile oil (possibly benzoic alcohol), and minute quantities of an odorous resin.

It is generally understood that, like the oil of bitter almonds, this oil does not exist ready-formed, but is produced by the action of water. The principles contained in the leaves, which are the cause of this reaction (which, as Mr. Umney, who made the investigation, points out, is *instantaneous*), are yet unknown and merit investigation.

* Pharm. Journ. [3] xviii. p. 170.

HELIOTROPE.

About 84 different species of these shrubs are known and botanically identified. They are found chiefly in tropical and subtropical regions and a few are acclimatized to the temperate countries of Europe, where they are valued and cultivated on account of their fragrant blossoms, being readily propagated from seeds and cuttings. In private gardens and for commercial cultivation, preference is generally given to the *Heliotropium Peruvianum* (Linn.) (syn. *H. odoratum*, Mœnch). This is a native of Peru and was introduced into Europe in 1757. It requires a rich soil and a sunny situation. In England, if grown in a conservatory and given free root-room by being planted in the ground, it will form a bush 8 or 10 feet high and flower throughout the year, but if grown out in the open air will be killed by the first winter. Its purple spikes of flowers are terminal and spirally revolute. There is a hybrid variety with larger flowers. In the South of France it will attain almost the size of the English " May tree " and form a hedge. It is grown to rather a large extent for its exquisitely perfumed flowers. To extract the perfume, the system of maceration is employed; the flowers must be used *immediately* after being gathered, and the melted purified grease kept at as low a temperature as possible, no more heat being applied than just sufficient to maintain the grease in a liquid state. The flowers are strained out and changed every 24 hours until the grease is saturated, it is then exhausted or washed with alcohol, as before described.

The plant called " Winter Heliotrope " is the *Tussilago fragrans* (syn. *Petasites fragrans* and *Nardosmia fragrans*), sometimes called " Fragrant Coltsfoot." It is a native of the South of Europe and a hardy plant in England, commencing to flower early in December. The fragrance of its purple flowers, which are produced in great abundance, is very similar to that of the heliotrope in a mild form. This plant will thrive in any situation, even in the shade, but it prefers a damp clay soil. It requires no cultivation at all, as it spreads so freely as to be practically irrepressible; in fact it becomes a nuisance and difficult to eradicate. If once established, its strong running roots rapidly make a tour of the garden and send up shoots in the paths and other places where not wanted, covering everything and crowding out all other small plants. There is a white-flowered variety (*Petasites alba*). It is a smaller plant and the flowers are not quite so highly scented as the purple-flowered.

Artificial Heliotrope.

Piperonal, $C_8H_6O_3$, commercially known as "Heliotropine," has a very agreeable odour very much like that of heliotrope. The starting-point in its manufacture is *Piperine*, $C_{17}H_{19}O_3$. Ground pepper, preferably the white Singapore pepper, as it contains the largest amount of alkaloid (9·15 per cent.), is mixed with twice its weight of slaked lime and a sufficient quantity of water; the solution is then evaporated to dryness on a water-bath and the powder exhausted with commercial ether, from which the piperine can be obtained nearly pure on evaporation, in large crystals of a faint straw-yellow colour. To obtain it perfectly pure, it must be dissolved in alcohol and re-crystallized. Another process of preparing piperine is to exhaust the pepper with alcohol of sp. gr. 0·833 and distil the tincture to the consistence of an extract. This extract is to be mixed with potash-lye, which dissolves the resin and leaves a green powder; by washing this in water, dissolving in alcohol, crystallizing and re-crystallizing, it is obtained colourless *.

Piperine is converted into potassium piperate by boiling it for 24 hours with its own weight of caustic potash and from 5 to 6 parts of alcohol in a large retort, using an inverted Liebig's condenser. On cooling, the potassium piperate crystallizes out in shining yellow laminæ. It is washed with cold alcohol and re-crystallized from hot water. If coloured, it is bleached by animal charcoal. As thus obtained, it is in nearly colourless crystals, which become yellow under the influence of light.

One part of potassium piperate is dissolved in from 40 to 50 parts of hot water, and a solution of 2 parts of potassium permanganate is gradually poured into the hot liquid with constant stirring. Each drop of the latter is almost instantly dissolved, and the solution acquires a very pleasant odour. A pasty mass of brown manganic hydrate separates, which is placed on a filter and washed with hot water until the washings cease to smell of heliotropine. These

---

* Poutet, Journ. de Chim. Med. i. p. 531.

washings are added together and the whole distilled over an open fire. The first portions of the distillate contain the largest proportion of piperonal, the greater part of which crystallizes out on cooling. The remainder may be obtained by agitation with ether *.

Piperonal crystallizes from water in colourless, transparent, highly lustrous prisms, an inch long. It is sparingly soluble in cold water, easily in cold alcohol, and in all proportions in boiling alcohol or ether. It melts exactly at 37° and boils without decomposition at 263°, forming a vapour which has a sp. gr. of 5·18 †. Tiemann and Haarmann state ‡ "that the odour of piperonal is possessed by 'vanillon,' a kind of vanilla, which forms thick, fleshy capsules and is obtained from the West Indies. This sort of vanilla is employed in perfumery for the preparation of essence of heliotrope; it contains no piperonal, but vanillin and an oil which is not yet identified. The perfumers, in preparing essence of heliotrope, add a little of this oil to the extract of vanillon. If a little be added to a solution of pure vanillin, both substances can be recognized by their smell for some time, but after standing for months the mixture acquires the smell of heliotrope."

The perfume of "Heliotropine" is completely destroyed by the action of direct sunlight; it is also injured by heat; it should therefore be stored in a cool place in the dark, such as a cool cellar, and be kept in yellow glass bottles, the yellow glass intercepting the chemical rays.

HAWTHORN.

Under the pseudonym "Aubepine" a preparation having the odour of Hawthorn (*Cratægus oxycantha*) has recently been brought out by a Paris firm as a novelty, but it appears to be a definite chemical body, long known to chemists as anisic aldehyde (Paramethoxybenzaldehyde), $C_8H_8O_2$. Cahours found § that it is formed, together with anisic acid, by the oxidation of oil of anise. It can be prepared by gently heating the oil of anise for about an hour with three times its volume of nitric acid of sp. gr. 1·106 (14° Beaumé). The heavy oil which is thus formed is washed with

---

* Chemiker Zeitung, Feb. 1884, and Ann. Ch. Pharm. clii. p. 35.
† Ber. Deutsch. chem. Ges. x. p. 1274.
‡ Ibid. ix. p. 1287.
§ Ann. Chim. Phys. [3] xiv. p. 484, and ibid. xxiii. p. 354.

dilute potash and distilled.   The distillate is agitated with a warm
solution of acid sodium sulphite of sp. gr. 1·25, and the crystalline
compound thus formed is collected on a funnel, thoroughly washed
with alcohol, dissolved in as little luke-warm water as possible, and
the solution heated with excess of strong sodic carbonate, when
the hydride of anisyl (anisic aldehyde) separates out and floats on
the surface.   It is then purified by redistillation.   It is a yellowish
liquid of sp. gr. 1·09 at 20° C.   Its boiling-point is 253° to 255° C.
It is almost insoluble in water, but soluble in all proportions in
alcohol and ether.   When exposed to the air, it gradually absorbs
oxygen and is converted into anisic acid.   It possesses the property,
peculiar to aldehydes, of forming crystalline compounds with the
acid sulphites of the alkali-metals.

Rossel prepared anisic aldehyde by oxidizing oil of anise with
chromic acid *.   300 grams of strong sulphuric acid and 850 grams
of water are poured upon 200 grams of potassium dichromate in a
capacious flask; 100 grams of oil of anise are added after the
liquid has completely cooled, and the whole is vigorously shaken,
the temperature of the mixture rising to 70°–80°.   When the
reaction is complete, the liquid is diluted with water to one and a
half times its volume and distilled, warm water being added through
a funnel-tube to replace that which distils off.   In this way a
quantity of anisic aldehyde is obtained equal to 50 per cent. of the
oil of anise used, together with 10 per cent. of anisic acid.   According to Rossel, pure anisic acid boils at 247°–248° C. under a
pressure of 733·5 mm. (?) and has a density of 1·228 at 18° C.

* Ann. Chem. Pharm. cli. p. 25.

CHAPTER X.

CINNAMON.—CASSIA.—CLOVE.

CINNAMON.

THE bark known as "Ceylon Cinnamon" is derived from the *Cinnamomum Zeylanicum*, Breyne, a native of Ceylon, where it is widely distributed in the forests at altitudes varying up to 3000 feet, and one of its varieties is found as high as 8000 feet. It is a small evergreen tree with beautiful shining leaves, bearing panicles of greenish flowers somewhat resembling mignonette, but of an unpleasant odour. The aspect, height, and dimensions of the tree are very variable. Many varieties differing greatly from each other have received distinctive specific names; the numerous intermediate forms merging one into the other.

Although formerly exclusively derived from Ceylon, the tree is now grown in India, Mauritius, Cayenne Mahé, (one of the Seychelle Islands), and the Antilles, from seed obtained from Ceylon, and these barks now to some extent rival those of Ceylon. It is cultivated extensively in Jamaica, where it was introduced from Ceylon about 1782.

In the humid forests of the South-west of India there are seven or eight clearly marked varieties which may be regarded as so many distinct species, but as they are so gradually linked by intermediate forms, it is impossible to distinguish them specifically by any sufficiently constant feature. As they are found from the sea-level up to very great altitudes, it has been thought that their difference in appearance may be due to local influences.

Several distinct varieties are known in Ceylon. The finest bark is produced by a choice cultivated variety of the tree, called *Rasse curunda* or "honey cinnamon." This is the true *C. zeylanicum*, Breyne, a tree of about 5 to 7 metres in height and 30 to 40 centimetres in diameter round the trunk. The leaves and leaf-stalks of the young branches are glabrous; the leaves are nearly opposite, oblong-ovate, obtuse, the largest being from 11 to 14 centimetres in length by 5 to 7 centimetres in width, but often much smaller, coriaceous, shining, and of a bright green above, glaucous beneath. Besides the middle vein of the leaf, there are two other veins on each side of it, also starting from the stalk, rounded to the shape of the edge of the leaf to nearly its extremity. The leaves on drying acquire a reddish-brown colour, due to the oxidation of the essential oil contained in them. The small flowers are disposed in terminal panicles, appearing in January and February; their strong perfume resembling a mixture of rose and lilac *. The berry is of a deep purplish-brown colour, shaped like an acorn, enclosed at the base by the calyx. It contains a soft green pulp and one seed. The berry ripens in August, and is gathered by the natives for the purpose of extracting the oil from the seed.

The locality most suitable to the cultivation of the finest variety of the cinnamon tree is situated in the south-west of the Island of Ceylon, between Negumbo, Colombo, and Matura. The principal plantations are in the immediate neighbourhood of Colombo, within little more than half a mile of the Fort, occupying a tract of country upwards of ten miles in length. The road, commencing at the west gate of the Fort, and returning by the south gate, makes a winding circuit through the woods.

The soil of these cinnamon gardens is mostly of a loose white sand over a rich sub-soil of sandy loam mixed with decayed vegetable matter. The situations most suitable to the growth of the cinnamon tree appear to be those which are fully exposed to the sun, but yet sheltered from the wind. Such shelter appears to contribute to its luxuriance, as it is found to grow with unusual vigour near to houses. When the ground is prepared for planting cinnamon, the low brushwood and young trees are cut down, but lofty trees are allowed to remain at intervals, as it is found that the tender plants thrive better under their shade than when exposed to the direct rays of the sun. The planting usually takes place at the end of autumn, when the seeds are ripe. A line is stretched across the ground, and guided by it the planter turns up about a foot square at intervals of 6 or 7 feet. The brushwood and branches having been previously burned, their ashes are then

* These flowers are not as yet utilized, but their perfume is well known.

spread upon the newly-dug spots, and into each of them 4 or 5 cinnamon berries are sown, in holes made with a dibble. They are then covered with earth, and branches are laid over the parts to prevent the earth from becoming parched and to protect the young plants as soon as they come up. This takes place in about 3 weeks. Sometimes the berries are sown in nurseries and the young plants are transplanted in October and November. In favourable situations plants attain the height of 5 or 6 feet in about 6 or 7 years, and a healthy bush will then afford 2 or 3 shoots fit for peeling, but in unfavourable situations there is no yield until the eighth or the twelfth year. In a good soil, from 4 to 7 shoots may be cut from one tree every second year; thriving shoots of 4 years' growth are sometimes fit for cutting, but they may be sometimes cut at the age of 2 years. A greyish corky appearance is an indication of their fitness.

As 4 or 5 seeds are usually sown in one spot, and in most seasons the greater part of them germinate, the plants grow in clusters, not unlike a hazel bush. If the season be unusually dry many of the seeds fail, and the want of moisture is often fatal to the young plants, so that it is sometimes necessary to plant a piece of ground several times successively. A plantation of cinnamon, even on good ground, cannot be expected to make much return until a lapse of 8 or 9 years. The plant is likewise sometimes propagated from shoots cut from large trees, or by layers, also by transplanting large stumps or divisions of a parent stump. The method of culture by seeds is considered the least advantageous, as it requires greater attention than other modes and the trees are longer before they arrive at perfection. If cultivated from shoots, the cuttings must be continually watered or they will not thrive. Those selected for the purpose should be very young, not having more than three leaves; if older they die. The method by layers has been recommended because the numerous side-branches which issue from the bottom of the trunk always furnish a plentiful supply well adapted to the purpose. Plants raised in this way or from cuttings soon require pruning to prevent them growing too tall, they then assume the form of stocks or pollards. The transplanting of divisions of old roots is a practice which is much approved, as they yield shoots of useful size 12 months after they have been placed in the ground. Great care is, however, necessary in their removal, for should any of the rootlets, even of one tenth of an inch diameter, receive injury, the whole root will certainly perish. Thunberg mentions a fifth method of cultivation, or rather a manner of obtaining cinnamon of superior quality :—" When the tree is cut down, and a fire kindled on the spot to consume the stumps, the roots afterwards throw out a number of long straight shoots which yield incomparably fine cinnamon. From these are cut the common walking-sticks, which, in appearance, resemble those of the hazel tree and retain the taste and smell of cinnamon.

In taking the harvest, the shoots are not all cut at once, but by degrees, as they arrive at the required maturity. The shoots which are cut are usually from a half to three-quarters of an inch in diameter and from 3 to 5 feet in length. When the cultivator perceives a shoot of proper growth, he strikes an instrument which resembles a small bill-hook into it, obliquely. He then gently opens the gash to discover whether the bark separates freely from the wood; should this not be the case, he leaves the sucker for a future time. Some shoots never arrive at a fit state for decortication. Plants of several years' growth sometimes bear numerous marks of " annual experiments " made for the purpose of ascertaining whether the bark was in a fit state for removal.

Two crops are gathered during the year ; the first, and largest, lasts from April till the end of August, the second commences in November and finishes in January. The reason of selecting these periods is that the sap is more active after the rains, and then the bark is more easily detached from the wood.

The branches are lopped off by means of a long knife in the shape of a hook or sickle. The leaves and outer bark are removed, and the inner bark is cut round at distances of about 30 centimetres. Two cuts are then made lengthways, one on either side of the branch, and when the branch is thick three cuts are made. The bark is then carefully removed by inserting a peeling-knife beneath it. When the bark adheres firmly the separation is facilitated by friction with the handle of the knife.

After 24 hours each tube of bark is placed on a small stick of convenient thickness and the outer epidermis and green pulpy matter from the inner surface carefully scraped off with a knife. After a few hours the smaller quills are introduced into the larger ones, so forming solid sticks often measuring 40 inches in length. In this state they are left for a day in the shade and are then placed on hurdles to dry in the sun. When sufficiently dry they are put up in bundles weighing about 30 lbs. each.

The cinnamon thus prepared appears in commerce in the form of long brittle sticks of a pale yellow-brown colour composed of numerous layers of bark, as thin as paper, rolled one over the other, the edges not overlapped but both edges rolled inwards, so forming a longitudinal groove the length of the stick. The taste is agreeably aromatic, warm, and sweet, and the odour very sweet. By distillation it yields about $\frac{1}{2}$ to 1 per cent. of a very sweet and powerful essential oil.

Cinnamon is re-baled on its arrival in London, and as the sticks are very brittle a quantity of chips and small pieces collect. These are collected and sold separately to druggists and distillers. They are often of excellent quality.

The tips of the branches and the trimmings which collect are carefully dried and shipped to Europe, where they are distilled and the oil sold as " Ceylon cinnamon oil." The export of " chips " from Colombo and Galle amounts to about 500,000 lbs. annually.

The leaves which are stripped from the branches are distilled in Ceylon, very seldom by the cinnamon growers themselves, but as a rule by persons who pay the proprietor of an estate fifty to one hundred rupees a year for permission to use as many leaves as may be required for a still. From 80,000 to 100,000 ounces of " cinnamon-leaf oil " are annually distilled in Ceylon.

Inspection and tasting are the methods resorted to for ascertaining the quality of cinnamon. The bark of Ceylon cinnamon is characterized by being cut obliquely at the bottom of the quill, whereas the other kinds are cut transversely. Inferior kinds are thicker, darker, browner, and have a more pungent taste, succeeded by a bitter taste. The most inferior quality of cinnamon bears such a resemblance to the best cassia that this last may be substituted for it or used as an adulterant to powdered cinnamon without being at once detected (of course when the bark is entire the difference is apparent). The following reactions are useful in examining powdered cinnamon :—Make a decoction of pulverized cinnamon of known purity, also a decoction of the suspected sample. Filter the decoctions when cold, and add to 30 grammes of each one or two drops of tincture of iodine. The decoction of pure cinnamon is but very slightly affected, but

o 2

that containing cassia immediately takes a blackish-blue coloration.
The cheap sorts of cassia known as *Cassia vera* can be distinguished
from China cassia and from cinnamon by their richness in
mucilage. This can be extracted by cold water as a thick glairy
liquid, which, on the addition of corrosive sublimate or neutral
acetate of lead (but not of alcohol), yields a dense, viscous precipitate.

When the Ceylon cinnamon trees become too old to produce
good growth they are cut down, and the bark of the larger branches
and of the trunk removed. This cinnamon is called *Mate*. The
odour and taste are agreeable, but feeble, and poor in essential oil.
An oil is also derived from the root; this is lighter than water and
smells of cinnamon and camphor mixed.

Although the finest bark is derived from the cultivated trees, all
forms of the tree yield a more or less odorous bark. The finest
of the uncultivated trees are distinguished by the large size of
their leaves, but yet the quality of the bark cannot always be
judged by this sign, so the bark-gatherers remove a piece of the
odourless, hazel-looking exterior bark, and taste the inner bark
before commencing operations; leaving those trees which are not
of the quality sought. Some varieties, such as the *C. multiflorum*
and *C. ovalifolium*, yield barks of such inferior quality that they
are rarely gathered except to adulterate a finer description.

Of Indian cinnamons there are the Tellicherry or Bombay cinnamon; in appearance it is equal to the Ceylon kind, but the
internal surface of the bark is more fibrous and the flavour inferior,
but it is superior to the Malabar variety which is grown on the
Coromandel coast. This Madras or Malabar cinnamon approx-
imates to *Cassia lignea* in thickness, but it is not the old Malabar
cinnamon, which was the product of the *Laurus Cassia*, Linn., and
which was destroyed by the Dutch. It is the Ceylon cinnamon
propagated in India by the English, and has nearly all the characters
and quality of the Ceylon; it is, however, distinguished by being
paler in colour and having a more feeble and less permanent
odour. It is made up in sticks as long as the Ceylon growth, but
the pieces of bark are in reality shorter and the length of the
sticks is due to the method of telescoping the strips of bark one
in the other. The layers of bark forming the sticks are not so
thin as those produced in Ceylon, and the sticks are thicker and
more cylindrical.

The Java cinnamon ranks between Ceylon and Tellicherry in flavour, and is imported almost exclusively into Holland. For recent observations on "China cinnamon" see article on *Cassia*.

The Cayenne cinnamon (also derived from *C. Zeylanicum*, Breyn.) is almost as thin and long in the bark as the Ceylon grown, but it is paler in colour, more feeble in flavour and odour, and its essential oil is more acrid. It is, however, frequently sold as Ceylon cinnamon.

The same tree is grown in Brazil and in the Antilles, but the resulting barks are very variable in quality and always inferior to the Ceylon growth. The Brazilian has the least flavour of any ; it is spongy and almost inodorous.

The cinnamon tree yields essential oils from its leaves, bark, and root, each oil differing in composition and value. The most valuable is the oil from the bark, consisting chiefly of cinnamic aldehyde, now called cinnamaldehyde, and a variable quantity of hydrocarbon, the nature of which has not yet been definitely determined. The oil of cinnamon bark is worth about eighteen times as much as the oil distilled from the leaf, which contains chiefly eugenol, a hydrocarbon having an odour of cymene, a little benzoic acid, and some cinnamaldehyde. The oil from the root contains cinnamaldehyde, hydrocarbon, and ordinary camphor. The oil of the root is lighter than water, that of the leaves and bark being heavier.

Oil of cinnamon is obtained in Ceylon by macerating the inferior pieces of the bark and broken quills which are incapable of being worked in with the usual quills, reduced to coarse powder, in a saturated solution of common salt for two days, and then submitting the whole to distillation. The yield of oil varies according to the quality of the bark, from $\frac{1}{2}$ to 1 per cent. This oil is largely distilled in Ceylon. As imported into London it varies somewhat in colour from yellow to cherry-red and very much in value, the paler varieties are the most esteemed ; hence London druggists frequently redistil the imported oil, by which they procure two pale yellow oils ; one lighter (amounting to about a quarter of the whole), the other heavier than water. The loss on this process being near 10 per cent.

The oil distilled from the finer sorts of cinnamon bark is of a golden colour when fresh, becoming red by age. The sp. gr. of the fresh oil is 1·035. The oil obtained from the coarser bark is darker and brownish. Cinnamon oil of fine quality when brought into contact with the tongue should produce a taste of intense sweetness, far exceeding the sweetness of sugar: with an inferior quality a clove-like taste is first developed, and after a time a slight sweetness,—such an oil has no higher value than cassia oil. Cinnamon oil and cassia oil are said to be of the same chemical composition, the marketable value of both being estimated by the amount of cinnamaldehyde contained in them ; methods of estimating this are also given under the article " Cassia."

As before remarked, the principal constituent of oil of cinnamon is *cinnamaldehyde*, it also contains a hydrocarbon which has not been thoroughly investigated. These may be separated by bringing the oil into contact with concentrated nitric acid; the crystals which separate out in long oblique rhombic prisms or small plates are then decomposed by water into nitric acid and the free aldehyde. Cinnamaldehyde may also be obtained pure by agitating oil of cinnamon with 3 or 4 volumes of a concentrated solution of acid potassium sulphite; the compound, which separates in scales, is dried, washed in alcohol, again dried, and finally decomposed with dilute sulphuric acid *. According to Peine,† an alcoholic solution of 50 parts of oil of cinnamon is agitated with 90 parts of a 50-per-cent. solution of sodium sulphite, and the compound, after washing with alcohol, decomposed with sulphuric acid ; 40 cubic centimetres of this, diluted with an equal volume of water, being employed for every 100 c. c. of the sulphite solution. The aldehyde is then distilled with steam, the distillate extracted with ether, and after the evaporation of the ether, fractionated under diminished pressure.

Cinnamaldehyde is a colourless, very pleasant smelling liquid, which decomposes on distillation at the ordinary pressure, but boils without alteration at 130° under a pressure of 30–40 mm.

Cinnamaldehyde may be very easily prepared synthetically by allowing a mixture of 10 parts of benzaldehyde, 15 parts of acetaldehyde, 900 parts of water, and 10 parts of a 10-per-cent. solution of caustic soda to stand for 8 or 10 days at a temperature of 30°, the whole being frequently agitated and finally extracted

---

* Ann. Chem. Pharm. lxxxv. p. 271.
† Ber. Deutsch. chem. Ges. xvii. p. 2109.

with ether *. Piria obtained it by the distillation of a mixture of calcium cinnamate and calcium formate †.

Cinnamic acid was so named by Dumas and Peligot in 1834, who found that oil of cinnamon bears the same relation to it as oil of bitter almonds or benzoyl hydride to benzoic acid, and therefore gave it the name of cinnamyl hydride ‡. It also occurs in the so-called " flowers " of cinnamon, the unripe fruits of " bastard cinnamon " (Cassia) §.

Cinnamic acid occurs in liquid styrax, partly in the free state and partly as styracin or cinnyl cinnamate. The balsams of Peru and Tolu contain this acid in the free state and accompanied by the benzyl ether of cinnamic and benzoic acids. Both these acids have also been found in Sumatra benzoin. Cinnamic acid also occurs in the leaves of the Japanese garden plant, *Enkyanthus Japonicus* ‖, and has been found in *Globularia alypum* and *Globularia vulgaris* ¶.

Cinnamic acid can be obtained synthetically by heating benzaldehyde with acetyl chloride to 120–130° **. It is also formed when benzaldehyde and glacial acetic acid are heated together to 16° in presence of hydrochloric acid or zinc chloride ††. Perkin found that the acids of the cinnamic series may be synthetically prepared by heating benzaldehyde with the anhydride of a fatty acid and the anhydrous sodium salt of the fatty acid. In order to prepare cinnamic acid in this way, a mixture of 1 part of sodium acetate, 2 parts of benzaldehyde, and 3 parts of acetic anhydride is boiled for a day, or heated for 5 to 6 hours in a sealed tube at 180°. The product is boiled with water to volatilize any unaltered benzaldehyde, and the impure cinnamic acid, which crystallizes out on cooling, is washed and dissolved in hot sodium-carbonate solution; the liquid is allowed to cool and is then filtered in order to remove any oily impurity, the cinnamic acid being precipitated with hydrochloric acid and finally crystallized

---

* Peine, Ber. Deutsch. chem. Ges. xvii. p. 2117.
† Ann. Chem. Pharm. c. p. 104.
‡ Ann. Chem. Pharm. xiv. p. 50, and Ann. Chim. Phys. lvii. p. 305.
§ Ann. Chem. Pharm. xxxiv. p. 147.
‖ Ber. Deutsch. chem. Ges. xx. ref. 66.
¶ Ann. Chim. Phys. [5] xxviii. p. 67.
** Ann. Chem. Pharm. c. p. 125.
†† Ber. Deutsch. chem. Ges. iii. p. 412.

from dilute alcohol. According to the reactions which take place in this operation, cinnamic anhydride is formed and is then decomposed by the boiling with water. When acetic anhydride and benzaldehyde are heated together no cinnamic acid is formed, benzidene diacetate being in this case the product. The sodium acetate therefore acts as a dehydrating agent, or to bring about a decomposition of the benzidene diacetate which is first formed. Perkin found that the sodium acetate can be replaced by butyrate or valerate without affecting the production of cinnamic acid *.

When benzaldehyde is heated with acetic anhydride, benzidene diacetate is formed, as above mentioned. If this be heated with sodium acetate, cinnamic acid is formed, and can be even more readily obtained by Caro's process of heating benzidene chloride with sodium acetate †, the diacetate being, however, the first product.

Cinnamic acid is also formed when benzaldehyde is heated to 140° with malonic acid. Cinnamic acid was formerly prepared from liquid styrax, but is now manufactured by the method discovered by Caro, which can also be used for the preparation of substituted cinnamic acids by the employment of substitution products of benzidene chloride ‡.

Cinnamic acid dissolves in 3500 parts of water at 17°; it is much more readily soluble in boiling water and crystallizes from it in lustrous plates, while it separates from alcohol in monosymmetric prisms, melting at 133°. It sublimes in a similar manner to benzoic acid, but somewhat less readily, is volatile in steam and boils at 300°–304°, but partially decomposes on continued boiling into styrolene and carbon dioxide. It yields benzene in considerable quantity on fusion with caustic soda §, while it is resolved into benzoic acid and acetic acid when caustic potash is employed ||. Oxidizing agents convert it first into benzaldehyde, so that it can in this way be readily distinguished from benzoic acid ¶. It is converted into hydrocinnamic acid by sodium amalgam and water.

---

* Journ. Chem. Soc. 1877, i. p. 838.
† Roscoe & Schorlemmer, Org. Chem. iii. pt. v. p. 214.
‡ Ber. Deutsch. chem. Ges. xv. p. 969.
§ Ibid. xiii. p. 1257.
|| Ann. Chem. Pharm. lxxxvi. p. 264, and cxlvii. p. 112.
¶ Ibid. lv. p. 1.

Methyl cinnamate forms crystals which have a very pleasant odour, melt at 33°·4 and boil at 263°.

Ethyl cinnamate is a pleasant smelling liquid boiling at 271°; it is contained in liquid storax.

Benzyl cinnamate (Cinnamein) is a constituent in Peru balsam, and is formed when sodium cinnamate is heated with benzyl chloride. It crystallizes in prisms which melt at 39°.

Hydrocinnamic acid is prepared by heating one part of cinnamic acid for an hour with four parts of hydriodic acid of boiling point 127° and a little amorphous phosphorus. The acid separates out on cooling in a solid cake, which is dissolved in ammonia and precipitated by hydrochloric acid in the filtered solution. The hydrocinnamic acid which separates is then distilled; the almost pure compound, containing only a small quantity of an oily substance, passes over at about 280° *. Hydrocinnamic acid possesses a characteristic *goat-like odour*, melts at 47°·5, and solidifies on cooling in long brittle needles; it boils at 280°, is readily volatile with steam, dissolves in 168 parts of water at 20°, more readily in hot water, and separates in oily drops when the solution is slightly cooled. It is readily soluble in alcohol, from which it separates in indistinct crystals †.

Ethyl hydrocinnamate is a powerfully refractive liquid which possesses an overpowering odour resembling that of *pineapple*, it boils at 247°–249°.

Oil of Cinnamon leaf is distilled in Ceylon from the leaves of the *Cinnamomum Zeylanicum*. Both in appearance and properties it very closely resembles the oils of cloves and pimento. It has a brownish colour and an aromatic penetrating odour. Its taste is exceedingly pungent. As examined by Stenhouse ‡ its specific gravity is 1·053. It has an acid reaction, and when treated with solutions of potash or ammonia it solidifies, forming a butyraceous crystalline magma. Like clove and pimento oil, it is essentially a mixture of eugenol and a neutral hydrocarbon of the formula $C_{10}H_{14}$. Cinnamon-leaf oil is remarkable, however, for containing a small quantity of benzoic acid. When the crude oil was distilled a colourless oil came over, while the residue in the retort became gradually darker. The last portions of the oil

---

* Ber. Deutsch. chem. Ges. xiii. p. 1680. † Ibid. xviii. p. 321.

‡ Pharm. Journ. Jan. 1855.

which passed over, on being set aside for some time, deposited
crystals which had all the appearance and properties of benzoic
acid. Stenhouse adds that the quantity of these crystals at his
disposal was so small that they could not be subjected to analysis.
He proved them not to be cinnamic acid, for on treating them
with a hot solution of hypochlorite of lime they did not give the
very characteristic reaction of cinnamic acid, viz., " the production of that singular chlorinated oil which is always produced
when cinnamic acid is treated in this way." They were proved
to be benzoic acid in the following way:—" When the last portions of the oil which had come over, together with the residue
remaining in the retort, were digested with strong nitric acid,
along with much oxalic acid, a small quantity of another kind of
crystals were obtained, which in appearance and properties agreed
perfectly with nitrobenzoic acid. The first portions of the colourless oil which distilled over were again rectified, when the boiling-
point was by no means constant, though the greater portion of
the oil came over at a temperature approaching the boiling-point
of eugenol *.

" The oil was treated with an excess of potash, and the hydrocarbon which did not combine with the alkali was drawn off by
a pipette. It was next treated with fused potash, then with
calcic chloride, and lastly rectified over potassium. In this way a
colourless, highly refractive liquid was obtained, the greater
portion of which distilled over between 160° and 165° C. Its
sp. gr. was 0·862 and its odour closely resembled that of cymene,
the result of its analysis agreeing exactly with the formula $C_{10}H_{14}$
(a body which is contained in several essential oils, or which can
at all events be withdrawn from them).

" The portion of the oil which dissolved in potash lye was
heated for a considerable time, with agitation, in order to drive off
the last portions of the hydrocarbon which might be adhering to
it. It was next saturated with sulphuric acid and the liberated
eugenol still further purified. It agreed perfectly in its characters with those ascribed to eugenol by Bonaster, Ettling, and

---

* In Stenhouse's original paper he calls this " eugenic acid,"—a term which
was formerly applied to eugenol because this liquid forms crystallizable compounds with the alkaline bases. (Although Stenhouse's researches may now
appear antique, they are yet of interest to instance his method of procedure.)

Boeckmann. Its boiling-point was found to be 242° C., and its sp. gr. 1·076 * ".

An examination of pure cinnamon-leaf oil from cinnamon trees cultivated in the Seychelle Islands has been made by Messrs. Schimmel (1890). The oil was found to correspond remarkably with the thin cinnamon-root oil from Ceylon. It consists chiefly of eugenol and its sp. gr. is 1·060.

CASSIA.

The dried bark known in commerce as " Cassia Bark," " Cassia lignea," and " China Cinnamon " is produced in the hot climates of Asia from several species of *Cinnamomum*. The trees differ considerably from each other in foliage, inflorescence, and aromatic properties, but the distinctive difference of several species has not yet been clearly established.

Cassia lignea or *China cinnamon* is annually brought in large quantities to Canton from the province of Kwangse in the south of China, whose principal city, Kwei Lin Foo (literally the city of the Forest of Cassia trees), derives its name from the forests of Cassia around it.

Until as lately as the year 1884 the exact botanical source of Chinese Cassia lignea was not known with certainty, although it was generally attributed to the tree now proved to yield it, viz., *Cinnamomum Cassia*, Blume, which is cultivated in China and is probably a native of Cochin China.

The investigations which led to the identification of the tree were made by Mr. Charles Ford, the Superintendent of the Botanical and Afforestation Department, Hong Kong, who, with the consent of his Government, proceeded to the Cassia plantations on the West River for the purpose of reporting on the cultivation and collection of Cassia lignea, as well as of bringing back for distribution from the Hong Kong Botanic garden living and dried botanical specimens of the authentic plant. He completely succeeded in the object of the expedition, and described his journey and its results in a Report to the Hong Kong Government. This Report was printed as a Government Notification (No. 339), and was republished by Mr. Thiselton Dyer, with

* This boiling-point differs slightly from the recorded observations of other chemists; see " Eugenol " in the Article on Cloves.

comments thereon, in the Journal of the Linnean Society, December 1882; the following details are abstracted therefrom:—

"There are three chief districts where the Cassia is cultivated, viz.:—Taiwu, in Lat. 23° 34' N., and Long. 110° 18' E., in the Kwangsi province; Lukpo, in Lat. 23° 6' N., and Long. 112° 24' E.; and Loting, in Lat. 22° 52' N., and Long. 111° 8' E., both in the Kwangtung province. These are the market towns of the district; but the Cassia is cultivated over a large area of country stretching to considerable distances from the towns, the extent of which could not be ascertained owing to the unreliable accounts given by the different people questioned, who either had very vague notions of area, or were disinclined, as they usually are, to give information to foreigners. Taiwu is about four or five miles from the West River; but the nearest Cassia plantations are situated 25 or 30 miles further in a southern or south-westerly direction, to which there is no communication by river. Taiwu is about 180 miles west of Canton. The Taiwu people said that the area of cultivation was not increasing. The next most important, if not the *most* important, district is the Loting one, commencing at about 8 or 10 miles distant from the city of Loting. After leaving the West River about 80 miles of the Loting River—the Nam Kong—has to be traversed before reaching the city, and from there the distance to the plantations has to be accomplished overland. One of the largest cultivators said that in this district there were about 1,000,000 *maus* (about 52,600 English acres) under cultivation, and that the area was greatly extending every year. The cultivation of Cassia has been carried on here for only about 25 years, *i. e.*, since the Tai-Ping rebellion, at which time, for the preservation of the plants from destruction by the rebels, they were transferred from a district further south, at which it is reported the cultivation of Cassia was abandoned when it was commenced at Loting. The next district is that of Lukpo, which is much less important than the other two. The city of Lukpo is situated on the northern bank of the West River, and the plantations are situated at about 15 miles between the nearest one and the city. In addition to these places, there are several small localities near the West River at intermediate places, where small patches of Cassia are grown; and as the quantities of bark obtained are too small to send to market towns, it is brought off by small boats and sold to larger boats which carry produce down the river. About six miles south-west of the small town of To Shing, which is situated on the southern bank of the river, about 25 miles above the confluence of the Loting and West rivers, there are some plantations, from which, however, no bark has been obtained for two years, and no new plantations made for ten, because the low prices which can now be obtained for the bark do not leave any profit to the producers. This was the only instance which came to my knowledge of the decrease of the trade in the Cassia-production, although it is said that the Java Cassia trade, in consequence of the lower prices at which the Cassia can be produced, is cutting out and crippling the China trade."

From each of the districts of Taiwu, Lukpo, and Loting Mr. Ford obtained and sent to Kew copious and excellent specimens, which were examined by Professor Oliver and all found to belong to the same species, which was undoubtedly the *Cinnamomum Cassia*, Blume. Mr. Thiselton Dyer adds that Mr. Ford in his Report further states that "this is the *only* tree from which Cassia-bark, 'buds,' or leaves of commerce in China (so far as he could ascertain from personal inspection and reports) are obtained. All the trees seen in the districts of Taiwu, Loting, and Lukpo, and intermediate localities where Cassia was grown in smaller quantities, were of this species, nor were there, apparently, distinct varieties of the species in cultivation. The cultivators and other natives were much interrogated as to whether they knew or had heard of any other tree which yielded the products under notice, and the invariable reply was that there was no other kind. There is therefore I think no doubt but writers who have named other kinds as Cassia-yielding trees of China have been mistaken or misinformed. .... *Cinnamomum Burmanni*, Blume, which it has been supposed may probably yield in part the Cassia-bark of the Canton market, does not, I feel sure, supply Cassia-bark to any extent. I did not see it anywhere cultivated; nor was it seen growing wild in any but very small quantities, and these wild trees bore no signs of having been cut, as had the Cassia trees; many natives were asked if it was ever used, but, with one exception, all denied that it afforded any Cassia-bark. The one exception said that its bark was sometimes, but rarely, used to adulterate the true Cassia-bark." "Mr. Ford on his return journey paid a visit to the well-known Chinese botanist,

Dr. Hance, H.M. Vice-Consul at Whampoa, who identified the specimens of the Cassia-lignea tree collected by Mr. Ford as belonging to *Cinnamomum Cassia*. There is, in fact, in the Kew Herbarium a specimen of the same species collected by Dr. Hance in 1876. This specimen is the material upon which the plate given by Bentley and Trimen (Med. Plants) is based, and represents no doubt the true plant."

" *Cinnamomum Cassia* was first described by Blume in 1825 *. The species was apparently founded on cultivated specimens from Java, where Blume states it was ' ex China introductum.'

" The Kew Herbarium possesses a cultivated Java specimen contributed by the Leyden Herbarium. This is no doubt an authentic type of the plant described by Blume, and Professor Oliver finds that it agrees precisely with the plant collected by Mr. Ford on the West River. It may therefore be considered finally settled on the one hand that the Chinese Cassia-lignea plant is really the *Cinnamomum Cassia*, Blume, and on the other hand that the plant cultivated in Java is identical with that now known to be the source of the spice in China.

" It is remarkable that though the cultivation of the Cassia-lignea tree has apparently been carried on in Southern China from time immemorial, it does not appear to be indigenous there. In Cochin China, however, there appears to be some probability of its being really wild."

The exact part of Cochin China where the Cassia-tree grows wild is mentioned in ' Pharmacographia,' 2nd ed. p. 320, as follows :—" The French expedition of Lieutenant Garnier for the exploration of the Me Kong and of Cochin China (1866–68) found Cassia growing in about N. Lat. 19° in the forests of the valley of the Se Ngum, one of the tributaries on the left bank of the Me Kong, near the frontiers of Annam. A part of this Cassia is carried by land into China, while another part is conveyed to Bangkok in Siam." The thick bark of the old uncultivated trees growing near the Annam frontier is very highly valued by the Chinese on account of its supposed medicinal properties, especially a bark called *Ching Fa Kwei* from trees growing on the Ching Fa mountain in Annam †.

In the Report of Mr. Ford above referred to the following account of the mode of collecting and preparing Cassia lignea is

---

* Bijdragen Fl. Nederl. Indië, ii. p. 570.   † Pharm. Journ. [3] xxi. p. 123.

given :—" When the trees are about six years old, the first cut of bark is obtained. The season for barking commences in March and continues until the end of May, after which the natives say the bark loses its aroma and is therefore not removed from the trees. The branches, which are about an inch thick, being cut to within a few inches of the ground, are carried to houses or sheds in the vicinity of the plantations.

" All the small twigs and leaves being cleared off, a large-bladed knife, with the cutting-edge something like the end of a budding-knife, is used to make two longitudinal slits, and three or four incisions, at sixteen inches apart, round the circumference through the bark; the bark is then loosened by passing underneath it a kind of slightly curved horn knife with the two edges slightly sharpened. Pieces of bark sixteen inches long and half the circumference are thus obtained. The bark, after its removal and while it is still moist with sap, is then laid with the concave side downwards, and a small plane passed over it, removing the epidermis. After this operation the bark is left to dry for about twenty-four hours and then tied up in bundles about 18 inches in diameter and sent into the merchants' houses in the market towns.

" The leaves which are cleared from the branches that are barked are carefully preserved and dried. They afford an oil by distillation. A large quantity of leaves are sent to Canton, where I was told the operation of distilling is performed. The twigs are removed from the cut branches at the same time as the leaves. They are a marketable commodity for native use."

The twigs and trimmings of the branches are imported into Europe as " chips " and distilled. In the European markets the Chinese Cassia lignea is the most esteemed of all the Cassia-barks, and is the one which most nearly approaches to cinnamon in its properties, but its substance is thicker, its appearance coarser, its colour darker, browner and duller; its flavour, though approaching to that of cinnamon, is much less sweet and fine ; it is more pungent and followed by a bitter after-taste. The pieces are not uniform in size or regular in shape, and are not enclosed one in the other like cinnamon. The epidermis has been removed with less care than is taken with cinnamon.

Cinnamomum Cassia is botanically very closely allied to *C. obtusifolium*, Nees, which is found on the Khasia Hills in the East of Bengal ; the Cassia lignea of this tree is taken to Calcutta for

shipment.   Three species of *Cinnamomum* are found in this region
growing at elevations of from 1000 to nearly 4000 feet and
yielding barks of a more or less cinnamon-like flavour; these are
the *C. obtusifolium*, Nees, *C. pauciflorum*, Nees, and *C. tamala*,
Nees and Eberm.

The *Cinnamomum iners*, Reinw., a very variable species found
in India, Ceylon, Tavoy, Java, Sumatra, and other islands of the
Indian Archipelago, is possibly a simple variety of *C. Zeylanicum*,
but is distinguished from it by its leaves being paler and thinner
and differently veined; also by its bark being of a different
flavour.   This tree probably furnishes the bark known as the
Cassia or Wild Cinnamon of Southern India.

A certain portion of the Cassia of Northern India is probably
yielded by the *C. Tamala*, Nees and Eberm., which, besides being
found on the Khasia Hills, grows in the regions of Silhet, Sikkim,
Nepaul, and Kumaon.   It has also been found in Australia.

Large quantities of a thick-bark Cassia, known to merchants as
" Cassia Vera," are occasionally shipped from Batavia; these are
probably derived from the *C. cassia*, Blume, and *C. Burmannii*,
Blume, which grow in Sumatra and Java; this last variety also
furnishing the " Cassia Vera " of Manilla, the epidermis of which
is sometimes very imperfectly removed, being warted and covered
with crustaceous lichen.

Cassia buds.   (*Flores cassiæ immaturæ* ; *Clavelli cinnamomi.*)—
The calyces of the immature flowers of the same tree which yields
Cassia lignea.   Cassia-buds bear some resemblance to cloves, but
are smaller; or to nails with round heads; they have the odour
and flavour of Cassia lignea or cinnamon.   They are gathered in
the Kwangtung Province in China when about one fourth of their
normal size, and shipped from Canton.

Buds and the seeds which are annually required for sowing are
obtained from the trees, ten years and upwards of age, that are
left standing at about 50 and 100 feet apart amongst the trees
which are cut down every six years for their bark.   These seed-
bearing trees are not cut unless there is a demand for the very
thick bark on their trunks, when some of the trees which can be
conveniently spared are sacrificed.   In the south of India the
Cassia-buds are gathered when in a more mature state from one
of the varieties of *Cinnamomum iners*, Reinw., but they are very

inferior to Chinese cassia-buds. Cassia-buds yield a volatile oil by distillation.

Oil of Cassia is considered to be exactly similar in a chemical point of view to oil of cinnamon, but not so fine in flavour.

Cassia oil distilled by the Chinese is frequently adulterated with colophony, and as this has a greater sp. gr. than the oil they add petroleum to regulate the consistence and sp. gr. With the view of checking the sale of this systematically adulterated oil, Messrs. Schimmel & Co. of Leipzig have published a Circular dated October 1889, giving very useful data respecting the pure oil, and describing methods of detecting the adulteration. As Messrs. S. & Co. have undertaken this trouble for the benefit of consumers, some extracts from the Circular may here be given :—

The specific gravity of " Extra pale Colophony " is . 1·070
The specific gravity of " Pale Colophony " is . . . 1·110

The darker sorts occurring in trade, to judge from the colour of the adulterated oil, are still heavier. Therefore, as the sp. gr. of pure Chinese Cassia oil amounts on an average to 1·060, any oil heavier than 1·070 must always be looked upon with suspicion. The table on p. 210 details the properties of *pure* Cassia oils and of *adulterated* Cassia oils examined by S. & Co.

The demands to be met by a good marketable cassia oil result, *ipso facto*, from the analyses. Briefly they are as follows :—

1. It should have at 15° C. a specific gravity of 1·050 to 1·070.
2. On distilling, about 90 per cent. of pure cassia oil should pass over. The residue must not become solid after cooling and take the character of a brittle resin, but must remain, at least, semi-fluid. It may amount to from 6 to 7 per cent., but is in no case to be more than 10 per cent.

No practical value for the detection of resin can be attached to testing the solubility of the oil in more or less diluted spirit, as is proved by the examination of oils which had been purposely mixed with resin and petroleum.

As a matter of course it is possible that adulteration with fatty oil will again occasionally be resorted to and that the fluid resin obtained by distilling will amount to more than 10 per cent. Any such oil must of course be rejected.

P

Origin.	Colour.	Sp. gr. at 20° C.	Rectification residue.	
			per cent.	
1. Cassia oil. Own distillation from Cassia chips ; age 4 months..............	Pale yellow.	1·035	5·4	Liquid.
2. Do. Do. from Cassia buds, age 4 months	Brownish.	1·026	4·4	,,
3. Do. Brand AYONG; age 60 to 80 years. Marks of the cases ⟨BV&Cº⟩ .	Do.	1·062	6.	,,
4. Do. Brand AYONG ; age 24 years. Marks of the cases ⟨T⟩ Season 1865.	Yellow.	1·060	8.	,,
5. Do. Brand AYONG ; age 22 years. Marks of the cases ⟨S⟩ E S & Co. Season 1867–8	Do.	1·060	7.	,,
6. Do. Brand HOP LEE A. CHIP ; very old. Marks of cases ⟨ES&Cº⟩ .	Do.	1·059	7.	,,
7. Do. Brand TAC FOONG ; age unknown. Cases marked Deetjen & Van Bergen.	Pale yellow.	1·060	5·5	,,
7a. Do. Brand YING CHONG ; age un-known.........................	Yellow.	1·055	7.	,,

The adulterated Cassia oils gave the following results :—

Origin.	Colour.	Sp. gr. at 20° C.	Rectification residue.	
			per cent.	
8. YAN LOONG	Reddish brown.	1·057	26.	Solid.
9. Do........................	Do.	1·059	23.	,,
10. CHEONG LOONG	Do.	1·056	24.	,,
11. Do........................	Do.	1·051	26.	,,
12. Do...	Dark brown.	1·061	33.	,,
13. LUEN TAI	Do.	1·060	38.	,,

The last two oils, taken from the most recent arrivals, were of a syrup-like consistence.

The following is a description of a simple practical process recommended for estimating the rectification residue of a sample :—

" Weigh out about 50 grammes of the oil into the small fraction retort *a*, connect it with the cooling-tube *b*, and place the thermometer *c*, by means of a perforated cork, in such a manner as the bulb of the thermometer stands about 5 or 10 centimètres above the fluid. The retort may not be more than half full.

Fig. 8.

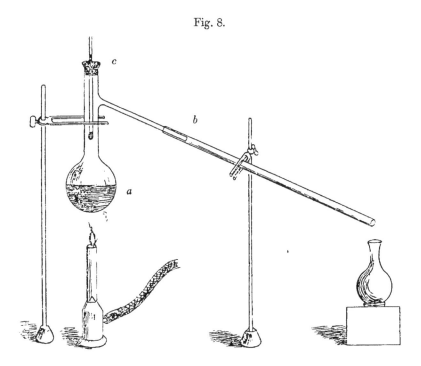

" To prevent bursting, the heat must be applied by degrees. For this purpose the flame of a Bunsen-burner or of a spirit-lamp must be moved to and fro underneath the retort. At first some water escapes with a crackling noise from the fluid. Usually the oil commences to boil at about 200°, and then the thermometer rises quickly to 240°. The bulk of the oil distils over between 240° and 260°. At last, white vapours develop in the retort, the thermometer rising at the same time from 280° to 290°. When this takes place, the distillation is to be interrupted; the residue is allowed to cool in the retort and is weighed with it. (The retort having, of course, been previously tared.)   If the residue becomes hard and solid after cooling, the oil is to be considered adulterated. Genuine non-adulterated oil also leaves a residue (up to about 10 per cent.), but it never becomes solid and remains in a semi-fluid state, even after having cooled off completely.

" It will be observed that samples 1 and 2 of the foregoing list show a materially lower sp. gr. than the pure Chinese oils.   This is probably in consequence of the more perfect distillatory apparatus employed by Messrs. Schimmel, so enabling more efficiently the removal of the highest boiling ingredients of the oil, than it is possible for the Chinese to do with their primitive apparatus. It may also be owing to the superiority of the raw material employed.

" The market value of commercial oil of Cassia is estimated by the percentage of cinnamaldehyde present in the sample.   The amount of cinnamaldehyde in the parcels above referred to was estimated by Messrs. S. & Co. as follows :—

Pure Oils.

No. 1 .... 88·9 per cent.
   2 .... 80·4   ,,
   3 .... 76·0   ,,
   4 .... 72·9   ,,
   5 .... 89·4   ,,
   6 .... 76·6   ,,
   7 .... 78·4   ,,

Adulterated Oils.

No. 8 .... 58·0 per cent.
   9 .... 63·2   ,,
 10 .... 58·7   ,,
 11 .... 52·9   ,,
 12 .... 57·8   ,,
 13 .... 47·1   ,,

" The amount of cinnamaldehyde was found by ascertaining the percentage of non-aldehydes after separating the cinnamaldehyde by bisulphite of sodium.

" The results show that every adulteration diminishes the amount of aldehyde. As to the normal amount of cinnamaldehyde in cassia oil, the matter up to the present has not been sufficiently investigated, but an oil containing less than 70 per cent. may be considered as adulterated, and probably an oil with less than 75 per cent. can be looked upon with suspicion.

" In earlier years cassia oil was also adulterated, especially with fixed oils ; but pure cassia oil prevailed in trade and the adulteration was carried on in rather more moderate limits. Different kinds of old oil examined by Messrs. S. & Co. gave the following results :—

> No. 14, age 9 years .... 79 per cent. cinnamaldehyde.
>   15, ,, unknown 70   ,,   ,,
>   16, ,,   ,, ... 73   ,,   ,,

" No. 14 left a residue from which a fatty oil to the amount of 10 per cent. was isolated. This oil would have shown, *before the adulteration*, about 90 per cent. of cinnamaldehyde."

This important Circular further adds :—" We are occupied with a thorough investigation of pure Cassia oils and have stated that the chief constituent of the non-aldehydes is the *Acetic ether of Cinnamyl*. In submitting the non-aldehydes to repeated fractional distillation *in vacuo*, we found the fraction boiling at 135°–145° (at 11 mm. atmospheric pressure) to be entirely the above-named ether. The cinnamic alcohol, obtained by saponification, crystallizes in ether, in white solid crystals, boils at 137° (at 11 mm. atmospheric pressure), and has a somewhat hyacinth-like odour.

" Besides this ether—if a conclusion from its boiling-point and the alcohol obtained is allowed—the presence of *acetic ether of phenylpropyl* is very probable.

" Terpenes of the constitution $C_{10}H_{16}$ are excluded. The presence of sesquiterpenes and polyterpenes is only presumed and requires further confirmation.

" Free cinnamic acid, formed by oxidation of the cinnamaldehyde when in contact with the open air, was found in both the old and the fresh distilled oils, but always in very small proportions."

A short time after this Circular was issued, the Chinese, finding they were detected, gave up the "resin-adulteration," and the samples of Cassia oil offered for sale in Hong Kong were fairly p ure so far as the resinous contents were concerned.

The oils had the normal sp. gr., were thin and beautifully clear, showing no adulteration with either resin, petroleum, or fatty oils, but they showed a striking lack of sweetness and left on the tongue an unpleasant bitter taste. Upon determining the proportion of cinnamaldehyde, it was found to vary from 49·4 to 69·8 per cent., as follows :—

Investigations from 4th November to 28th December, 1889.

Brand.

Brand		Sp. gr.	Resin			Cinnamaldehyde	
CHEONG LOONG.	Sp. gr.	1·060.	Resin 7·7 p. c., soft.			Cinnamaldehyde 55	p. c.
Do.	„	1·060.	„	6·0 „	„	„	58·9 „
Do.	„	1·058.	„	8·0 „	„	„	49·4 „
YAN LOONG.	„	1·061.	„	6·0 „	„	„	69·8 „
CHEON LOONG.	„	1·060.	„	6·0 „	„	„	61·5 „

" Notwithstanding the low percentage of cinnamaldehyde, no direct adulteration can be proved in these oils. The non-aldehydes consist almost exclusively of acetic ether of cinnamyl $(CH_3 . CO . OC_9H_9)$. These oils were worth hardly as much as those adulterated with 40 per cent. of resin. On this fraud being unmasked, the following parcels were offered.

" Investigations from 12th to 26th March, 1890.

Brand.

Brand		Sp. gr.	Resin			Cinnamaldehyde	
CHEONG LOONG.	Sp. gr.	1·061.	Resin 4·6 p. c., soft.			Cinnamaldehyde 77·7	p. c.
Do.	„	1·061.	„	4·5 „	„	„	76·1 „
YAN LOONG.	„	1·060.	„	5·0 „	„	„	77·1 „
YE TAC.	„	1·061.	„	6·6 „	„	„	70·9 „
Do.	„	1·061.	„	6·6 „	„	„	73·5 „

" Since, as before stated, the quality and value of oil of cassia depend entirely upon the amount of cinnamaldehyde it contains, a scientific estimate of the worth of an oil is to be reached only by determining the proportion of aldehyde present. One method of doing so is as follows :—

" 75 grammes of the oil are mixed in a good-sized flask with 300 grammes of a boiling 30-per-cent. solution of bisulphite of sodium. The acid sulphite compound $(C_6H_5 . CH : CH . COH . HNaSO_2)$ separates in the form of a coagulating precipitate. It is shaken vigorously and allowed to stand a short time. (Oils rich in aldehyde usually develop considerable heat which must be reduced by addition of cold water.) Then about 200 grammes of hot water are added, and the whole, with frequent shaking, is warmed in a water-bath until the combination of aldehyde with the bisulphite of sodium is entirely dissolved and the non-aldehydes, as an oily layer, float on the solution of the aldehyde salt. It is now allowed to cool, is shaken up twice with ether, first with about 200 c. cm. and then with 100 c. cm. The ethereal extracts of non-aldehydes are separated off, brought together and filtered into a good-sized beaker previously weighed, in which is placed a spiral-shaped platinum wire. The ether is evaporated as quickly as possible, by putting the beaker in hot water. As soon as the liquid ceases to foam up when moved around, it is allowed to cool and is weighed. Then the beaker is again placed in the water-bath for 10 minutes, weighed again after cooling, and so on until the difference between two weighings is not more than 0·3 gram. The last but one is then taken as the correct weight. (The method of expelling the ether has a great influence upon the accuracy of the result. The non-aldehydes can be volatilized, but not easily. The ether must therefore be boiled rapidly and the beaker must not be allowed to stand in the water-bath any longer than is necessary for the evaporation of the ether.) The weight of non-aldehydes thus obtained being subtracted from that of the cassia oil used, the remainder gives the amount of cinnamaldehyde.

" *Example.*—Oil used, 79·71 grams.

1st weight of glass, after expulsion of the ether, 147·55 grams.
2nd   „   „   „   146·84   „
3rd   „   „   „   146·58   „

From 146·84
Deduct 128·34 Tare.

Non-aldehydes in the oil   18·50 grams, or 23·1 per cent.

Therefore 100 less 23·1 = 76·9 per cent. of cinnamaldehyde.

"By repeating the experiment upon a second portion, the variation in result (if the directions are closely followed) generally amounts to a few tenths of 1 per cent., rarely to 1 per cent., which in practice is more than sufficient.

"The other constituents of cassia oil, including eventual adulterations (resin, petroleum, fatty oil, etc.), are not affected by the reagent used, and, by shaking up with ether, can with ease be entirely separated from the watery solution. (Except the cinnamic acid.)

" The cinnamic acid found in every cassia oil is determined in this process as cinnamaldehyde. As, however, even in very old oil, not so much as one per cent. of cinnamic acid was found, the error is insignificant."

Cinnamic acid may be removed by shaking the oil with hot sodium-carbonate solution, and the amount found deducted from that of the aldehyde found. Cinnamic acid can be identified by its melting-point 131° C. (the melting-point of pure cinnamic acid is 133°), and by the formation of benzaldehyde by oxidation with potassium permanganate. When the oil of cassia is heated with a solution of bisulphite of soda for the purpose of determining the aldehyde, the cinnamic acid is dissolved with it, forming, according to Valet *, phenylsulphopropionate of soda. The pure acid is easily soluble in water, and cannot be extracted with ether from the solution acidified with sulphuric acid, therefore, according to the manner of determining cinnamaldehyde, it is reckoned as aldehyde.

A later Circular issued by the same Firm describes another process for the estimation of cinnamaldehyde as follows :—A specially manufactured glass flask is required, of about 100 c. cm. capacity, with a neck about 13 cm. long and 8 mm. internal width, which is divided into tenths up to 6 c. cm.; also a pipette of 10 c. cm. capacity. Ten c. cm. of the oil to be examined are measured with the pipette and allowed to run into the flask; the pipette is drained for a minute or two, and the last drops blown out with the mouth. The flask is then about three-fourths filled with a 30-per-cent. solution of bisulphite of sodium, and the curd that forms immediately shaken a few times, powerfully but carefully, without allowing a drop to spurt out. The flask is then placed in a hot-water bath. The whole is kept hot for several hours and occasionally agitated. It is sometimes kept hot for half a day and at least till the curd is completely dissolved and there floats on the surface a layer of clear oil, sharply defined against the solution. [A still later Circular, dated October 1891, says the results may be arrived at more quickly by a slight modification of the process :—After 10 c. cm. of the oil have

* Annalen der Chemie, cliv. p. 62.

been allowed to run into the flask from a pipette, and have been warmed on a water-bath, the entire quantity of sodium bisulphite solution is not at once added, but small portions at a time, waiting after each addition until the solid mass first formed has become partly fluid. Proceeding in this way, the formation of the liquid compound of the cinnamaldehyde with sodium bisulphite is complete in 10 to 15 minutes if the proportion of aldehyde is high. In the presence of large quantities of non-aldehydous contents the estimation requires a correspondingly longer time.] On observing the supernatant layer of clear oil above referred to, the flask is removed from the water-bath, allowed to cool, and filled up with the bisulphite solution (towards the end with great care drop by drop) till the oil has entirely risen into the neck and its lower limit accurately coincides with the lowest mark on the neck of the flask. This oil consists of the non-aldehydes, whose volume has to be subtracted from the 10 c. cm. of cassia oil taken. The difference shows the amount of cinnamaldehyde in 10 c. cm. of oil. Assuming that 2·7 c. cm. of oil were read off in the neck of the flask, then there were 2·7 c. cm. of non-aldehydes in 10 c. cm. of cassia oil, or, in other words, the oil contained 27 per cent. of non-aldehydes and 75 per cent. of cinnamaldehyde. Strictly speaking these are volume and not weight percentages ; as, however, the sp. gr. of the non-aldehyde in cassia oil (1·060 at 20° C.) almost exactly accords with that of cassia oil (1·059–1·061 at 20° C.), the actual difference is only small and of no practical importance.

In carrying out the process some patience must be exercised and the operator must not be content with only a nearly complete solution of the curd. Not a single flock ought to be perceivable either in the oil or adhering to the sides of the flask, as otherwise an accurate reading off of the quantity of oil is impossible and an error of several per cent. may arise. If the bisulphite solution is turbid, it must be previously filtered *. If the curd will not dissolve, although the flask has stood an entire day in the boiling water, an extraordinarily heavy adulteration of the cassia oil with a hard resin may be assumed. In this case a volumetric estimation of the cinnamaldehyde is impossible. When the oil is adulterated both with resin and mineral oil (the most frequent and almost exclusive kind of adulteration) the separated non-aldehydes

---

* It is necessary that the sodium bisulphite should have a sp. gr. of not under 1·03 and that it be kept in well-stoppered bottles.

are still liquid in the presence of a resin-content of 30 per cent. Two estimates of every oil ought to be made, of which the average should be taken. Both should accurately agree, or at most show a difference of 1, 2, or 4 per cent.

The following tests are given by E. Hirschsohn in the ' Journal of the Chemical Society ' (and reprinted in the Pharm. Journal, May 2, 1891). One volume of the oil should give a clear or merely opaline solution with three volumes of 70-per-cent. alcohol at 15°; a turbidity or sediment indicates the presence of petroleum or of foreign ethereal or fatty oils, or of a large excess of colophony. Also, when the alcoholic solution is added drop by drop into half its volume of a cold saturated solution (saturated at the tempera-ture of the room) of lead acetate in 70-per-cent. alcohol, no pre-cipitate should be produced, otherwise colophony or a similar resin is present. Oils without any addition of resin do not become turbid on applying this test. Even old cassia oils with their slightly increased content of cinnamic acid remain quite clear. However, some old cassia oils that have been kept in badly closed vessels may possibly precipitate cinnamate of lead. This test with acetate of lead is particularly commended when little material is available for the investigation.

To detect the adulteration of oil of cassia by oil of cloves, a drop of the oil should be heated on a watch-glass. Genuine cassia evolves a fragrant vapour possessing but little acridity ; when, however, clove oil is present, the vapour is very acrid and excites coughing. With fuming nitric acid, cassia merely crystallizes ; if clove be present it swells up, evolves a large quantity of red vapour and yields a thick reddish-brown oil. Cassia oil when pure solidifies with concentrated potash, but not when mixed with clove oil.

It has been noticed that oils of cassia and cinnamon may be highly adulterated with resin oils and still pass the tests of the German Pharmacopœia *.

With nitric acid sp. gr. 1·45 at 15°, or with 1·50 acid at 6°, both the pure and impure oils give crystals without development of heat ; however, with the 1·50 acid at 15°, both react violently, with development of heat and without the formation of crystals. Therefore the P. G. test, as neither the sp. gr. nor the temperature

* Chem. Zeit. xiii. pp. 1406-7, and Journ. Chem. Soc. Apr. 1890.

of the acid is stated, may lead to the condemnation of a pure oil, and *vice versâ.* By determining the " acid number," the adulteration can be detected, as the following numbers show :—

	Acid number.
Genuine oil of Cassia (with 6 per cent. non-volatile residue)	13
Ditto, after 40 hours' aëration	13
Genuine Ceylon oil of Cinnamon (2 per cent. residue)	9
Ditto (2½ per cent. residue)	10
Adulterated oil of Cassia (28 per cent. residue)	47
Ditto (prepared from pure oil of Cassia by inter-mixing 20 per cent. of colophony)	40
Colophony	150

CLOVES.

The spice commonly known by this name consists of the dried, unexpanded flower-buds of the *Caryophyllus aromaticus,* Linn. Sp. Pl. 735, and Gærtner, De Fruct. i. p. 167, tab. 33 ; DC. Prodr. iii. 262.

Syn. *Eugenia caryophyllata,* Thunberg, Dis. p. 1.
Myrtus caryophyllus, Sprengel.
Caryophyllus, Rumph. Amb. ii. tab. i. figs. 2, 3.
Bentley & Trimen, Med. Plants, t. 112.
Curtis, Bot. Mag. t. 2749.
Woodville, Med. Bot. t. 135.

In Sanskrit, *Lavunga.*
Persian, *Meykuk.*
Arabic, *Kerunpul.*
Bengalee, *Lung.*
Hindee, *Laung.*
German, *Gewurzuelken.*

The Clove-tree is a beautiful evergreen, growing to a height of about 20 or 30 feet. Its trunk is straight and rises four or five feet before it throws out branches ; having the appearance of a gigantic myrtle, with long ovate-oblong leaves. It bears quantities of flowers arranged in trichotomous terminal cymes. The bark is smooth, adhering closely to the wood, which is hard and strong, but of an ugly grey colour, therefore not suitable for cabinet work.

The name *clove* is derived from the French *clou*, a nail, in allusion to the shape of the bud, the corolla forming a ball, the four petals being adherent at the points, and this knob, between the four teeth of the calyx, surmounting the long tube of the calyx looks like a nail. For the same reason the Dutch call it *Naghel*, the Spaniards *Clavo*, and the Italians *Chiodo*.

The seed-berry of the clove is oblong, 1- or 2-celled and as many seeded.

A peculiarity of the clove-tree is that every part is aromatic, owing to essential oil contained in minute glands, but the greatest strength is found in the bud.

The clove-tree is indigenous to five of the Molucca islands and was originally confined to them, viz. Tarnati, Tidori, Mortir, Bachian, and Machain, chiefly this last. These islands constitute a string of islands westward of the large island of Gilolo, where, strange to say, the tree does not appear to grow in the wild state. According to Rumphius it was introduced into Amboyna a short time before the arrival of the Portuguese, and it is still largely cultivated there as well as in the adjacent islands of Haruku, Saparna, and Nusalant.

The portion of Amboyna called Leytimeer and the Uliasser islands produced no cloves until the arrival of the Dutch, by whom the cultivation was restricted to Amboyna, every effort being made to extirpate the plant elsewhere.

It is also grown on a large scale at Sumatra, Penang, Malacca, Madagascar, the islands of Zanzibar and Pemba on the east coast of Africa, and in the East Indies. Several of the West Indian islands possess it, as St. Vincent, Trinidad, Martinique, St. Kitts. The French also introduced it into Bourbon and the Mauritius. Small parcels of exceedingly fine quality have recently been imported into France from St. Marie, Madagascar. The tree furnishing the spice is a cultivated variety, of smaller growth than the wild tree but more aromatic.

There appear to be five varieties of the clove, viz. :—the ordinary cultivated clove; the clove called the " female clove " by the natives, which has a pale stem ; the *Keri* or *leory* clove ; the *"Royal Clove,"* which is very scarce ; and the wild clove, which has hardly any aromatic flavour and is consequently of little value. The first three descriptions are equally valuable as spices, the "female" being considered fittest for the distillation of essential oil. The "Royal Clove" is a curious monstrosity which formerly had a great reputation as the "*Caryophyllum Regium*" by reason of its rarity and the curious observations which were made respecting it[*]. It is a very small clove, distinguished by an abnormal number of sepals and by large bracts at the base of the tube of the calyx; the corolla and internal organs being imperfectly developed.

The soil most suitable to the clove-tree is a dark loam, having a substratum of dark yellow earth intermixed with gravel. A sandy soil, a hard clay, and a wet ground in which sedges grow are to be avoided. The tree may be propagated either by setting the seeds, or by transplanting the young plants found in the clove gardens which have come up from self-sown seed. The plants raised by the first method, although luxuriant, are not thought to be so fruitful as the self-sown plants.

In Amboyna it is thought best to set the young plants amongst other trees which shade them from the sun, and as the clove-trees grow up the other trees are removed, leaving here and there a few fruit-trees, such as the kanari and the cocoanut. The clove-trees must be kept pruned and care be taken that they are not choked with weeds; failing these precautions the plants languish or degenerate into wild cloves. The health of the tree much depends on the nature of the soil and ground.

"In Amboyna the harvest begins when the cloves begin to turn red. The ground beneath the trees is swept clean. The nearest clusters are taken off with the hand, and the more distant with the assistance of crooked sticks. As the boughs are tender, great care should be taken not to handle them roughly, as an injury would prevent them bearing for years. The curing of the cloves consists in placing them for some days on hurdles, where they are smoked by a slow wood-fire, which gives them a brown colour, and afterwards drying them in the sun, when they turn black. In some places they are scalded with hot water before being smoked, but this practice is not common. Such cloves as casually fall to the ground and are picked up in small quantities, the cultivators do

---

[*] Rumphius, Hort. Amb. ii. xi. t. 2; also Hasskarl, Neuer Schlussel zu Rumph's Hort. Amb. Halle, p. 166; Berg, in Linnæa, 1854, p. 137; and Valmont de Bomare, Dict. d'Hist. Nat. 1775, iii. p. 70.

not think it worth while to subject to the process of smoking, and they are merely dried in the sun; they are discoverable by their shrivelled appearance and are of inferior value. The period of harvest is here from October to December, and the average production of all trees in a plantation may be taken at above five pounds, this is allowing that only two-thirds of the trees are in bearing condition, the remaining third being either barren or young trees. According to these figures, the produce of an acre will be 375 lbs. avoirdupois, and deducting one-eighth for young trees under ten years, 328 lbs." *

On the Moluccas the clove-tree begins to blossom when about seven or eight years old; the average produce is about 4½ lbs. of cloves from each tree yearly. The harvest takes place twice a year, in June and in December; the buds are partly collected by hand, and partly knocked off by bamboo canes and collected in cloths spread beneath the trees.

As cultivated in the Bencoolen districts in Sumatra, the seeds are planted in rich mould at the distance of 12 inches from each other, screened from the sun and duly watered. They germinate within five weeks, and when four feet high are transplanted to distances of thirty feet, with a small admixture of sand with the red mould peculiar to these districts, so as to reduce its tenacity; they are then cultivated in the same way as the nutmeg-trees, only that when full grown they require less manure in the proportion of one-third. They yield generally at the age of six years, and at the age of twelve years are in their highest state of bearing; the average produce may then be estimated at six or seven pounds of marketable fruit from each tree during the harvest, which takes place in the rainy seasons, i. e. twice a year.

At the commencement of the wet season the tree throws out a profusion of new leaves. Soon after, the germs of the fruit are discovered at the extremities of the young shoots, and in the four following months the cloves are completely formed. The buds, at first of a pale green colour, assume in time a pale yellow and ultimately a blood-red colour. They are then ripe for gathering for the purpose of drying for the market, but this is not the actual period of maturity of the fruit, i. e. the time when the seed is developed and ripe for purposes of propagation. For this another three weeks are required, and in this short period the fruit swells

* 'Indian Agriculturalist,' reprinted in Pharm. Journ. Aug. 20, 1887.

to an extraordinary size, loses much of its spicy quality, and contains a hard nucleus like the seed of the Bay. This state of the fruit is what Europeans call the "mother clove" and the natives *poleng*.

"The sight of a young plantation just coming into bearing is very pretty. The leaves of various shades of green tinged with red serve to set off the clusters of dull red clove-buds.

"The buds are plucked by hand, so that the process of gathering is tedious. They are then dried for several days on mats in the sun, until they break easily between the fingers and assume a dark brown colour. The loss of weight in drying is about 60 per cent. When past its prime the clove-tree has a rugged and straggling appearance. Its term of existence is from 20 to 24 years; hence it is necessary to plant a succession of seedlings, to take the place of worn out trees." *

When speaking of Zanzibar cloves we include the products of the islands of Zanzibar and Pemba, three fourths of the entire crop being produced in Pemba, about 25 miles distant. Those grown on the island of Zanzibar are reckoned of superior quality and command a higher price, but this is probably due to the fact that the owners reside at Zanzibar and can thus give their affairs the benefit of direct supervision. Certainly the conditions for their successful cultivation are most favourable at Pemba, where the rainfall exceeds that of Zanzibar, but the management being left to careless overseers, the result is the cloves are imperfectly cured and (being carelessly handled) are frequently marketed in an inferior condition.

The clove-tree was first introduced into Zanzibar about the year 1830 by the Sultan of that period, since which its cultivation has gradually extended, until it is now the chief industry of the islands.

The industry received a check in 1872, the date of the great hurricane. At least nine-tenths of the trees were destroyed, so that the larger part of those now standing are of new growth.

The method of cultivation at Zanzibar is as follows :—The seeds are planted in long trenches and kept well watered until after sprouting. In the course of forty days the shoots appear above ground; they are then watered when necessary and the ground

* 'Journal of the Indian Archipelago,' v. p. 78.

well kept for the space of two years, when they should be about three feet high. They are then transplanted, being set about thirty feet apart and kept well watered till the tender roots are established. From this time they only require ordinary care, though the best results are obtained when the ground about the trees is well worked over and kept free from weeds.

The growth of the tree is very slow, and five or six years are required for it to come into bearing, at which time it is about the size of an ordinary pear-tree and is usually very shapely.

" As soon as the buds are fully formed and assume this reddish colour, the harvesting commences, and is continued for fully six months at intervals, as the buds do not form simultaneously but at odd times throughout the whole period. The limbs of the tree being very brittle, a peculiar four-sided ladder is used for gathering the harvest. As fast as collected the buds are spread out in the sun, until they assume a brownish colour, when they are put in the store-house and are ready for the market.

" A ten-year old plantation should produce an average of 20 lbs. of cloves to a tree. Trees of twenty years frequently produce upwards of 100 lbs. each " (?)*.

The Consular Report above quoted from adds :—" The present season, commencing July 1889, is very favourable and the crop will exceed that of any previous season. It will, in all probability, amount to 13,000,000 lbs., averaging a local value of 10 cents per lb. The Sultan derives no inconsiderable portion of his revenue from this source, since the duty is levied at 30 per cent. *ad valorem,* thus placing to the Sultan's credit for the present year (1891) nearly, if not quite, 400,000 dollars."

(Further restrictions have since been placed on the export of cloves by the Sultan, and the market will probably be controlled by an arrangement of fortnightly auctions held under his authority.)

Consul Pratt's Report continues as follows :—

" Besides the clove-buds the flower-stalks are also gathered and form an article of commerce, commanding about one-fifth of the price of cloves and having about the same percentage of strength. To this circumstance is due the fact that ground cloves can frequently be purchased in the market at a lower price than whole cloves.

" For the past 15 years the cultivation of cloves has been the chief occupation of the Arab planters and has always netted good

* Report of Consul Pratt, Zanzibar.

returns. It seems probable that it will continue to be a profitable crop, as the consumption of the article appears to keep pace with the inevitable increase of production.

" Up to the present time the plantations have been worked with slave labour at comparatively small expense; but with stoppage of slave supplies from the mainland great difficulty will be experienced by the planters during harvest-time. One result will be an increase in expenses; but what the planters have most to fear is that the curtailment of the labour-supply will entail a direct loss by rendering it impossible to harvest the crop until after it has bloomed, when it would be unfit for the uses of commerce."

From a still more recent Report of Consul G. H. Portal of Zanzibar, we learn that " four-fifths of the world's crop of cloves is produced in Zanzibar and Pemba, and this harvest forms the staple item upon which the country may be said to depend. The cultivation has been so remunerative that almost every available acre of (suitable) ground is devoted to them. But the average price has now (1892) gradually declined to about one-third of what it was * ; the market is overstocked and the demand fails to keep pace with the supply. Rather more than half the crop reaches Bombay and New York, whilst London, Hamburg, Marseilles, etc. take the rest. Clove-stems are also exported in quantity . . . . . The Northern and Western portions of the Island are extremely fertile, being covered with clove plantations and cocoa-nut palms . . . . Wherever the ground is not cleared for cloves it is usually overrun by a luxuriant growth of aloes and common pine-apples."

The finest quality of cloves are dark brown in colour, with full perfect heads, free from moisture. The varieties of cloves met with in commerce are structurally similar in appearance. The inferior sorts are somewhat smaller, of inferior colour, and poorer in essential oil. In the London " Price Currents " cloves are quoted, according to their relative value, as " Penang," " Bencoolen," " Amboyna," and " Zanzibar."

(The large yield of cloves per tree mentioned above by Consul Pratt seems rather exaggerated, and his estimation of the percentage of strength of the flower-stalks is also wrong, the yield of oil from the stalks being only one-fourth that from the cloves.)

* [The price in 1892 is less than a third of that in 1888 and only a fifth of that in 1879. The present price of the oil is less than half the price it was in 1889.]

The flower-stalks of cloves were formerly an important article of commerce in Europe; they were known in France as " *Griffes de Giroflé*," " *Pédoncules de Giroflé;*" in Italy as " *Fusti* " and " *Bastaroni* " ; and in Latin as " *Stiptes Caryophylli.*" The leaves of the clove-tree were also commercially dealt in. There is still a market in the " stalks "; they are called by the natives of Zanzibar " *Vikunia,*" and they are imported into England for the purpose of adulterating powdered cloves ; they may be detected by the microscope by the fact of their containing thick-walled cells, which are not present in cloves. The flavour of the stalks is moderately aromatic. They yield by distillation 4 or 5 per cent. of essential oil rich in hydrocarbon ; for this reason the action of the oil is lævogyre on polarized light.

Powdered cloves are also adulterated with powdered Jamaica pimento, and this adulteration can be equally discovered by the microscope by reason of the thick ligneous walls of the cells. Pimento also contains a quantity of starch granules, which are not visible in pure powdered cloves. The microscope also reveals a great number of large starch granules in the *seed* of the clove, which is also imported into London for the same purpose of adulteration. The seed is much poorer in essential oil than the flower of the clove.

It is a common practice to falsify this spice with spent cloves from which the oil has been distilled. They are dried and rubbed between the hands, previously moistened with a little sweet oil to brighten their colour, after which they are mixed up with fresh cloves for sale.

Oil of Cloves.—It is estimated that the flower-buds of the Clove-tree yield on distillation as follows :—

Amboyna, 19 per cent.; Bourbon, 18 per cent.; St. Marie (Madagascar), 18 per cent.; Zanzibar, 17·5 per cent.

Clove stalks 6 per cent.

" Clove root " (*Geum urbanum*) 0·04 per cent.

Few plants possess organs so rich in oil as the clove. This oil is at first colourless or yellow, darkening by age and exposure to the air. It is optically inactive. It consists of a mixture in variable proportions of a sesquiterpene and an oxygenated oil. The first constituent, sometimes termed " Light oil of cloves," passes over with the vapour of water when the crude oil is distilled with strong potash lye ; its composition is $C_{15}H_{24}$; its sp. gr. is 0·910 at 15° C. ;

its boiling-point 251°–254° C.; its optical power is very slightly lævorotatory. The other, and the chief, constituent is *Eugenol*, $C_{10}H_{12}O_2$, which exists to the extent of 76 to 85 per cent. in the oil. It has been found that very fine samples may contain as much as 90·64 per cent. of eugenol.

Good oil of cloves should have a sp gr. of 1·067 at 15° C., and be freely soluble in alcohol of 90 per cent. An adulteration of turpentine would lower the sp. gr. and diminish the solubility in alcohol.

Eugenol is a strongly refractive liquid with the characteristic smell and burning taste of cloves; by exposure to the air it becomes brown. On fusion with caustic potash it yields protocatechuic acid, and is convertible into vanillin by the action of potassium permanganate (see Vanillin).

Besides forming the chief constituent of oil of cloves, it is found to a large extent in Allspice (*Eugenia Pimenta*), in the leaves of the Cinnamon-tree (*Cinnamomum Zeylanicum*), in Canella Bark (*Canella alba*), and probably in the Brazilian Cinnamon (*Dycipellium caryophyllatum*). It also occurs in the leaves of *Illicium religiosum* and of *Laurus nobilis*. It has been produced artificially by the action of sodium amalgam on coniferyl alcohol *.

Pure eugenol has a sp. gr. of 1·072 at 15° C. Its boiling-point is 253°–254° C., and it forms a clear solution in a 1 per cent. caustic-potash solution.

The market value of an oil of cloves being dependent on the amount of eugenol contained in it, it becomes necessary to quantitatively estimate that amount.

The usual method of separating eugenol from oil of cloves is by shaking up three parts of the oil with a solution composed of one part of caustic potash or soda in ten parts of water, pressing the crystalline paste of eugenol alkali that forms, taking up the press-residue with water, decomposing with hydrochloric acid, washing the liberated eugenol with water, drying it with calcium chloride, and then rectifying.

A more simple process for determining with accuracy (at least to within 1 per cent.) the amount of eugenol, has been recommended by Thoms, in a paper read at the meeting of the " Society of Naturalists and Physicians " in Halle †. The oil is converted

---

* Ber. Deutsch. chem. Ges. ix. p. 418. † Pharm. Centralhalle, Oct. 8, 1891.

into benzoyl-eugenol, $C_{17}H_{16}O_3$, by first agitating with caustic potash or soda solution, and then adding an equivalent of benzoyl chloride, C_7H_5OCl. Upon shaking, a considerable heating of the mass takes place, which is sufficient to complete the formation of the benzoyl-eugenol which is found in a crystalline mass on cooling. In order to remove a small quantity of sesquiterpene adherent, the crystalline mass, after being washed several times, is heated on a water-bath in a definite quantity of 90 per cent. by weight of alcohol (25 c. c. where 5 grammes of the oil has been taken). The mixture is stirred until complete solution takes place, then cooled to 17° C. and filtered, sufficient alcohol being added to make up for any loss. The filtrate at this temperature contains all the sesquiterpene, together with a small quantity of benzoyl-eugenol, amounting to 25 c. c. to 0·55 gramme. The crystalline mass is then placed, with the still moist filter, in a weighing-bottle, the weight of which, together with that of the filter, dried at 101° C., has previously been taken, and dried at 101° C. until of constant weight. From the weight of this benzoyl-eugenol, after an addition has been made for the quantity remaining dissolved in the filtrate, the quantity of eugenol present in the oil taken can be calculated in the usual way; the molecular weight of the former compound being 164 and the latter 268.

Experiments made by Messrs. Schimmel to test the accuracy of this process resulted as follows :—

1. Eugenol prepared by saponification of benzoyl-eugenol, sp. gr. 1·072 at 15° C. Boiling-point 253°–254° C. Forms a clear solution in 1-per-cent. caustic-potash solution. 5 grams yielded 8·22 gram benzoyl-eugenol, equivalent to 100·6 per cent. eugenol. Thus

$$\frac{8\cdot22 \times 100 \times 164}{5 \times 268} = 100\cdot60.$$

2. Eugenol obtained from clove-oil by shaking with alkali. Sp. gr. and boiling-point exactly the same as No. 1. 5 grams yielded 8·20 grams benzoyl-eugenol, equivalent to 100·35 per cent. eugenol :

$$\frac{8\cdot20 \times 100 \times 164}{5 \times 268} = 100\cdot35.$$

3. Eugenol prepared as No. 2, but a less pure product. Formed a clear solution with 10-per-cent. caustic-potash solution, but turbid

with 5-per-cent. solution. 5 grams yielded 7·9 grams benzoyl-eugenol, equivalent to 96·6 per cent. eugenol:

$$\frac{7·9 \times 100 \times 164}{5 \times 268} = 96·67.$$

4. A mixture of 8 parts eugenol (No. 1) with 2 parts sequi-terpene from clove-oil. 5 grams yielded 6·45 grams benzoyl-eugenol, equivalent to 78·94 per cent. eugenol:—

$$\frac{6·45 \times 100 \times 164}{5 \times 268} = 78·94.$$

Oil of cloves is frequently adulterated with phenol. This may be recognized by the following test, recommended by Flückiger[*]. The suspected oil is shaken up with fifty times its volume of hot water; after cooling, it is decanted, and concentrated at a gentle heat to a small bulk. A drop of liquid ammonia and a pinch of chloride are then dropped on the surface. If phenol be present, the liquid will assume a green colour, changing to blue, which will remain for several days. If the oil be pure, no coloration will be produced. This process is based on Berthelot's discovery that ammoniacal phenol becomes blue on the addition of hypochlorite of lime.

The above test is certain in its action, but has since been modi-fied by Jacquemin, in a way to test very small samples :—One drop of the suspected oil is mixed with a small trace of solution of aniline by means of a glass rod and then shaken with 5 or 6 c.c. of distilled water. By the addition of a few drops of sodium hypochlorite to the mixture, the characteristic blue coloration due to phenol will be developed in a few minutes, whereas with the oil nothing but the purplish-violet colour of aniline will be perceived. Stirring or shaking must be avoided after the addition of the hypochlorite, the reaction will take place without it. This is a very delicate test. A drop of oil containing 1 per cent. of adulteration distinctly develops the blue colour.

In 1863, Scheuch detected the presence of *salicylic acid* in oil of cloves, existing probably in the form of a compound ether. It can be isolated by agitating the oil with a solution of carbonate of ammonium. It is probably the presence of this acid in the oil

---

[*] Schweizerig Woschenschrift für Pharmacie, 1870, p. 200.

which causes the greenish-blue coloration when brought in
contact with an alcoholic solution of perchloride of iron, and
produces the intense violet colour when the oil is agitated with
metallic " reduced iron."

Caryophylline, $C_{10}H_{18}O$, is a neutral, tasteless, inodorous substance, isomeric with common camphor, crystallizable in prismatic
needles.   It was obtained by Flückiger and Hanbury in small
quantity by extracting with ether cloves previously deprived of the
greater part of their essential oil by a little alcohol.   In 1873,
Mylius obtained from cloves by nitric acid, crystals of Caryo-
phyllinic acid.

Cloves also contain 16 per cent. of a peculiar tannic acid, 13
per cent. of gum, and about 18 per cent. of water and extractive
matter.

The perfume of cloves is strongly developed in flowers of certain
plants, notably in those of a clove-scented Convolvulus found in
the forests of Midnapore in Bengal, called *Lettsomia Bona-nox*,
Roxburgh *, synonymous with *Argyreia Bona-nox*, Sweet †.   The
native name of this plant in Bengal is *Kulni-luta*.   The flowers,
which are produced during the rainy season, are large, pure white,
expanding at sunset and perfuming the air to a considerable distance
with a fragrance resembling that of the finest cloves.   At sunrise
these flowers wither ; a peculiarity which is noticeable in another
scented Convolvulus, the *Ipomœa grandiflora*, Roxburgh, called
Doodiya-Kulmi in Bengalee, *Munda Valli* by Rheede ‡.   This
plant will twine up to a height of 20 feet and bears white flowers
4 to 6 inches in diameter.   It is common in hedges near Samul-
cota and on the banks of watercourses amongst bushes.   It has
been considered very nearly allied to, if not identical with, the
Lettsomia Bona-nox above-mentioned.

The perfume of cloves blended with a trace of that of methyl-
salicylate or a compound organic ether, is conspicuous in several
species of *Dianthus* or " Pink " ; plants belonging to the extensive
order Caryophyllaceæ.   Most of the species are natives of Europe,
temperate Asia, and North Africa.   *Dianthus Caryophyllus* or
" Clove Pink " is the original of the garden Carnation.

* Flor. Ind. ii. p. 494.   † Hort. Brit. p. 289.
‡ Rheede, Hort. Mal. xi. p. 103, t. 50, and ' Asiatic Researches,' iv. p. 257.

It is stated that when balsam Tolu is dissolved in a small quantity of solution of potash it loses its own characteristic odour and acquires that of the " *clove-pink.*"

The odour of cloves is strongly developed in the bark of *Laurus Culilawan,* L. * (*Laurus caryophyllata,* R.) and in *Cinnamomum sintoc,* Blume; also, in combination with the odour of nutmegs, in *Culilawanus Papuanus* †. The bark of *Cinnamomum Culilawan,* a native of Amboyna, is called " Clove Bark " by reason of its strong flavour of cloves. " Clove Cassia " is the name applied to the bark of *Dicypellium Caryophyllatum.* It is also called " Brazilian Clove Bark." The tree producing it is found in Para and Rio Negro. Doubtless all these barks owe their aromatic quality to the presence of eugenol. The " Madagascar Clove-Nutmeg " is the fruit of *Agathophyllum aromaticum,* a laurel; its leaves are used by the natives as a condiment.

Analogous products, such as Pimento, " Wild Clove," West Indian Bay and European Bay, will be fully described in Series 2 of this work.

* Sp. Pl. ed. Willd. ii. p. 478.
† Martini, Ency. i. p. 436.

CHAPTER XI.

BENZOIN.—STORAX.—BALSAM PERU.—BALSAM TOLU.

BENZOIN.

IN the earliest known mention of gum benzoin the drug is called "Incense of Java," *Lubán Jáwi.* The word Java was applied to the island of Sumatra, and was in fact used indiscriminately by the Arabs to designate in a general way the islands of the Archipelago. The Arab name Lubán Jáwi became successively corrupted into Banjawi, Benjui, Benzui, Benzoi, Benzoïn, and ultimately into the common English name Benjamin.

The Sumatra and the Java benzoin are produced by the *Styrax Benzoin* tree (Dryander). The botanical source of the Sumatra benzoin was determined by Dryander, and an account and figure of the plant were published by him in the ' Philosophical Transactions ' for the year 1787, lxxvii. p. 303. According to recent investigations made by Mr. Holmes *, the Penang benzoin is similar in appearance to the Sumatra kind ; but it has an odour quite distinct and so strongly resembling storax that it is probably produced by a different species, possibly the *S. subdenticulata,* Miq., which occurs in West Sumatra and has the same vernacular name, " Kajoe Kéminjan." That these two species should receive the same native name is not surprising, since the leaves are very similar in shape and appearance and the fruit of *S. subdenticulata* apparently only differs from that of *S. Benzoin* in being obovate instead of globular and depressed. Mr. Holmes adds :—" The Palembang benzoin is probably derived from *S. Benzoin*; it has the same odour as ordinary benzoin from Sumatra and Java, but

---

* Pharm. Journ. [3] xiv. p. 354, and xx. p. 519.

is more translucent, contains a larger percentage of benzoic acid, and appears to contain a considerable amount of moisture, very likely owing to some difference in the mode of preparation or melting into blocks. The Siam benzoin comes from Luang Prabang in the Laos States; it has a distinct odour of vanilla, and there is sufficient botanical evidence of its being the product of a different species of *Styrax*.

In Sumatra, benzoin is collected in the northern and eastern parts of the island, especially in the district of Batta *, situate in the south of the State of Achin. The tree also grows abundantly in the high parts of Palembang in the south of the island. Large plantations are generally established in the vicinity of the coast. It is also cultivated along the Batang Leko river, where the trees attain a height of about 15 feet. The benzoin produced in the interior is generally obtained from wild trees which grow at the foot of the mountains at an altitude of 300 to 900 feet. The tree is a rapid grower; being planted along the banks of rivers, the only attention they require for the first six years is to prevent them being overgrown by other plants. At that age the trunks will have attained a thickness of 6 to 8 inches, and are old enough to yield resin. An incision being made in the trunk, a thick white resinous sap flows out, which rapidly hardens by exposure to the air; this is carefully scraped off with a knife. Each tree will continue to yield during 10 or 12 years about 3 lbs. of resin per year. The tree is then felled. The resin exuded during the first 3 years is the most rich in the white tears, consequently superior to that which is afterwards produced; it is designated by the Malays "Head Benzoin." The resin produced during the 7 or 8 succeeding years is browner and not so fine, this is termed "Belly benzoin." The third quality, called "Foot Benzoin," is obtained by cutting down the tree and scraping the wood. The benzoin is carried down to the port in large cakes called "Tampangs," enveloped in rush mats. They are then broken up and re-melted, either by the heat of the sun or by boiling water, and then filled into square boxes.

The only information known respecting the method of harvest in *Siam* was furnished a few years ago by the English Consul at

* Miquel, 'Prodromus Floræ Sumatranæ,' p. 72, and Flückiger, 'Hist. des Drogues,' ii. p. 12.

Bangkok. It is to the effect that the whole surface of the bark is incised, and that the resin which exudes accumulates and hardens between the wood and the bark, which is then stripped off. This information is confirmed by the aspect of the Siam benzoin of commerce, but evidently the entire supply of Siam benzoin is not obtained in this way. The Consul adds that the resin is much damaged and broken in transport to the navigable places on the Menam river, whence it is taken down to Bangkok.

The finest *Siam* benzoin consists entirely of a mass of flattened tears of opaque milk-white resin tightly agglutinated together; these tears are sometimes of from 1 to 2 inches in length. Generally, commercial benzoin is a compact mass, consisting of a quantity of white tears about the size of an almond, embedded in a translucid dark-brown resin. Sometimes the translucid resin predominates, and the white tears are few or almost entirely absent. In some consignments the tears of white resin are very small, and the mass presents an aspect of reddish-brown granite. There is always found a certain admixture of fragments of wood, bark, and other accidental impurities. The white tears when broken exhibit a certain stratified formation. The lumps of resin, which were exteriorly of a milky whiteness, gradually become brown and transparent on the surface. On the authority of Professor Flückiger, this opacity does not appear to be due to enclosed moisture, but rather to a particular molecular (semi-crystalline?) condition of the resin.

Siam benzoin is very brittle; the fracture of the opaque tears is rather waxy; the translucent part being of a glassy fracture. It softens quickly in the mouth and can be masticated. Its fragrance is very delicate, balsamic, and recalling perceptibly that of vanilla, but its taste is very feeble. On being heated, it exhales a very strong odour and disengages irritating vapours of benzoic acid. It melts at 75° C. Siam benzoin is imported in cubic blocks, having the form of the boxes into which it was packed when in a soft state.

The *Sumatra* benzoin is imported in the same shaped cubic blocks as the Siamese, but the aspect of it differs by being generally of a greyer tint. When of fine quality it contains numerous opaque trees embedded in a translucent greyish-brown resin, sometimes containing fragments of wood and bark. In inferior qualities the white tears are not present and the proportion of impurities is larger, sometimes very large. The odour of Sumatra benzoin is weaker and less agreeable than that of the Siam variety. The Sumatra benzoin is inferior in appearance to the Siam and is not generally so pure, hence its marketable price is lower. The greyish-brown part melts at 95° C., and the tears at 85° C.

As before observed, the Penang benzoin is markedly different in odour to that of Siam or Sumatra.

Another description of benzoin is yielded by the *Terminalia angustifolia*, Jacq. * ; syn. *T. Benzoin*, Linn. † ; *Catappa Benzoin*, Gærtner ‡. This is a tree of 30 or 40 feet in height ; a native of the East Indies. The gum is procured by wounding the tree; it is composed of large white and light brown pieces, breaking very easily between the hands. It contains benzoic acid. When gently dried it forms a white powder, formerly in great request as a cosmetic, its perfume being very agreeable.

Benzoic acid has a sharp acid taste and produces a peculiar irritation in the throat. It melts at 121°·4 and boils at 249°, but volatilizes at 100°, and sublimes rapidly at 140°. It also volatilizes with steam, one gramme passing over with two litres of water. Its vapour has an aromatic penetrating odour, produces coughing and attacks the eyes violently, more mildly when it is mixed with steam.

Benzoic acid crystallizes in lustrous, flat, monoclinic plates or needles ; by the gradual evaporation of its solution it is obtained in larger tablets, which, however, are always thin.

1000 parts of water dissolve

at

10°	20°	30°	40°	50°	60°	70°	80°	90°	100°
2·10	2·90	4·10	5·55	7·75	11·55	17·75	27·15	40·75	58·75

parts of this acid §.

100 parts of absolute ether at 15° dissolve 31·35 parts ; 100 parts of 40 % alcohol 41·62 parts ; and 100 parts of absolute alcohol, 46·48 parts of benzoic acid ‖. It also readily dissolves

* Hort. Vind. iii. t. 100.
† Sup. Lam. Dict. i. p. 349.
‡ Fruct. ii. p. 206, t. 127.
§ Ann. Chim. Phys. [5] xv. p. 168.
‖ Bull. Soc. Chim. xxix. p. 242.

in chloroform, carbon disulphide, volatile and fatty oils, and concentrated sulphuric acid.

It is characteristic of benzoic acid that certain impurities, even
when they are present in extremely small quantities, alter its
physical properties to a very considerable extent.

Many varieties of benzoin contain cinnamic acid in addition to
benzoic acid, and frequently only the former *. Both these acids
occur either free or in the form of ethereal salts, together with
other aromatic compounds, in Tolu balsam, Peru balsam, Mecca
balsam (Balsamodendron Opobalsamum et Gileadensis), Myrrh
(B. Myrrha), liquid Styrax, acaroïd resin (Xanthorrhœa hostilis),
Dragon's blood, and other resins. Benzoic acid has also been
found in the oil of Ylang-Ylang (Cananga odorata), in plums
(Prunus domestica chlorocarpa), and the cranberry. It also occurs
in vanilla, the fruit of the clove-tree, the seeds of the spindle-tree
(Euonymus Europæus), and the root of the sweet flag (Acorus
calamus).

Fourcroy and Vauquelin found, in 1797, that the urine of grami-
nivora contains benzoic acid; and Liebig, in 1829, showed that a
new nitrogenous acid, which he named hippuric acid, splits up
when the urine is allowed to stand, yielding benzoic acid. According to some observers, however, benzoic acid frequently occurs
along with hippuric acid in the urine, and it has been found in a
gland in the beaver †, and in the kidneys of the ox ‡. It is
probable that in all cases the acid is formed by the decomposition
of hippuric acid. It also occurs in coal-tar.

Benzoic acid, $C_6H_5 . COOH$, was formerly prepared exclusively
from gum benzoin by sublimation; thus obtained it always
contains a small amount of an ethereal oil, which gives it its
peculiar smell. In order to prepare it in this way, the coarsely
powdered resin is heated to about 170° in a flat iron vessel; this is
covered with filter-paper, and fitted with a conical cap of strong
paper in which the acid sublimes. This acid is commercially
called "English benzoic acid" or "Benzoin-benzoic acid."

A second process for procuring it from its natural source
(Scheele's process) consists in boiling a mixture of equal weights

* Kolbe, Ann. Chem. Pharm. cxix. p. 136.
† Ann. Chem. Pharm. lxvii. p. 360.
‡ Chem. Centralbl. 1861, p. 241.

of the powdered resin and slaked lime, evaporating the filtrate to one-sixth of its bulk, treating with bleaching-powder solution, and then boiling with hydrochloric acid until all the chlorine has been removed. The acid separates out on cooling, and is re-crystallized from hot water.

The first process yields the most fragrant acid, the second yields the largest result.

Benzoic acid is prepared from the urine of cows or horses by allowing it to stand for several days, clarifying with milk of lime, evaporating the liquid to one-fourth of its bulk, and precipitating the benzoic acid with hydrochloric acid. As the evaporation produces a very unpleasant smell, it is better to precipitate the excess of lime by carbonic acid and add ferric chloride to precipitate ferric benzoate, which is then decomposed by hydrochloric acid. The acid thus obtained is purified by being re-dissolved in milk of lime with the addition of a little bleaching-powder solution, separated by hydrochloric acid and re-crystallized from hot water. The final product (*acidum benzoicum ex urina*) is not only devoid of the fragrance of " Benzoin-benzoic acid," but it still smells of urine, and is consequently unfit for purposes of perfumery. About 2 lbs. of acid are obtained from 1000 lbs. of urine. This variety is commercially known as " urine-benzoic acid " and " German " benzoic acid ; it is probably used as an adulterant to the sublimed acid. When the " urine acid " is sublimed, it may, as averred, lose its offensive smell, but it does not contain the adherent volatile oils which add to the fine odour of the natural product.

Artificial benzoic acid is now prepared readily and cheaply from toluene, which is a bye-product in the manufacture of nitro-benzene. Toluene is first converted into benzyl chloride (as previously described). 100 parts of this are then boiled with 300 parts of nitric acid sp. gr. 1·313 and 200 parts of water for about 10 hours in an apparatus connected with an inverted condenser, until the smell of benzyl chloride and benzaldehyde has disappeared, and the liquid solidifies on cooling to a crystalline mass, no oily drops being formed.

This form of acid retains a peculiar odour of nitrobenzene, which is more noticeable when the package containing it is first opened. It is known as " Toluene-benzoic acid," and is largely made in Germany. As neither the " hippuric " nor the " toluene-benzoic acid." has the peculiar fragrance of the "benzoin-benzoic acid," it is said that the latter acid is added to the other varieties to give them a more "genuine" odour, which is probable, as these are produced at a very much lower cost than the genuine article.

The difference in the behaviour of the different benzoic acids towards potassium permanganate is much more characteristic when in alkaline solution than when in acid solution. If 0·1 gram of the benzoic acid be dissolved in 3 c. c. of potash solution sp. gr. 1·177 at 15° C., the solution diluted with 3 c. c. of distilled water, 5 drops of ½-per-cent. permanganate solution added, and the whole heated to boiling, all the benzoic acids, with the exception of those obtained by sublimation and in the nascent way from Siam benzoin, yield deep dark-green liquids, in which gradually a brown precipitate forms, whilst the two kinds mentioned give immediately decolorized liquids over brown precipitates. The different specimens present the same appearances after standing several hours. In this way therefore it can be ascertained with ease and certainty whether a genuine resin benzoic acid is present or not [*].

A "Toluene-benzoic acid" is found in commerce which has been sublimed with one-fifth and sometimes with one-tenth of its weight of Siam benzoin to disguise its origin. This acid forms in white silky loose crystals, smelling strongly and very agreeably of benzoin, but, when treated with potassium permanganate in alkaline solution in the above-mentioned proportions and with heat, does *not* behave like true "gum" benzoic acid.

There are various odorous compounds artificially prepared from benzoic acid, besides those mentioned elsewhere under their appropriate headings, e. g. :—

Methyl benzoate. Two parts of benzoic acid, one part of methylic alcohol, and two parts of strong sulphuric acid are distilled together. The residue is redistilled two or three times with fresh portions of methylic alcohol, and all the distillates are mixed together with water. The crude methyl benzoate sinks to the bottom. In this state it is sometimes sold as "Niobe essence." To purify it, it is washed two or three times with water, dried by agitation with calcic chloride, and rectified over dry oxide of lead, the portion which comes over above 198° C. being collected apart. It is a colourless, oily fluid, with a pleasant

* Schacht, Pharm. Jnl. [3] xii. p. 518.

balsamic smell; insoluble in water, soluble in alcohol and ether; boils at 198°·5 C. at 761 millim., and at 199°·2 at 746 millim.; its sp. gr. is 1·10 at 17° and 1·0876 at 16° C. (Kopp). Its constitution is represented by the formula $C_8H_8O_2$. According to another authority methyl benzoate is conveniently prepared by passing hydrochloric acid into a solution of benzoic acid in methyl alcohol, distilling, and then precipitating the ether with water *.

Ethyl benzoate was prepared by Scheele as long ago as 1785, by the distillation of a mixture of alcohol, benzoic acid, and hydrochloric acid. It is not formed when an alcoholic solution of the acid is allowed to stand in the cold, but the ether is gradually formed if a little hydrochloric acid be added, or if the liquid be heated to 100°. The method adopted for the preparation of methyl benzoate may be followed, or it may be prepared by distilling four parts of alcohol with two parts of benzoic acid and one part of fuming hydrochloric acid till two-thirds have passed over, and pouring back the distillate into the retort two or three times, a little of the ether passes over, but the greater part remains in the residue, whence it is separated by the addition of water. It is also obtained when a solution of three parts of benzoic acid in two parts of boiling alcohol of 80 per cent. is heated for some time in a vessel provided with a reflux condenser. It is represented by the formula, $C_9H_{10}O_2$.

It is also readily formed by the action of benzoyl chloride (C_7H_5OCl) on alcohol, which is very susceptible of its presence. The presence of alcohol even in very dilute aqueous solution can therefore be detected by warming it with a little benzoyl chloride, and removing the excess of this acid by caustic soda; even when only 0·1 per cent. of alcohol is present, the characteristic odour of ethyl benzoate can be distinctly recognized (*Berthelot*).

Isonitrosobenzyl ether, $C_6H_5CH_2ON$, is prepared by heating benzyl chloride (C_7H_7Cl) with a solution of sodium and isonitroso-acetone in absolute alcohol. It crystallizes from petroleum ether in colourless tablets, which have a pleasant smell of flowers. It melts at 45° to 46° †.

Benzoylacetone is formed together with acetophenone and benzoic acid when benzoylacetic acid is boiled with water ‡. It is

* Ann. Chem. Pharm. cx. p. 210.
† Ber. Deutsch. chem. Ges. xv. p. 3071.
‡ Ibid. xvi. p. 2239.

also a product of the action of sodium ethylate on a mixture of acetone and ethyl benzoate * ; but it is best prepared by covering sodium ethylate, free from alcohol, with an excess of acetic ether, and gradually adding the calculated quantity of acetophenone, the liquid being cooled with ice. The liquid product solidifies after a short time to a thick mass of light yellow crystals of sodium benzoylacetone. The mass is triturated with ether and filtered, the residue being dissolved in water and decomposed with acetic acid, which precipitates the benzoylacetone in the form of small prisms †. This substance melts at 60°–61°, boils almost without decomposition at 260°–262°, and is readily volatile with steam. It has a very pleasant penetrating odour, is only slightly soluble in cold, more readily in hot water, and is readily soluble in alcohol and caustic-soda solution. It gives an intense claret-red coloration with ferric chloride, and is decomposed by boiling with alkalies into acetic acid and acetophenone.

Amylbenzene has been obtained by treating a solution of amyl bromide and bromobenzene in benzene with sodium ‡. A better yield is obtained when ether is used as a diluent instead of benzene. It is also formed by the action of amyl chloride on a mixture of benzene and aluminium chloride §. It is a pleasant smelling liquid, boils at 193°, and has a sp. gr. of 0·859 at 12°.

Pentylbenzene, $C_6H_5C_5H_{11}$, has been obtained by the action of sodium on a mixture of benzyl bromide and butyl bromide. It is a very pleasant smelling liquid, which boils at 200°·5–201°·5, and has a sp. gr. of 0·8602 at 22°||.

Butylbenzene is formed by the action of sodium on a mixture of propyl bromide and benzyl bromide, or of butyl bromide on bromobenzene. It is a liquid of a very pleasant odour, boils at 180°, and has a sp. gr. of 0·875 at 0° and of 0·8622 at 16° ¶.

Isobutylbenzene is obtained by treating a mixture of bromobenzene and isobutyl bromide or isobutyl iodide, or of benzyl chloride and isopropyl iodide with sodium **. It is also formed by heating

* Ber. Deutsch. chem. Ges. xx. p. 655.
† Ibid. xx. p. 2078.
‡ Ann. Chem. Pharm. cxxix. p. 369 ; cxxxi. p. 313 ; and cxli. p. 160.
§ Ann. Chim. Phys. [6] i. p. 454.
|| Ann. Chem. Pharm. ccxviii. p. 383.
¶ Ber. Deutsch. Chem. Ges. ix. p. 260, and x. p. 296.
** Ibid. iii. p. 779 ix. p. 1606; and viii. p. 500.

isobutyl alcohol with benzene and zinc chloride *, and by the action of aluminium chloride on a mixture of isobutyl chloride and benzene †. It is a pleasant smelling liquid, which boils at 167°·5, and has a sp. gr. of 0·890 at 15°.

STORAX.

The words *Storax* and *Styrax* have been used by some authors to distinguish between the solid and the liquid varieties of this drug. The solid kind usually referred to by Dioscorides and ancient classical writers as Στύραξ is yielded by the *Styrax officinalis*, Linn., a small tree nearly allied to the tree yielding benzoin. It is a native of Greece, Asia Minor, and Syria, and is cultivated in Italy and some parts of the south of France. It is figured and botanically described in Woodville's 'Medical Botany,' tab. 71, in Churchill and Stevenson's 'Medical Botany,' i. tab. 47 ; in Andrews's 'Botanical Repository,' 631, and Loddiges's 'Botanical Cabinet,' 928. It is said that this tree, when allowed to grow freely, will attain 15 or 20 feet in height ; but now, in most localities, it is stunted down to a mere bush through bad cultivation and cutting the tree periodically for fuel ; in such state it does not yield the odoriferous product, except possibly in the district of Alexandretta.

The *Styrax officinalis*, Linn., is indigenous in the mountainous woods on the east side of Toulon, in the direction of Cuers. It there grows in abundance, but as it is cut periodically for fuel in common with the other trees growing near it, it can seldom attain any considerable size. Hanbury states ‡ that at the time of his visit to this district, May 17, 1854, he did not observe any trees exceeding eight or nine feet in height ; . . . that the Styrax trees presented a beautiful appearance by their abundance of orange-flower-like blossoms, but that no trace of resinous exudation could be observed upon any of the trunks, nor did the fresh bark possess the least odour of storax.

This gum, which used to be known as " True Storax " (derived from the Arabic word *Assthirak*), is very fragrant, and appears in the form of separate or more or less agglutinated tears, exuding either spontaneously or after incision made in the trunk of the

* Ber. Deutsch. chem. Ges. xv. pp. 1066 and 1425.
† Bull. Soc. Chim. xli. p. 446.
‡ Pharm. Journ. [3] xiv. p. 12.

tree. This fine kind of storax, always extremely scarce, was called *amygdaloid*, from the small white almond-like tears of which it partially consisted. It also bore the name of *Styrax calamites*, a term derived from the ancient method of packing it in reeds (*calami*). This description of storax, which was undoubtedly the Storax of ancient times, has now wholly disappeared from commerce, the name *Styrax calamites* or *calamita* being retained to designate a manufactured compound, hereafter described, but not produced from *S. officinalis*, Linn.

Liquid Storax.—This balsam was mentioned at a very early date by the Arabian physicians and was exported by the Arabs to India and China, which countries still receive the larger portion of the annual production. The botanical origin of the drug was, until a comparatively recent date, obscure, but we know now with certainty that it is exclusively obtained from *Liquidambar Orientalis*, Miller (*L. imberbe*, Aiton), a tree whose geographical distribution is very restricted, but which forms large forests in the south-west of Asia Minor. These forests are described by Hanbury* as being found in the district of Sighala, near Melasso ; near Moughla, and near Giova and Ullà, in the Gulf of Giova; also near Marmorizza and Isgengak, opposite Rhodes. The trees are described as resembling the plane tree, although with a smaller leaf and being much denser in foliage than the plane; the height being twenty to sixty feet, and being especially large in the immediate vicinity of streams of water, and where they find sufficient air and space.

The tree is figured in Hooker's 'Icones Plantarum,' 3rd series, 1867, tab. 1019 ; Hanbury's ' Science Papers,' 1876 ; and Bentley and Trimen's Med. Plants, tab. 107.

The methods of extracting the Liquid Storax have been described as follows :—In June and July, the outer bark is stripped off on one side of the tree and reserved for purposes of fumigation. The inner bark is then scraped off with a semicircular or sickle-shaped knife, and when a sufficient quantity has been collected, it is packed in strong horsehair bags and subjected to pressure in a wooden lever press. Upon removal from the press, hot water is thrown over the bags and they are pressed a second time, after which the greater portion of the resin will have

* Pharm. Journ. [1] xvi. p. 461.

been extracted. Another account says the resin is chiefly collected by a tribe of wandering Turcomans called *Yuruks,* who boil the inner bark in a large copper, the liquid resin rising to the surface is skimmed off. The boiled bark is next put into horsehair bags and pressed, the extracted resin being added to the portion first obtained.

The result of these processes is an opaque grey or greyish-brown semi-fluid resin of about the consistence of honey, which is exported in casks to Constantinople, Smyrna, Syra, and Alexandria. Some is also packed with a certain proportion of water in goat-skins and sent to Smyrna, where it is transferred to casks and shipped mostly to Trieste. This balsam is known to the Turks by names which mean *" Black Frankincense oil," Incense-oil,"* and *" Sighala oil "* (from the district between Melasso and Macri, where much of it is collected). The Greeks often designate it by the first-mentioned Turkish name. As imported into Bombay from the Red Sea ports, this balsam is known as *Rose Malloes,* being evidently confused with the resin obtained from the *" Liquidambar Altingiana "* of the Indian Archipelago, a tree which bears the Malay name of *Ras-Sama-la.*

The residual bark, after the extraction of the liquid storax, is emptied out of the bags and exposed in the sun to dry. It has the appearance of brownish-red cakes consisting of thin, narrow reddish strips, tightly pressed together and having a sweet balsamic odour. These cakes are known as *Red Storax* and *Black Frankincense leaf,* and in pharmacy as *Cortex Thymiamatis, Cortex Thuris, Thus Judæorum, Narcaphthum.* In modern Greek it is known by the simple name " Storax."

The semi-fluid resin of *Liquidambar orientalis* always contains a certain amount of water which, by degrees, floats to the surface. By age the resin becomes more transparent and of a dark brown colour. It also becomes transparent and more fluid on the application of heat, parting with the water mechanically held in solution and depositing solid impurities at the bottom of the flask. On being spread out very thinly it partly dries, but does not quite lose its stickiness. After being separated from the water contained in it, it reddens litmus. It dissolves in alcohol, chloroform, ether, acetic acid, bisulphide of carbon, and in most essential oils, but *not* in petroleum benzene. In coal-tar benzene it dissolves with ease, and this forms the best menstruum for freeing it from impurity, of which there is usually present about 10 per cent. It has been suggested that this peculiarity of solubility may form a useful means of distinguishing between the two benzenes, or even detecting their admixture *. The odour of liquid storax is very pleasant and balsamic, especially when matured by age; the recent balsam having rather an odour of bitumen and naphthaline.

The presence of turpentines of the various species of *Larix* and *Pinus* as adulterants of liquid storax may, according to Hager, be determined by melting a small quantity (say 5 grams) in a test-tube, and exhausting it by agitating three times successively with petroleum ether. Upon evaporating the solution, the residue should consist of styrolene (formerly called styrol) and styracin only, should be colourless, bluish opalescent, and of pleasant odour. If a turpentine is present the residue is yellowish, and has the characteristic odour of turpentine. Good storax yields to petroleum ether about 50 per cent. ; if more is taken up, the presence of adulterants may be inferred †.

For the detection of crude turpentine in liquid storax, Hager also recommends the following test:—Fuse the sample in a test-tube by means of a water-bath, then agitate it with half its volume of absolute alcohol until dissolved. This is then to be thrice shaken up with several times its volume of benzene. The decanted benzene solutions are united and evaporated in a water-bath from a tared vessel. The residue should weigh 45 to 55 per cent. of the sample; it should have a bluish opalescence and an agreeable odour. If turpentine is present the residue will be yellowish, with an odour of turpentine, and its weight will be greater than that stated above ‡.

Liquid storax contains styrolene, cinnamic acid, styracin, phenyl-propyl cinnamate, ethyl cinnamate, and a small quantity of a pleasant smelling substance which is probably ethyl vanillin. The chief constituent, however, is the resinous styracin and its cinnamic ether §.

Styracin, or Cinnyl cinnamate, $C_{36}H_{55}(OH)_3$, can be isolated from liquid storax by filtering hot through a cloth and triturating

* Pharm. Journ. [3] xi. p. 431.
† Ph. Centralhalle, 1874, p. 161.
‡ Ibid. 1874, no. 21.
§ Miller, Ann. Chem. Pharm. clxxxviii. p. 184, and clxxxix. p. 338.

the filtrate with cold petroleum spirit. One half of the liquid is then distilled off, and the solution filtered from the precipitate, which consists of the ethyl cinnamate and a portion of the styracin. The clear liquid deposits the styracin on standing, in dazzling white fascicular crystals, which melt at 44°.

Styrolene, C_8H_8, formerly called *Styrol* or *Cinnamol,* is chemically *phenyl ethylene.* It is a volatile oil, and was formerly obtained by distilling liquid storax with water, to which sodium carbonate was added to prevent any cinnamic acid passing over ; it is strongly refractive, colourless, very aromatic, boils at 144°·5 and has a sp. gr. of 0·925 at 0°. The amount of styrolene present varies according to the age of the balsam, also by reason of the varying methods employed to extract the balsam. On oxidation it yields benzoic acid. When styrolene is heated for some time to 100° or for a short time to 200°, it is converted without changing its composition into Meta-styrolene, a vitreous, strongly refractive mass, insoluble either in alcohol or ether. Its sp. gr. is 1·054, and it is too hard to be cut with a knife.

When cinnamic acid is distilled with four times its weight of baryta, a body is formed which has been considered identical with styrolene *, and Hempel found that a similar hydrocarbon is formed when the vapour of cinnamic acid is passed through a red-hot tube, as well as by the distillation of copper cinnamate †, and the slow distillation of free cinnamic acid.

Styrolene has been obtained in other ways. Botsch found that when Dragon's blood is distilled with zinc dust, 66 per cent. of styrolene are formed, in addition to ethylbenzene, a little toluene, and a smaller quantity of higher boiling substance ‡. Berthelot obtained it, together with benzene and other hydrocarbons, by heating acetylene to the softening point of glass §, and by passing a mixture of ethylene and benzene through a red-hot tube ‖ ; he also discovered it in coal-tar ¶.

Styrolene is now obtained by the slow distillation of cinnamic acid, which is manufactured synthetically on the large scale (see Cinnamon). When cinnamic acid is allowed to stand for some

* Compt. Rend. xxi. p. 1376. † Ibid. lix. p. 316.
‡ Monatshefte f. chem. i. p. 609.
§ Ann. Chem. Pharm. cxli. p. 181.
‖ Ibid. cxlii. p. 257. ¶ Ibid. Suppl. iii. p. 368.

days in contact with the most concentrated hydrobromic acid, phenylbromo-propionic acid is formed, and this is decomposed by sodium-carbonate solution with formation of styrolene. A still better yield is obtained when phenyliodopropionic acid, which is prepared in a similar manner, is boiled with sodium-carbonate solution *.

The artificial, or synthetically prepared styrolene, which was termed by Herzog †, who prepared it by distilling cinnamic acid with slaked lime, *Cinnamol*, also the styrolene obtained artificially by Berthelot and other chemists, is optically inactive, whilst styrolene obtained by distillation from liquid storax is lævorotatory. This appeared remarkable ; but Van t'Hoff proved that the hydrocarbon obtained from storax contains varying quantities of an impurity to which it owes its rotatory power ‡.

Styrolene chloride is formed by the direct combination of styrolene with chlorine, but it is difficult to purify, since substitution products are instantaneously formed. It is a thick oily liquid which decomposes on distillation and is scarcely soluble in water, but imparts to it a very characteristic penetrating smell, resembling those of the oils of lemons and juniper berries.

Styrolene bromide has the same characteristic smell and taste as the chloride. It is best obtained by mixing styrolene with an equal volume of chloroform § or with two volumes of ether ||, and adding the calculated quantity of bromine to the well-cooled liquid. It may also be prepared by adding bromine gradually to hot ethylbenzene ¶, and purifying the product by re-crystallization. It crystallizes from alcohol in small plates or broad needles melting at $74°–74°·5$.

Cinnamic acid occurs in liquid storax, partly in the free state, partly as styracin. The uncombined cinnamic acid is easily extracted by boiling the drug in water with carbonate of soda or of lime. Cinnamate of soda is thereby formed, and can readily be decomposed by acids. By this process from 6 to 12 per cent., and

---

* Ann. Chem. Pharm. cxcv. p. 137.

† Compt. Rend. liii. p. 323.

‡ Ber. Deutsch. ix. pp. 5 & 1339, and xi. p. 1260; also Ann. Chem. Pharm. cxli. p. 378.

§ Ann. Chem. Pharm. cliv. p. 154.

|| Ibid. ccxvi. p. 288.

¶ Ber. Deutsch. chem. Ges. vi. p. 493, and Bull. Soc. Chim. xxxv. p. 55.

even more, of crystallized cinnamic acid can be obtained. The properties of this acid are described under the article "Cinnamon."

Styracin can be isolated by ether, benzene, or alcohol after the styrolene and cinnamic acid have been separated out from the resin; it being insoluble in water and only volatile at a temperature above red heat.

It was found in 1839 that the *Styracin* discovered by Bonastre in liquid storax is converted, by boiling with caustic-soda solution, into cinnamic acid and an oily liquid which was termed *Styracone.*

Later investigations * have shown that this is a mixture of phenylpropyl alcohol and phenylallyl alcohol, which last is also called Cinnamyl alcohol, or *Styryl alcohol,* or more conveniently *Cinnyl alcohol.* In order to prepare it, styrax is distilled with sodium-carbonate solution until styrolene no longer passes over, the aqueous solution containing sodium cinnamate removed, the residue distilled with caustic potash, and the cinnyl alcohol separated from phenylpropyl alcohol by the fractional distillation of the oily distillate. Pure styracin may also be distilled with caustic potash; the cinnyl alcohol crystallizes out of the distillate, while any remaining in the solution is precipitated with salt or extracted with ether. Cinnyl alcohol crystallizes in long thin needles which melt at 33°, and have a pleasant smell of *hyacinths*; it boils at 250°, is tolerably soluble in water, readily in alcohol and ether, and is oxidized to *cinnamaldehyde* by platinum black †. It is converted into phenylpropyl alcohol by the action of sodium amalgam on its warm aqueous solution, a little allyl-benzene being simultaneously formed.

Styrax Calamita.

The substance which bears this name is not the Storax Calamita of the Ancients, but a composition artificially made up of the pressed residue of the bark of *Liquidambar Orientalis,* from which the Liquid Styrax has been extracted. This cake of bark, called " *Cortex Thymiamatis,*" is coarsely pulverized and mixed with liquid storax in the proportion of three of the former to two of the latter. When first made it presents the appearance of a viscid mass, which after a few weeks becomes coated with small silver needles of styracin. It has a very sweet odour.

* Ann. Chem. Pharm. cxxxii. p. 22, and cxxxviii. p. 184.
† Ibid. xciii. p. 370.

When the bark residue is scarce common sawdust is used as a substitute, and inferior qualities are made up of olibanum, honey, red earth, and other substances. This drug is manufactured at Trieste, Venice, and Marseilles.

Other odoriferous substances are compounded from liquid storax, amongst which may be mentioned the "*Black Storax*" which Pereira notices [*]; it is a brownish-black substance which by degrees moulds itself to the shape of the vessel in which it is enclosed. In Hanbury's opinion it is this *Black Storax*, composed of olibanum and liquid storax made into cakes, which is sold for incense under the name of ψευδομοσχολιβανον or στοράκι by the Greek monks, particularly those of the island of Symi, and it is also this substance which constitutes the "precious incense" used at Easter in the Church of the Holy Sepulchre at Jerusalem, and of which small pieces are sold to the pilgrims at an enormous price,—not for burning, but chiefly to be used as a charm.

" *Cake Storax* " is met with in large blocks of 50 or 60 lbs. weight packed in canvas; it is brownish red, easily pulverized to coarse powder which can be again united into a mass by pressure. This is probably made of the bark coarsely ground in a mill and deprived of the bulk of its resin.

Resin of *Liquidambar Styraciflua*, Linnæus, is the produce of the "Liquidambar tree" of Louisiana, Florida, Mexico, and Guatemala. In the United States it forms a very large tree, known as the "Sweet Gum tree" which in southern latitudes attains an immense size, the balsamic exudation being much more abundant. It yields two sorts of balsam, physically different in character. The one is a thick, transparent, yellow oily fluid, which by age or exposure to the air becomes darker in colour and concretes. This is called "*Liquidambar liquid.*" It has a strong, agreeably balsamic odour, and an aromatic very bitter taste. It contains a large quantity of benzoic or cinnamic acid, a drop of it on litmus paper producing a deep red coloration. On being treated with boiling alcohol, a small quantity of white residue is precipitated. This balsam is obtained by making incisions in the bark of the tree, and immediately receiving the liquid into bottles to protect it from the action of the air. It is afterwards decanted, to separate an opaque portion which deposits at the bottom. The

---

[*] Mat. Med. ii. pt. i. p. 680.

second product of this tree, known as " *Soft* or *White Liquidambar*," consists partly of the opaque deposit above-mentioned, and partly of that portion of the balsam which has flowed from the tree and thickened by exposure to the air ; these are probably melted together. It is of the consistence of a very thick turpentine or soft pitch ; opaque, whitish in colour, and less odorous than the preceding. It contains benzoic acid, which frequently effloresces on the surface. By exposure to the air it solidifies and becomes almost transparent, but retains its odour. It is frequently used to adulterate the White Balsam Peru, but is distinguishable from it by taste and by a bitterness which it acquires by exposure to the air. Pereira says it is quite different from a genuine sample of the White Balsam of Peru received by him from Guatemala, and equally different from genuine "liquid storax."

A thick dark-coloured opaque substance is obtained from the young branches of this species by boiling them in water and skimming off the fluid balsam which rises to the surface. This also has been confounded with liquid storax.

The balsam previously referred to as " Rose Malloes," from the Indian Archipelago, is produced by the *Liquidambar Altingia,* Blume (*Altingia excelsa,* Noronha), a native of those islands and of Burmah and Assam. In Java this tree is said to attain an immense size. The word Ras-Sama-la is variously corrupted into Rosmal, Rosum Aloes, and Rose Malloes. The odoriferous semi-fluid resin is not abundant, and does not resemble "liquid storax." Two sorts of balsam are obtained from this tree in Burmah ; the one is pellucid, of a clear yellow colour, obtained by simple incision of the bark ; the other thick, dark, opaque, and of terebenaceous odour, obtained by applying heat to the tree after wounding it.

Resin of *Liquidambar Formosana,* Hance.—The tree furnishing this resin is supposed to be identical with the *Styrax liquida folio minore,* mentioned by Ray [*] as being amongst a collection of plants from Amoy. It is a native of Formosa and the south of China. It yields a hard resin which has an agreeable odour when warmed. The tree is figured in Hooker's ' Icones Plantarum,' series 3, i. tab. 1020.

---

[*] Hist. Plant. iii. Append. p. 233.

BALSAM PERU.

This fragrant balsam is extracted from the trunk of the *Myroxylon Pereiræ*, Klotzsch (*Myrospermum Pereiræ*, Royle), a handsome tree of about 45 feet in height, the trunk rising to a height of 6 to 9 feet before putting forth its branches. It inhabits the western part of the State of San Salvador, on the Pacific coast of Central America, known as the "Balsam Coast," extending between 13° 35' and 14° 10' N. lat., and 89° and 89° 40' W. long., a tract of land which formerly belonged to Guatemala. In this locality it is found growing wild in dense forests, but each tree has its particular owner; those which grow in clusters together are sometimes enclosed and those which grow singly are simply marked. They are sometimes let out for a certain number of years, or the produce of a certain number of trees may be contracted for. The principal towns and villages in the neighbourhood of the Balsam region are:—Juisnagua, Topecoyo or Coyo, Tamanique, Chiltiuapan, Talnique, Jicalapa, Jayaque, Teotepeque, and Comasagua.

The season for collecting the balsam commences in November. A portion of the bark on four sides of the tree is loosened by being well bruised with a hammer or the back of an axe, leaving between the parts so bruised four strips of similar size uninjured, so as to preserve the vitality of the tree. Then, to excite an abundant flow of balsam and facilitate the removal of the bark, it is usual in about five or six days after thus bruising it to char its outer surface by means of lighted torches, care being taken not to let them come in contact with the sap, the inflammability of which might cause the complete destruction of the tree. The torches are generally made of "chemaliate," a kind of resinous cane burning like a candle. After the lapse of about a week, the charred bark either falls off of its own accord, or is easily detached. The trunk then commences to exude the balsam, which is collected by wrapping rags round the naked wood and so soaking it up. After a few days, the rags being thoroughly saturated with balsam are thrown into an earthen vessel three parts full of boiling water, stirring them meanwhile. In a few hours all the balsam will be extracted from the rags and sink to the bottom of the vessel. From time to time the spent rags are withdrawn from the boiler, and replaced by others saturated by balsam. As the rags are withdrawn from the boiler they are submitted to strong pressure, and the balsam which is extracted from them returned to the boiler. When the decoction is cold the water is decanted and the balsam poured into gourds ready for transport to the coast. The next year the Indians again visit the same trees and perform the same operation on the portion of bark which was left untouched the year before.

The tree begins to be productive in its fifth year, and continues to yield for 30 years or more [*]. In 1861 the tree was introduced in Ceylon, with complete success.

As the tree is said to be capable of reproducing its bark in two years, a harvest can be gathered for many years, provided that from time to time it be allowed a few years of rest. Sometimes the naked wood is covered up with clay as a protection.

When Balsam Peru arrives at Acajutla and La Libertad, the ports on the " Balsam Coast " from which it is chiefly shipped, it is in a crude state, usually of a grey-green to a dirty yellow colour, and requires to be purified before it is fit for exportation. A first clarification is effected by allowing the crude balsam to stand in a large iron vessel capable of holding six or seven hundred pounds during a week or a fortnight, by which time the heavier impurities sink to the bottom and the lighter ones float as a scum on the surface. The clear balsam, which has already attained its characteristic black-brown colour, is then drawn off through a tap fixed about four inches from the bottom of the vessel and run into a tinned iron boiler set over an open fire and boiled moderately for two or three hours. All scum is removed as it makes its appearance, and the boiling is continued as long as any continues to be formed. It can easily be understood that the physical properties of the balsam will differ according to the temperature to which it is submitted during this boiling, and it is alleged that the lower specific gravity observed in balsam of Peru during recent years is attributable to a modification it undergoes in this operation, and is quite consistent with the genuineness of a given sample [†]; this may be so, but it would seem preferable to refine the balsam in Europe by a more careful process.

The sp. gr. of pure balsam Peru at $15°$ C. varies from $1·140$ to

[*] Am. Journ. Pharm. xxxii. p. 303.
[†] Gehe & Co.'s Market Reports, 1884.

1·145, according to the proportionate amounts of the bodies which constitute it. It is of a dark brown colour and thick consistence, somewhat like treacle, but not so sticky, and when pure it does not drop with the thread-like attenuated drops observable with treacle. In the bulk it appears to be black, but pressed into a thin film between two plates of glass it appears to be of a dark orange-brown and quite transparent.

It possesses a smoky balsamic odour, which becomes very agreeable when dropped on paper and warmed. Its fragrance is increased and somewhat changed when dropped on a red-hot coal, by reason of the decomposition of the inodorous resin it contains.

After long exposure to the air it remains unaltered and does not deposit crystals. It is insoluble in water, but yields up to it a small quantity of cinnamic acid and a trace of benzoic acid. Six or eight parts of crystallized carbonate of sodium are required to neutralize 100 parts of the balsam. It is only slightly soluble in dilute alcohol, benzene, ether, essential or fixed oils, and quite insoluble in petroleum ether. It dissolves easily in cold acetic acid, anhydrous acetone, absolute alcohol, and chloroform. The peculiarity of the process employed in the preparation of Balsam Peru accounts for its containing substances which are not found in the Balsam of Tolu, which is extracted in a more natural way from *Myroxylon Toluifera* ; the result being that these two drugs possess very different properties although produced by trees so very nearly akin that Professor Baillon * considered them specifically identical.

Pure balsam Peru does not diminish in volume when shaken with an equal bulk of water. Three parts of balsam mix easily with one part of carbon disulphide, but further addition of this last causes a brown flocculent precipitate of resin. When thrice its weight of carbon disulphide is added, a black coherent mass of resin is precipitated, amounting sometimes to 38 per cent. of the balsam, and attaches itself firmly to the glass. The carbon solution then appears as a perfectly transparent, slightly brown liquid which, when decanted and evaporated, leaves a brown thick aromatic liquid having a sp. gr. of 1·1 ; this is cinnameïn (or benzyl

---

* " Sur les caractères spécifiques des Toluifera," Bull. de la Soc. Linn. de Paris, 1874, p. 7, also Rép. de Pharm. n. s. i. p. 566.

cinnamate), $C_{16}H_{14}O_2$, as may be proved by the fact of its conversion into benzyl alcohol, $C_7H_{14}O_2$, and cinnamic acid by the action of caustic alkalies. It amounts to about 60 per cent. of the balsam. Cinnameïn can also be separated from the balsam by distillation, but with more difficulty, owing to its high boiling-point, about 305° C. Cinnameïn is a thick liquid, miscible in alcohol and ether, and not congealing at $-12°$ C. By boiling it suffers slight decomposition. By exposure to the air it slowly acquires an acid reaction ; submitted to the prolonged action of potash, especially in alcoholic solution, potassium cinnamate crystallizes out, and the oily liquid remaining consists of a mixture of benzyl alcohol and toluene called *Peruvin*,.C_7H_8O (so named by Frémy). Benzyl cinnamate can be prepared artificially by heating sodium cinnamate with benzyl chloride, and so obtained it forms crystals which melt at 39° C. and boil between 225° and 235°, so differing considerably from cinnameïn ; in fact it has been thought that cinnameïn contains, besides benzyl-cinnamate, benzyl-benzoate, cinnamyl-cinnamate, and some free benzyl-alcohol *. Both benzyl-benzoate and benzyl-cinnamate also occur in Tolu balsam, and benzyl-alcohol is found in liquid storax † and in oil of cherry-laurel ‡. Cinnamyl-cinnamate is also called cinnyl-cinnamate, and is identical with styracin $C_{36}H_{32}O_4$ or meta-cinnameïn $C_{18}H_{16}O_2$, which is contained in liquid storax, and which, although crystallizing in long rectangular prisms which melt at 38° C., frequently solidifies in a form which is not crystalline, or only crystallizes after remaining for a considerable time in a liquid, oily state.

By treatment with a concentrated solution of potash, styracin is decomposed into a cinnamate of potash and *Styrone*, $C_9H_{10}O$ §, having an odour of hyacinths.

The resin separated from the balsam as above mentioned by carbon disulphide consists of an amorphous, black, brittle mass which does not possess the characteristic odour of the balsam. It is soluble in caustic alkalies and in alcohol. It can be purified from its alcoholic solution by charcoal. It reddens litmus and yields an abundant precipitate on the addition of an alcoholic

* Ann. Chem. Pharm. clii. p. 131.
† Ibid. clxiv. p. 289.
‡ Pharm. Journ. [3] v. p. 761.
§ See p. 247.

solution of neutral acetate of lead. Kachler, in 1869, found that by melting this resin with potash, about $\frac{2}{3}$ of its weight of proto-catechuic acid was obtained. By destructive distillation it yields benzoic acid, styrol, and toluene.

From the researches above recorded it may be concluded that the balsam contains one-third more of a resin and probably nearly two-thirds of benzyl cinnamate, which undergoes some modification consequent on the method employed to obtain the balsam—a method which is certainly the cause of the presence of free acids and a black colouring-matter in the balsam. The balsam also contains 3 or 4 per cent. of cinnamic acid. The proportions of these constituents vary somewhat (but only to a small extent) in different parcels of balsam, and are probably accompanied by small quantities of other bodies such as benzylic alcohol, styracin, and benzoic acid.

The thick consistence and dark brown colour of the balsam render sophistication very easy. The principal adulterants are castor oil, purified storax, copaiba balsam, an alcoholic solution of benzoin brought to the consistence of a balsam, and a similar solution of colophony. The specific gravity is a very important criterion of unsophisticated balsam. All the substances above mentioned are lighter than the true balsam, the sp. gr. of which at 15° C. varies, as above stated, between 1·140 and 1·145, or perhaps 1·138 may be taken as the extreme minimum and 1·147 as the extreme maximum; therefore, as soon as the sp. gr. of a sample is found to be below 1·140, and certainly when below 1·138, the article becomes suspicious. The sp. gr. of castor oil varies between 0·95 and 0·97, and that of copaiba balsam between 0·94 and 0·99. The sp. gr. of purified storax, obtained as a brown transparent balsam by extracting liquid storax with alcoholic ether and evaporating the clear filtrate, was determined by Schlickum * as 1·093, that of the colophony solution as 1·016, and that of the benzoin solution as 1·080; therefore an addition of one of them would markedly lower its sp. gr. The plan re-commended by Hager, of observing whether a drop of balsam floats or sinks in a saline solution of known density, is considered to be defective, as the behaviour of the drop is affected by the conditions under which it reaches the surface of the solution.

In estimating the purity of this balsam, Professor Flückiger

* Archiv. der Pharm. [3] xx. p. 498.

remarks * that such physical observations as those above mentioned (sp. gr., consistency, stickiness, etc.) should first be made, and draws particular attention to the fact that adulterated samples exhibit a *thread-like attenuated drop*, which is presumed to mean a similarity to the manner in which treacle will drop. The chemical properties of the chief constituents should then be considered, and remarks on the adaptation of cinnameïn to this purpose, which may be obtained with the greatest facility, though not perfectly pure, when the balsam is shaken with three times its weight of carbon disulphide. The latter becomes only slightly coloured when a pure balsam is employed, while the adulterated often yields a very dark coloured solution, but he yet considers the amount of cinnameïn present altogether too variable or its relation to carbon disulphide, and too much affected by the possible admixtures to admit of its quantitative estimation. The cinnameïn may be obtained more pure by means of the lower-boiling fractions of petroleum. This solution is almost entirely colourless, and leaves, after the evaporation of the petroleum, as is known, a very fragrant cinnameïn, amounting to about half the weight of the balsam, thus far less than by the application of carbon disulphide. A petroleum boiling at 50° to 70° C. may be very well adapted; and the yield must be determined by testing numerous samples.

Besides the amount of cinnameïn, the amount of *resin* of Peru balsam may be also employed as a test. This can be separated, as above mentioned, by means of carbon disulphide or petroleum, and amounts in a pure balsam to more than one-third, or to about two-fifths; most of the adulterants will have the effect of decreasing the weight of the resin separated by the solvent, and inversely to increase the amount of the portion taken up by that solvent; *i. e.*, apparently to furnish more cinnameïn. As the cinnameïn and resin are determined in the same operation, the same objections apply to the latter as to the quantitative estimation of the cinnameïn. In a like manner the property of this resin, on the other hand, of not being rapidly attacked by alkalies is of value.

The important article by Professor Flückiger above quoted from adds that the free acid which occurs in the balsam, chiefly cinnamic acid, offers a point of attack which the German Pharmacopœia has already made use of, although in a manner which

---

* Pharm. Zeitung, 1881, p. 222.

leaves room for doubt; for in what manner can it be determined that " 1000 parts of balsam are neutralized by 75 parts of sodium carbonate " ? The execution of this experiment is not quite so simple as it would appear in this laconic requirement. Should the balsam be boiled with the finely-powdered carbonate, the action of the same aided by means of water, or must, inversely, the balsam be diluted with alcohol ? Good Peru balsam was boiled for a day with an excess of sodium carbonate and 10 times its weight of alcohol (sp. gr. 0·830) in a flask provided with an inverted condenser; 91·6 parts of sodium carbonate were required for 1000 parts of balsam. It is probable that hereby, finally, not only the free acid is combined with the sodium, but also that a decomposition of the cinnamic ether or cinnameïn begins. It appears therefore more advisable to extract the free cinnamic acid by means of lime, in that it may be accepted that the latter is without action on the compound cinnamic ether. If, for example, 50 parts of balsam are boiled for two hours with a mixture consisting of 20 parts of lime and 500 parts of water, the evaporated water restored, the boiling mixture filtered, and the mass twice washed, employing each time 200 parts of hot water, the cinnamate of calcium is thus obtained in solution. This is evaporated to 200 parts (whereby it becomes more and more of a yellowish colour, developing a coumarin odour, which resembles the odour of the Balsam Peru pods), and, after super-saturation with hydrochloric acid, is placed for some hours in the cold, whereupon the separated cinnamic acid is collected, after draining pressed between bibulous paper, first dried by exposure to the air and finally on the water-bath. When prepared from pure balsam the acid consists of loose, not smeary, somewhat brownish crystals, the weight of which amounts from 2 to 4 per cent. Adulterated balsams yield, according to the nature of the admixture, a much less pure cinnamic acid, or they give much more or less than 3 to 4 per cent. of acid. That cinnamic acid is obtained is manifest from the fact that it requires for solution 100 parts of boiling water, while benzoic acid dissolves at 100° C. in 15 parts of water. Upon cooling the hot, saturated, aqueous solution, the cinnamic acid is, for the most part, again separated. If 2 parts of the crystals, purified in this manner, are shaken in a flask with 1 part of potassium permanganate and 20 parts of luke-warm water, a strong odour of bitter-almond oil is developed, the cinnamic acid yielding benzaldehyde.

The free cinnamic acid can thus be employed qualitatively and quantitatively as a criterion of the purity of Peru balsam, but too much importance should not be attached thereto. This acid is not to be regarded like cinnameïn as an active constituent, and, on the other hand, does not occur to such an amount in the balsam as to be regarded, like the black resin, as a peculiarly indicative portion of the mixture. As the free acid always amounts to but a few per cent., the percentage amount would be but little changed even by a large adulteration, except in so far as benzoin is concerned, in which case readily-perceptible large amounts of benzoic or cinnamic acids would be introduced. A large admixture of storax, on the contrary, produced no correspondingly increased yield of cinnamic acid. In comparison with most materials which are adapted to its adulteration, the somewhat slighter tendency of Peru balsam to decomposition by the action of alkalies appears to be of service. This peculiarity was indicated by Dr. Grote, as he * recommended 3 to 6 drops of the balsam (about a quarter cubic centimetre) to be shaken with 2 to 3 cubic centimetres of ammonia, sp. gr. 0·960, or, according to the relations by weight, 2 parts of balsam with about 17 parts of ammonia. The free acid passes into solution, and of the remaining constituents only a small amount is emulsionized, while the chief portion is not at all further changed. From the pure balsam, after one day, a turbid liquid may be decanted, while the residue remains semi-liquid, or very soft. But little is here dependent upon the proportions, as it was found that the balsam shows still the same behaviour when it is shaken with only half its weight of ammonia. The action of ammonia upon adulterated balsams, however, is quite different; they solidify after a short time, as Dr. Grote has shown, to a stiff jelly from which no liquid can be decanted, or they become perfectly hard. Dr. Grote has, however, already indicated that it is chiefly colophony which may be detected in this manner, and that other admixtures, on the contrary, such as benzoin, storax, copaiba, and "gardschan" balsam ("wood oil") cannot be recognized by means of ammonia. The same is applicable to the fatty oils. If a balsam containing fat is shaken with carbon disulphide the latter is taken up by the solvent, whereby an (apparent) increase of the cinnameïn will appear. If this is saponified with alcoholic soda and carbonic

---

* Pharm. Centralh. May, 1880, p. 179.

acid gas then passed through the liquid in order to remove the excess of soda as carbonate, the filtrate will contain, besides sodium cinnamate, also the sodium salts of the fatty acids, and these latter would then be separated by boiling water from the acidulated solution of their salts, and from the cinnamic and benzoic acids.

By boiling the balsam with milk of lime there is extracted therefrom, as the above experiments show, for the most part simply cinnamic acid, and upon the filter there remains a soft friable mass. Slaked lime in a dry condition exerts the same slight action. If 2 parts of the balsam are triturated with 1 part of slaked lime, the properties of the mixture are changed no more than would be expected; a smeary, or, at all events, a *soft, kneadable or somewhat friable, readily divisible mass* is obtained, which, even after long exposure in the water-bath, does not harden. Adulterated specimens, on the contrary, furnish *very hard, no longer kneadable masses* when rubbed with half their weight of slaked lime. The same behaviour is shown by specimens of balsam to which storax, evaporated alcoholic solution of benzoin, colophony and copaiba are added in amounts of 10 per cent. or more. In every case the adulterated balsam solidifies with the lime. This lime test appears therefore to be of constant value, and in its simplicity leaves nothing to be desired. In concluding his observations on this test, Professor Flückiger remarks that Dr. Grote, who has treated the subject minutely and successfully, quite agrees with him as regards the action of slaked lime, and that, as a test, it may be exacted that " 10 *drops of Peru balsam shall furnish with ·4 gram of slaked lime a mixture which remains soft.*" This test, however, is not effectual if castor oil be present (or other fatty oil), but on warming such a mixture with the lime the fatty odour is plainly perceptible, if not, a very small amount of fat is added, and upon ignition decomposition products of the castor oil are formed, which possess a peculiar odour.

The results of the examination of Peru balsam made by Schlickum[*] differ materially in several respects from those of

[*] Archiv. der Pharm. [3] xx. p. 498 ; and Pharm. Journ. [3] xiii. p. 321.

Professor Flückiger; for instance, in the 'Pharmacographia' and in the Professor's article above quoted from, 38 per cent. of resin is stated as the amount contained in the balsam. Schlickum states the resinous residue obtainable by the same method (treatment with three times its volume of carbon disulphide) to be 16 per cent. at most, and that an admixture of benzoin increases the insoluble proportion. Mac Ewan, in a paper read before the British Pharmaceutical Conference, 1884*, corroborates Schlickum's observation as regards the maximum being 16 per cent., and adds, " It is apparent that the United States Pharmacopœia has fixed upon too high a maximum which might be profitably amended, since the test affords a good indication of the presence of benzoin." The maximum fixed by the U.S.P. is 40 *per cent.* Muter states the percentage at 38 ; Attfield 33 as maximum. The results of experiments made by Mac Ewan differ in some respects in a marked manner from those of Schlickum, and they also disagree with Professor Flückiger's " Lime test " † and Dr. Grote's modification thereof; consequently great doubt arises as to the purity of the samples operated upon by Messrs. Mac Ewan and Schlickum, and the mind has a tendency to rely on the investigations of the author of the 'Pharmacographia,' at least until further researches are made with samples of undoubted purity and known age.

The following process is recommended by Messrs. Gehe & Co. as convenient for determining with certainty the presence or absence of benzoin or storax in Peru balsam :—

Five grams of balsam, five grams of soda solution (sp. gr. 1·160), and ten grams of water mixed together are shaken with two successive quantities of ether of 15 grams each, and the ether poured off as completely as possible. The residue is heated to boiling and acidulated with hydrochloric acid; cold water is then added, and the resin which separates is removed, dissolved in about 3 grams of soda solution, diluted with 20 grams of water, heated to boiling, and precipitated with solution of barium chloride. This precipitate is drained on a filter, dried on a water-bath, extracted with alcohol, the alcohol extract evaporated, then taken up with

* Pharm. Journ. [3] xv. p. 238.
† Pharm. Zeitung, 1881, p. 222 ; and Pharm. Journ. [3] xii. p. 45.

concentrated sulphuric acid, chloroform added and the whole shaken. In the presence of benzoin or storax, the chloroform is coloured violet to blue. It is stated that by this method even small admixtures of these substances can be detected with certainty, and it is affirmed that although, like the petroleum spirit and nitric acid test of the German Pharmacopœia, it is dependent upon a colour reaction, it is more certain in its indications.

The method of detecting castor oil proposed by Wagner is to expose a small portion of the suspected balsam to distillation until somewhat more than one half has passed; to shake the distillate with baryta water, to remove by means of a pipette the layer of oil floating on the surface, and to shake this with a concentrated solution of sodium bisulphite. If castor oil be present the liquid will immediately become a crystalline mass *.

The United States Dispensatory gives the following test :—If one volume of the balsam be triturated with two volumes of sulphuric acid, a tough, homogeneous cherry-red mixture should result. If this be washed after a few minutes with cold water, it should be converted into a resinous mass which is brittle when cold.

A mixture of three parts of the balsam with one part of carbon disulphide remains clear, but a mixture of one part of the balsam with three parts of carbon disulphide separates from the balsam about 40 per cent. of resin. The liquid poured off from the latter should be transparent, should not have a deeper colour than light brownish, and should not exhibit more than a faint fluorescence. When distilled with 200 times its weight of water, no volatile oil should pass over.

There sometimes exudes spontaneously from the trees a gum resin of a feebly bitter taste and entirely devoid of balsamic odour; Attfield's analysis of this substance shows it to contain 77·4 per cent. of non-aromatic resin, without a trace of cinnamic acid, and entirely different from the Balsam Peru. The leaves of the tree contain a fragrant oil.

" White Balsam Peru," " Balsamo Blanco," or " Baume Blanc de Son Sonaté " is obtained by pressure from the fruit-pods of *Myroxylon Pereiræ*. It has no similarity whatever to the balsam obtained from the trunk of the tree. In appearance it resembles

* Am. Journ. Pharm. xxx. p. 570.

turbid honey, of a yellowish-white colour and of an odour of melilot or coumarin. It is but slightly soluble in cold alcohol, but more so in ether, which leaves on evaporation a matter which is more of the nature of a wax than of a resin. By treating this balsam with hot alcohol, Stenhouse * extracted a neutral, resinous, colourless substance, easily crystallizable, to which he gave the name of Myroxocarpine $C_{24}H_{34}O_3$.

These fruit-pods of the Balsam Peru tree also yield by distillation with water an almost colourless essential oil, with a sweet odour that recalls the fragrance of a field of beans in blossom †. By exposure to the air the odour becomes slightly altered, probably by oxidation of the oil, and approaches that of cedar wood. This oil is not entirely soluble in rectified spirit, a white precipitate settling down after a few days and leaving the supernatant liquid quite clear. The odour is quite different from that of either Peru or Tolu balsam, and is not exactly like any known perfume.

Other balsamic exudations from trees belonging to the genus *Myroxylon* are noticed by pharmacological writers; but they are not dealt in commercially in Europe, are rarely imported, and very imperfectly understood, although well deserving thorough investigation. Amongst such odorous resins may be mentioned that obtained from the *Myroxylon peruiferum*, Mutis & Linn. fil., identical with *Myrospermum peruiferum*, D.C., and *Myrospermum pedicellatum*, Lam. Dict. This tree is figured in Guibourt's Histoire des Drogues, 7ᵉ ed. iii. 472. It grows in Peru as a large tree of 65 centimetres diameter in the trunk, and is locally known as *Quino-quino*. This balsam exudes from incisions made in the trunk during the rainy season. It is preserved in bottles and remains fluid for some years; in this condition it passes under the name of "white liquid balsam," but when it is put up in calabashes, which is a common practice at Carthagena, it soon dries up into a solid resin and is then known as "Dry white balsam" or "Balsam Tolu" by the local druggists. This balsam is quite distinct from the *White Balsam of Son Sonate* above-described.

Balsam Tolu.

A balsam obtained by exudation after incision in the bark of *Myroxylon Toluifera*, syn. *Toluifera Balsamum*, Miller, *Myrosper-*

* Pharm. Journ. [1] x. p. 290.　　　† Ibid. [3] xv. p. 483.

mum Toluiferum, A. Rich., a lofty evergreen tree with a straight
trunk, sometimes rising to a height of 40 feet before branching
(evidencing in this respect a marked difference from the *Myroxylon Pereiræ* yielding Balsam Peru, with which Professor
Baillon argues it to be identical *), and attaining an average height
of about 70 feet.

The Balsam Tolu tree is found in the district of Plata on the
right bank of the Magdalena in New Granada, also in Venezuela,
Equador, and Brazil.

The balsam is obtained by cutting several V-shaped notches into
the bark, at the base of which the wood is a little hollowed out,
so as to allow calabashes of the size and shape of tea-cups to be
fixed. Sometimes as many as twenty calabashes may be seen on
various parts of the same trunk, and the *bleeding* is allowed to go
on for eight months in the year, so that ultimately the trees
become much exhausted and thin in foliage †.

As freshly imported, Balsam Tolu appears as a light brown
resin, and although not fluid or of a sticky surface, is yet sufficiently soft to receive the impression of the finger. It hardens
gradually by age, becoming brittle in cold weather, but softens by
the warmth of the hand. In an attenuated state, spread out in a
thin layer, it is quite transparent and yellowish-brown in colour.
Its perfume is very delicate and agreeable, recalling that of
benzoin and vanilla, and is very diffusible on the application of
warmth or when its alcoholic solution is left to evaporate on paper.
Its taste is feebly aromatic and of scarcely perceptible acidity,
although its alcoholic solution distinctly reddens litmus. Very
old samples, such as those of the last century which are imported
into Europe in small calabashes of the size and shape of an
orange, have become hard, brittle, and pulverulent; showing a
brilliant crystalline fracture of beautiful dark amber colour, and
having a more delicate perfume than recent balsam.

When Balsam Tolu is pressed between two slips of warmed
glass and the film examined with a magnifying-glass, a quantity
of crystals of cinnamic acid are observable.

The balsam is completely soluble in cold acetic acid, alcohol,
chloroform, and solution of caustic potash; it is only imperfectly
soluble in ether, scarcely at all in volatile oils, and quite insoluble

* Pharm. Journ. [3] iv. p. 382.
† Journ. of the Royal Hort. Soc. May 1864.

in benzene or carbon disulphide. Its solution in acetone is inactive to polarized light.

Balsam Tolu is partly constituted of an amorphous resin which is insoluble in carbon disulphide, and which is apparently identical with the black resin precipitated from Balsam Peru by carbon disulphide. This resin yields by distillation a hydrocarbon which was termed by Berzelius Toluol, a name which is still in use on the Continent, but which in England has been changed into Toluene, C_7H_8. It is a strongly refractive liquid smelling like benzene; it boils at 110°·3, and does not solidify at $-20°$; oxidizing agents convert it into benzoic acid.

Balsam Tolu contains a large quantity of cinnamic acid, which may be separated by boiling with water, also with carbon disulphide, and identified by heating with potassium dichromate and sulphuric acid, when benzaldehyde will be given off in quantity. By distillation with water, the balsam yields one per cent. of a volatile oil of sp. gr. 0·945, which consists of a terpene $C_{10}H_{16}$ called Tolene, boiling at about 170° C., and which rapidly absorbs oxygen from the air. This volatile oil of Tolu also contains compound ethers of cinnamic and benzoic acids, and is of very fine perfume.

By destructive distillation, Balsam Tolu yields the same products as Balsam Peru, amongst which Phenol and Styrol have been observed.

This balsam does not contain either cinnameïn or styracin (cinnamyl cinnamate), both of which are found in Balsam Peru.

According to the researches of Kopp there are two distinct resins in Tolu balsam, one $C_{18}H_{19}O_4$, and another sparingly soluble, $C_{18}H_{20}O_5$. According to Deville, however, there is only one resin, to which the second formula belongs. Trommsdorff obtained 88 per cent. of resin, 12 per cent. of free acid, and only 0·2 per cent. of volatile oil.

The officinal description given in the United States Dispensatory states that Balsam Tolu is " a yellowish or brownish yellow, semi-fluid or nearly solid mass, transparent in thin layers, and brittle when cold. . . . It is almost insoluble in water. . . . Warm carbon disulphide removes from the balsam scarcely anything but cinnamic and benzoic acids. On evaporating the disulphide, no substance having the properties of resin should be left behind."

It is stated that when Balsam Tolu is dissolved in the smallest

quantity of solution of potash, it loses its own characteristic odour
and acquires that of the Clove-Pink; also (by Alix) that the
balsam, if heated with sulphuric acid, dissolves without disengage-
ment of sulphurous acid, and yields a cherry-red liquid.

To detect storax or colophony in balsam Tolu, the following
process is recommended by Cripps *:—About 30 grams of the
sample are digested in carbon disulphide for about 15 minutes,
keeping it gently warm by occasional immersion in hot water.
The clear liquid is poured off, evaporated to dryness, and when
cold, sulphuric acid added to dissolve the resinous extract. A
bright red-rose coloration is produced, which in the case of
genuine tolu remains of a distinctly rose hue for some considerable
time. If, however, the sample be adulterated with either storax
or ordinary resin, the rose colour rapidly becomes more brown in
tint. The best way to apply the test is by performing the operation upon a genuine sample by the side of the suspected one. In
this way, a distinct difference in tint can be observed if only one
per cent. of the adulterant be present; with 4 per cent. of colophony, or rather more of storax, the difference in tint can be
readily distinguished without the blank experiment. If to the
sulphuric acid solution a fluid ounce of water be rapidly added, the
colour of the resulting liquid is much duller and paler when
colophony is present than with the pure balsam.

* Pharm. Journ. [3] xix. p. 422.

CHAPTER XII.

OPOPANAX.—BDELLIUM.—MYRRH.—BALSAM OF MECCA.—FRANKINCENSE.

OPOPANAX, BDELLIUM, AND MYRRH.

The gum-resin formerly sold under this name has almost entirely disappeared from commerce. Its botanical origin is doubtful. It has been ascribed to *Opopanax Chironium*, Koch[*], an Umbelliferous plant indigenous to the European shores of the Mediterranean; to *Opopanax Persicum*, Boissier[†], found on Mount Elbrus, in the north of Persia near Passgala, and on the rocky mountains in the district of Kuhkilouyed in Eastern Persia; also to *Diplotaemia cachrydifolia*, which occurs in high mountains extending northwards of Teheran, particularly near Azadbar, a plant used by the Persians as a culinary vegetable and called by them *Djaw-chive*.

Very extensive details of the gum-resins Sagapenum and Opopanax are given in the theses by Przeciszewski, 'Pharmacologische Untersuchungen über Ammoniacum, Sagapenum und Opopanax,' Dorpat, 1861; and Viguier, 'Gommes résines des Ombellifères,' Paris, 1869.

In any case it is a mistake to suppose that the opopanax now used in perfumery is the article which until the last few years was known as opopanax, and which unquestionably may be ranged among the fetid gum-resins, it having an odour like bruised ivy leaves but even more disagreeable[‡].

The oil of opopanax, now extensively used in perfumery, is distilled from the fragrant gum-resin which exudes from the *Balsamodendron Káfal*, Kunth[§]; Syn. *Protium Káfal* (Lindl. Fl.

---

[*] N. Act. Nat. Curies. xiii. p. 96.
[†] Diagn. sér. 2, fasc. 10, p. 36.
[‡] Pharm. Journ. [3] xviii. p. 624.
[§] Gen. Tereb. 16; De Cand. Prodr. ii. p. 76.

Med. p. 169); *Amyris Kafal*, Forsk. The term " perfumed bdellium " would be more appropriate than " opopanax " to distinguish it *.

The *Balsamodendron Káfal* is a native of Arabia, where it is called " Káfal." The tree attains the height of 20 feet, and a very fragrant balsam is obtained from its fruit.

In Dr. Dymock's valuable ' Notes on Myrrh and its allied gum-resins ' †, mention is made of " a perfumed bdellium " being found in small quantities in the bales of " ordinary bdellium " which are shipped from Berbera, on the Somali coast of the Gulf of Aden. It seems very likely that this is the produce of the *Balsamodendron Káfal* of Southern Arabia. Dr. Dymock describes this perfumed bdellium as a kind of " Bissa-bôl," occurring in irregularly shaped pieces, more or less flat, some of them having fragments of thick bark adherent, but not the birch-like bark which adheres to common Bdellium. The colour of the gum is dark reddish brown ; opaque, yellowish-white streaks are frequently met with in the semi-transparent reddish mass which forms the bulk of the drug. The odour, on fresh fracture, is powerful and pleasant. The Arabic name in use by the Somalis is Hábāk-hádee. In Bombay vernacular it is " Bysabōl," and in Sanskrit " Mhaisabol."

Byssabāl is a name also applied in India to ordinary African Bdellium, the produce of *Balsamodendron Kataf*, Kunth, syn. *Amyris Kataf*, Forskal, a native of Arabia Felix. This tree does not attain the height of the " Kafal tree " ; the shape of its leaves is the same, viz., palmately trifoliate and serrated at the apex, but its berry is globose, that of " Kafal " being compressed.

The ordinary African Bissa-bōl bdellium very much resembles myrrh, with which it has been confused by many authors, but it is darker and more reddish than true myrrh, has a stronger acrid taste, and differs also in its peculiar odour ‡. It is but sparingly soluble in bisulphide of carbon, and the solution does not assume the violet shade characteristic of myrrh on the addition of bromine.

The Indian bdellium or false myrrh, from Coromandel, is said

* Pharm. Journ. [3] xxi. p. 838.

† Ibid. [3] vi. p. 641.

‡ Kew Report, 1880, p. 50; Pharmacographia, p. 146; Bentley & Trimen, Med. Pl. p. 60; Dymock, Mat. Med. of Western India, p. 128.

to be the produce of *Boswellia glabra*, and that from the Western Himalaya of the *Boswellia serrata**. This kind of bdellium softens in the hand, and has an acrid taste without the aroma of myrrh. The odour has a faint resemblance to that of cedar. The surface of the pieces frequently has hairs or fragments of a papery bark attached to it. Dr. Roxburgh says that the trunk of this tree is covered with a light coloured pellicle, as in the common birch, which peels off from time to time, exposing to view a smooth green coat, which in succession supplies other similar exfoliations †. It is identical with *Balsamodendron Mukul*, Hooker, which grows in Scinde.

Another bdellium termed "*Googul*," from Bengal, has been attributed to *Balsamodendron Mukul*, Hooker ‡ ; it somewhat resembles Indian bdellium in appearance, but (in the specimen in the Museum of the Pharmaceutical Society) the odour is different and recalls that of Burgundy pitch or castor. From a paper by Dr. Stocks in Hooker's 'Journal of Botany,' i. p. 257, it would appear that this tree is not identical with the *B. Mukul*, which, from the similarity of its native name "Googul," has been mistaken for it.

Dr. Roxburgh § observes, regarding the Bdellium googul, which he attributes to *Amyris comiphora*, that the whole tree, while growing, is very odoriferous, and if broken in any part diffuses around a grateful fragrance like that of the finest myrrh; yet that the juice never congeals, but is carried off by evaporation, leaving little or nothing behind. The googul is collected in the cold season by making incisions in the tree and letting the resin fall to the ground. Googul is used as incense. It is now many years since Dr. Royle remarked that "all the species of this genus require to be examined from good and authentic specimens, accompanied by their respective products, as so much doubt still remains in the opinions of botanists regarding the trees producing these substances." Even now, in 1892, very little seems to have been done to clear up the doubts complained of by Dr. Royle.

There is an "*opaque bdellium*" said to be the produce of *Balsamodendron Playfairii*, Hooker, met with in North-East India

---

* Hall's 'Dict. of the Economic Products of India,' Calcutta, 1889, p. 426.
† Flor. Ind. ii. p. 245.
‡ Cooke, 'Report on Gum Resins in the India Museum,' 1874, p. 72.
§ Fl. Ind. ii. p. 244.

and called in the Somali vernacular "*Hotai.*" It is an opaque, whitish gum-resin, called in Arabic "*Dukh,*" and in the Bombay dialect "*Meena kärma.*" Chemical examination has shown that opaque bdellium and *Hotai* are far from being identical *.

The word Myrrh is derived from a Hebrew word signifying *amer* or *bitter*, the Greek word σμύρνα being derived from the Arabic *Mur.* The ancient Egyptian appellation *Bola* or *Bol,* and the Sanskrit equivalent *Vola,* are still retained in the modern Persian and Indian languages as *Bol* and *Bola,* the true Myrrh being called *Heera-bol.* This gum-resin is secreted between the cortical layers of the *Balsamodendron Myrrha,* Nees. It is a small tree or shrub of low stature, unattractive aspect, rigid, often spiny, with scanty foliage, minute flowers, and small, oval, dry berries. It is a native of the hot and dry countries around the southern extremity of the Red Sea, viz., the country about Ghizan, on the eastern shore of the Red Sea; the southern Arabian coast eastward of Aden; the Somali country south and west of Gardafui; and the region lying between Tajúra and Shoa, including Harar to the south-east.

There are probably at least three distinct species of myrrh tree. Myrrh trees abound on the hills near to Sureea in the territory of the Fadhli tribe, east of Aden. The myrrh there produced is gathered by the Somalis, who cross over from the opposite coast and pay a tribute for the right of gathering it, also the gums bdellium and olibanum. The produce is sold at an annual fair held at Berbera in November, December, and January, and shipped to Bombay, where it is sorted. From this source is derived the "Turkey myrrh" of commerce or "myrrh of the first quality."

The appearance of myrrh as it exudes from the trees is that of an oily liquid, yellowish-white in colour, and of buttery consistence; the colour changes to golden and reddish in the process of drying, and it darkens and loses value by age. On arrival in London it appears in pieces of irregular forms and of variable sizes, either distinct or agglomerated, usually covered with a fine powder or dust. The colour varies from pale reddish-yellow to red and reddish-brown. The pieces are fragile, semi-transparent, with a dull, in part splintery, fatty kind of fracture. The odour of myrrh is aromatic and balsamic, peculiar, but to most persons

* Hall's 'Dict. of Economic Products of India,' i. p. 426.

pleasant. The purest, palest, and most odorous pieces are sold as *picked myrrh*. Sometimes the same chest contains myrrh of all qualities ; it is then termed *myrrh in sorts*.

Myrrh cannot be reduced to fine powder until a part of the essential oil and water which it contains has been dried out of it. On being heated it does not liquefy like resin. Water disintegrates it, forming a light brown emulsion which, under the microscope, is found to be composed of colourless drops mingled with granules of yellow resin. Alcohol dissolves the resin of myrrh and leaves the non-crystalline gummy matter and fragments of bark. On treating myrrh with water, about 40 or 50 per cent. of gum is dissolved out, sometimes as much as 67 per cent.; part of this can be precipitated by neutral acetate of lead, thus differing from gum arabic ; but a part of it, about a quarter, resembles that gum as regards the action of acetate of lead upon it *. African bdellium, which is frequently found in imported parcels of " unpicked myrrh," contains a much smaller proportion of gum soluble in water; according to the analyses of Parker (Pharm. Journ. [3] xi. p. 41) the composition of African bdellium is as follows :—

Soluble in alcohol	15·4
Gum soluble in water	33·2
Gum insoluble in water	37·8
Moisture	13·6

Details of this investigation and comparative results obtained from other varieties of bdellium are given in the paper referred to.

Flückiger and Hanbury found that myrrh yields on distillation 3·4 per cent. of a thick, yellow, neutral oil, sp. gr. 0·988 at 13°. In a tube of 50 millim. it deviated the ray 30·1 to the left. This oil commenced to boil at 266°, and distilled between 270° and 290° C.

The Stacte (σтαктή) often mentioned by the ancients is, according to Pliny, a liquid which exudes spontaneously from the myrrh tree †. Theophrastus ‡ mentions two sorts of myrrh, one liquid and one solid, but no modern drug has been identified with Stacte or the liquid myrrh of the ancients ; whatever it was, it was

---

* Flückiger & Hanbury, Hist. des Drogues, i. p. 272.

† Vincent, 'Commerce of the Ancients,' ii. p. 316.

‡ Lib. ix. c. 4.

obtainable in quantity, as 150 lbs. of it are said to have been offered by an Egyptian city to St. Sylvester at Rome*. It may have been " liquid storax," *Liquidambar orientalis,* Miller.

Balsam of Gilead and Balsam of Mecca.

These oleo-resins were considered to be the produce of two distinct trees, but it is now believed that they are derived from one and the same tree, or that the *Balsamodendron Gileadense,* Kunth †, is only a variety of *Balsamodendron Opobalsamum,* Kunth & Brandis ‡.

The synonymous names under which the trees have been described by other authorities are *Balsamodendron Ehrenbergianum,* Berg § ; *Amyris Gileadensis,* Linn. ‖ ; *Amyris Opobalsamum,* Forsk. ¶ ; *Balessan,* Bruce ** ; *Balsamea,* Gled. †† ; *Protium Gileadense,* Wright & Arnott ‡‡ ; *Balsamea Meccanense,* D.C. §§.

The French translator of the ' Pharmacographia ' remarks :— " To judge by the description and figure given by Berg‖‖ , it was indeed difficult to distinguish between the plants, and our opinion that they were identical was fully confirmed on examining a specimen of the plant sent by Berg to the Museum in Paris, labelled *Balsamodendron Ehrenbergianum,* Berg, and comparing that specimen with that of *B. Opobalsamum* in the same Herbarium."

There are slight variations in the analyses of this plant made by various botanists, but generally it is described as a small tree of 14 to 20 feet high, with very short, slender, divergent, purplish branches ; sparsely furnished with small, alternate, palmately trifoliate, glabrous leaves, and blunt entire leaflets. The small whitish flowers appear three together on separate stalks, which are shorter than the leaf-stalks. The flowers are of separate sexes, with a 4-toothed permanent calyx ; the corolla has 4 fleshy petals inserted under an annular disc, provided with 8 glands. The 8

---

* Vignolius, ' Liber Pontificalis,' 1724, i. p. 95.
† Gen. Tereb. p. 16; D.C. Prodr. ii. p. 76.
‡ For. Fl. p. 65; and Ann. Sc. Nat. ser. 1, 1824, ii. p. 348.
§ Reg. Bot. Zeit. 1862, p. 163.
‖ Mant. p. 6; and Vahl. Symb. i. p. 28, t. 11.
¶ Descr. p. 79 ; and Linn. Amn. vii. p. 68.
** Travels, Fr. ed. t. 25.
†† Act. Soc. Cur. Nat. Berl. iii. p. 127.
‡‡ Prodr. i. p. 177.　　　　　　　　　　§§ Prodr. ii. p. 76.
‖‖ Darstell. und Beschreibung . . . offizin. Gewächse, iv. t. 32.

stamens are inserted under the annular, cup-shaped disc. The ovary is sessile, surmounted by a very short blunt style and a quadrilobed stigma. The berry is oval, smooth, containing a solitary seed.

Guibourt finds that the second variety (*B. Opobalsamum*) only differs from the first named (*B. Gileadense*) in that its leaves are composed of one or two pairs of sessile leaflets, with an odd one *.

These trees are very rare and difficult to cultivate. They have gradually disappeared from the countries where they were formerly known to have grown. Thus, from Judea, where according to Theophrastus, Dioscorides, Pliny, Justin, and Strabo, they were in ancient times grown, they have, since many years, completely disappeared. In Egypt, where they were either obtained from Judea or Arabia, they were cultivated between the 11th and 16th or 17th centuries, in a place near Cairo called Matriya. This garden of Matriya was 7 " feddans," or more than 9 arpents, in extent †; it was enclosed by walls and guarded by Janisaries. Abd-ul-Latif, who lived from 1161 to 1231, described the extraction of the balsam at the garden of Matriya near Cairo. He says that incisions are made through the bark down to the wood, the juice is scraped from the tree and preserved in bottles, which are buried in the earth for a time, and afterwards exposed to the sun until the balsam has separated from the impurities; it is then subjected to some secret process, after which it is stored in the King's Treasury.

At the time of the visit of Bélon to Cairo in 1550, notwithstanding that the trees in this garden had many times been renewed by importations from Mecca, there remained only 9 or 10 trees, almost leafless, and no longer yielding balsam. The last tree died in 1615, through an inundation of the Nile. Therefore it is no longer Egypt or Judea which furnishes balsam of Mecca, but rather Arabia Felix, in the environs of Medina and Mecca, where the tree grows naturally, and where it has never ceased to exist. It is stated that the tree grows near Bederhunin, a village between Mecca and Medina, in a sandy, rocky soil, confined to a tract about a mile in length.

Strabo alone, of all the ancient writers, has given an account of

---

* Hist. des Drogues, iii. p. 506.

† 'Relation de l'Egypte,' traduite par Sylvestre de Sacy, Paris, 1810.

the place of its origin. "Near to this," he says, "is the most happy land of Sabeans, Saba, Sheba, or Arabia Felix, and they are a very great people. Among them frankincense, myrrh, and cinnamon grow, and in the coast that is about Saba, the balsam also, among the myrrh trees behind Azab; all along the coasts, to the Straits of Babelmandeb, is its native country." We need not doubt that it was early transplanted into Arabia, that is, into the south part of Arabia Felix, immediately fronting Azab. The first plantation that succeeded seems to have been at Petra, the ancient metropolis of Arabia, now called Beder or Beder .Huncin. (Hall states, 'Economic Plants of India,' Calcutta, 1889, that it is found on both sides of the Red Sea, south of 22° N. lat., and is also recorded from several places on the Nubian coast and in Abyssinia. It is met with on the Asiatic side of Ghizandad in Arabia, at Aden and Yemen.) Afterwards, being transplanted into Palestine, it obtained the name of *Balsamum Judaicum* and *Balm of Gilead*, and became an article of commerce there.

According to Bruce, the tree is only 5 or 6 feet high, branching much, with the aspect of a standard cherry tree, having red branches and white flowers, but according to other observers the tree attains a height of 14 to 20 feet.

In the beginning of April the trees drop their juice from gashes which are made in the smaller branches into vessels set under them to receive it. A tree will not yield more than 10 to 15 drachms in one season. Another account says the wound in the tree is made in July or August, when the juice is in its strongest circulation; it is then received into small earthen bottles, and every day's produce is poured into a larger vessel, which is kept closely corked. The Balsam of Judea appears to be the same balsam adulterated. An inferior sort of balsam is prepared by boiling the twigs in a quantity of water; the balsamic matter rises to the surface and is skimmed off. After they have procured all they can, it is said that they increase the fire, and a large quantity of thicker balsam rises, which is preserved separately and is principally the description which is sent to Europe. The other can only be obtained as presents, and that which naturally distils from the trees hardly being sufficient to supply the Seraglio and great officers of State, there is none of it sent out of the country. This in Europe is never obtained genuine. Guibourt mentions having obtained it pure, but says it is of great rarity, and that it is put up in square leaden bottles containing about 250 grammes, but is, as a rule, adulterated, and many chemists even sell Chian turpentine or Canada balsam instead of it. Oil of sesamum is likewise used as an adulterant.

According to Prosper Alpinius * the true balsam is at first turbid and white, of very strong pungent smell, like that of turpentine, but much sweeter, with a bitter, acrid, astringent taste ; upon being kept for some time it becomes thin and limpid, of a greenish hue, then of a golden yellow, and at length of the colour of honey. A drop of the balsam, when let fall into a vessel containing water, at first sinks to a certain depth, and then rising to the surface, *instantly* and *completely* spreads out into a thin nebulous film, which under a magnifying-glass presents the appearance of an infinite number of little globules uniformly spread over the entire surface of the water. After about a quarter of an hour this film solidifies, by reason of the rapid evaporation of its volatile oil; it can then be lifted out entire on the point of a pencil. This invariable property was first observed by Prosper Alpinius † ; it is of great exactness and is one of the best tests of the purity of the balsam. It was confirmed by Guibourt on experimenting with a sample which he had reason to believe was quite pure, and with which sample he made the following further observations :—The balsam thickens with age, becomes more brown in colour, and does not rise so quickly and spread out on the surface of water; a natural result of resinification and loss of volatile oil. When rubbed on the hand, the balsam rapidly loses its volatile oil and becomes very sticky. Dropped on paper it spreads a little, but does not penetrate or render the paper transparent. After twelve hours of exposure to the air, the drop of balsam becomes so adhesive that on doubling the paper in two it becomes difficult to separate the sheet without tearing it. On being triturated with an eighth of its weight of calcined magnesia it does not solidify as do the turpentines of pines and several balsams. Five grammes of balsam mixed with 30 grammes of 90 per cent. alcohol form a liquid white as milk, which only becomes transparent after being left standing for 8 or 10 days ; a glutinous matter being then deposited which consists of insoluble resin. This resin

* ‘ Dialogue du Baume,’ traduction d’Antoine Collin, Lyon, 1819.
† Ibid. p. 61.

T

readily dries on paper without penetrating it and without rendering it transparent.

In Bombay the balsam is known as *Balsán-Ka-tel*. It is imported from Arabia under the Arabic designation *Duhnul-balasán*. When freshly imported it is a greenish-yellow turbid oleo-resin of the consistency of honey, of powerful pleasant odour, somewhat like rosemary. The balsam is in high esteem among the Eastern nations as a medicine, as an odoriferous unguent, and as a cosmetic.

In the dialects of Persia and Bombay the wood of this tree is called *Ude-i-balasán,* the word *ūde* being pronounced broadly as *aood*, which approaches the English word " wood " when pronounced slowly. This wood, known to the druggists as Xylobalsamum, consists of small branches about 16 centimètres in length and as thick as a pen, marked with small alternate excrescences which are the remnants of the secondary branches which bore the flowers. The bark is of a reddish brown, marked with regular longitudinal streaks; the wood is white, hard, of very feeble perfume, and devoid of taste by reason of age. Guibourt mentions having also met with, at the native druggists, small tips of branches 11 to 14 millimètres in length and of not more than 2 millimètres in thickness, covered with a rough reddish bark transversely striated; this substance was of an aromatic, rather bitter taste, and of a sweet agreeable odour when perceived in bulk. On being pinched by the hand it developed a strong odour similar to that of rosemary.

The fruit or berries of the tree, called in Persian *Tukhme-i-Balasán,* and in the Bombay and Arabic dialects *Habul-Balasán,* are also imported into India and kept by the native druggists. They are known as *Carpobalsamum.* They are of a greyish-red colour, about the size of a pea, pointed at the ends. The kernel is oily and of agreeable aromatic taste. Dr. Dymock reports * having compared these berries with the figures and descriptions given in Bentley and Trimen's ' Medicinal Plants ' (t. 59) and considers there can be no doubt of their identity. If soaked in water they soften, so that they can be easily dissected, and the remarkable form of the pulpy layer within the epicarp is seen. Sections of the epicarp show very large ramifying balsam cells, which appear to communicate one with another.

* Pharm. Journ. [3] viii. p. 104.

Both the wood and the fruit are used medicinally by Yunani Hakims of India.

FRANKINCENSE OR OLIBANUM.

Records of the employment of Frankincense in the celebration of religious services date back into the remotest antiquity, and it is remarkable that a controversy has been going on for ages concerning the identity of the trees yielding it.

" Frankincense " is largely imported into London under the name of Gum Olibanum, and is used principally for compounding incense for burning in the Roman Catholic and Greek Churches. The Greek word λίβανος, the Latin *Olibanum*, the Arabic Lubân, and analogous words in other languages are all derived from the Hebrew *Lebonah*, which signifies *milk*, in allusion to the sap of the trees, which, before becoming dry by exposure to the air, has the appearance of milk. Under the name of *Ju-siang*, meaning *perfume of milk*, this drug was imported into China from Arabia as far back as the tenth century, and is still imported to a large extent at Shanghai to this day.

Even at present the trees yielding the olibanum of commerce are still imperfectly known, but it is believed that this gum-resin is obtained by incision and exudation from the stems of various species of *Boswellia*, natives of hot, arid districts on the mountains of Hadramaut along the south-eastern coast of Arabia, and of the opposite shore of North-eastern Africa, on the limestone mountains which extend westward from Cape Gardafui through the country of the Soumalis. The bulk of the commercial olibanum is probably derived from *Boswellia Carterii*, Birdwood, called *Mohr Madow* by the Soumalis, and *Maghrayt d' Sheehaz* by the Maharas in Arabia, this species being presumed to be found in both countries. Also from *Boswellia Bhau Dhajiana*, Birdwood, called *Mohr-Add* by the Soumalis. The arguments on this subject are gone into at considerable length by Dr. Birdwood in the ' Transactions of the Linnean Society,' xxvii. p. 143, botanical descriptions and figures of the trees being given, also those of the tree yielding a variety called Lubân Meyeti *, which is not found in European commerce. In the London Drug Market olibanum is

* Sometimes spelt Maitee.

characterized as "African" and "Indian," which are misleading
terms, as both are produced from the above-mentioned trees in
Southern Arabia and North-eastern Africa, but as it is partly
shipped direct, and partly shipped first to Bombay before being
forwarded to London, the mistaken idea may have arisen that
some of it was produced in India, and was thought to be yielded
by the *Boswellia serrata* of Colebrooke, but it is now known that
although this tree furnishes an aromatic gum-resin used as incense
in India, it is not found in any notable quantity in any European
market. Therefore the "Indian" and the "African" olibanum
do not differ in their origin and nature, but only in their quality.

The description given by Vaughan * on the subject of African
Olibanum is as follows :—"The Lubân tree is a native of the
eastern coast of Africa and flourishes on the highlands which
intersect the whole of the Soumali country, where I have had an
opportunity of seeing it not far from Cape Gardafui. The hill-
ranges on the eastern coast of Africa are composed entirely of
white limestone, in some parts so compact as to resemble alabaster ;
this appears to be the soil most genial to the tree, and in no
instance did I find it growing in sand or loam. The tree is first
met with at a few miles inland from the coast, and at an altitude of
about 300 feet above the level of the sea. Its appearance is strikingly
singular, seeming at first to be destitute of roots and clinging to
the hard uncreviced rock by masses of a rhomboidal and fantas-
tically shaped wood, with the most obstinate adherence. The
stem is nearly at right angles with this substance, ascending
almost invariably in an upright direction, and attaining the height
of from 12 to 15 feet. At the base the circumference is equal to
that of a man's thigh, gradually tapering towards the top, where
it shoots off its branches and leaves. The wood is white, fibrous,
and somewhat soft, the bark is of a light brown colour, very
succulent, and covered with a glossy cuticle. This usually bursts
or cracks with the natural increment of the tree, and may then be
removed in cutaneous flakes, when it presents an appearance not
unlike that of prepared oil-paper, and something akin to a similar
coating observable on the English birch. The old and decayed
portions of the tree assume a cinereous hue, whereby they are
easily distinguishable from the younger and more healthy plants.
At the proper season incisions are made in the stem, from which

* Pharm. Journ. 1853, xii. p. 228.

the juice flows forthwith in a copious stream (frequently covering the entire stem), until the wounds are closed by the desiccation of the fluid into a gum. In this state the trees glisten in their rich investiture, and, as if vexed at being prevented from pouring forth all their store, the bark distends from the abundance of the sap within. After the juice is inspissated and dried by the action of the atmosphere and the sun, it is scraped off the trees and the ground beneath and collected by the natives, who store it in large loose heaps at particular places on the sea-coast. It is then packed in sheep and goat skins and transported on camels to the great fair held in Berbera, from whence it is shipped in native vessels to Aden and other parts on the Arabian coast. It is, however, sometimes purchased by the Banians and sent direct to the Bombay market. The olibanum of commerce appears in oblong, pear-shaped, or rounded tears, of a yellow or reddish colour; always covered externally with a fine white dust, and even where that is wiped off the tears appear translucent, milky, and semi-opaque, but the finest quality is almost colourless and of a greenish tint. When rubbed in a mortar with water it forms a white emulsion. It softens between the teeth, producing an aromatic, slightly rough taste. Its odour is agreeably aromatic, but is only developed by exposing the substance to a high temper - ature. It is but partially soluble in water and alcohol. It melts with difficulty and imperfectly when heated. At 100° C. it softens without melting, and on the temperature being raised it begins to decompose. It burns with a bright white flame on the approach of a taper."

According to Braconnet * it contains 56 per cent. resin soluble in alcohol (but insoluble in alkalies), and 8 per cent. of essential oil. The chemical constitution of the resin agrees nearly with the formula $C_{40}H_{30}O_6$. The residue, about 33 per cent., was found by Hekmejer † to consist of a gum identical with gum arabic.

By distilling this gum resin with water in a cast-iron retort, an oil is obtained constituting about 7 per cent. of the resin taken alcohol extracts from the residue about 72 per cent. of resin, and the remainder is gum. The crude oily distillate boils between 160° and 170°, and contains oxygen, as Stenhouse previously

* Ann. Chim. Phys. [2] lviii. p. 60.
† Jahresb. 1858, p. 482.

observed *. By fractional distillation a terpene, known as *olibene*, boiling at 156°–158° and sp. gr. 0·862 at 12°, is obtained, together with a small quantity of an oxidized substance boiling above 175° and not yielding a hydrocarbon when treated with sodium. Olibene is identical with pinene; it is soluble in alcohol and ether, and is resinified by nitric acid; it absorbs 1 molecule of hydrogen chloride, giving on standing a crystalline camphor-like body, $C_{10}H_{16}HCl$, which melts at 127°.

The resin, exhausted as above by alcohol, melts rapidly and yields by dry distillation traces of an organic acid, and an oil which boils above 360°, and contains less oxygen than the resin itself †.

The *Boswellia serrata* (*B. glabra*), Roxburgh ‡, synonymous with *B. thurifera*, Colebrooke §, above referred to as being a native of mountainous parts of India, and producing a gum-resin consumed there, is figured by Roxburgh ‖ and described by Royle as found at Oude and Rohilcund; also at Behar by Hooker, and at Kattyawar by Birdwood. Colebrooke found it between Sône and Nagpur, on the route by which he travelled to Berar.

The natives of India recognize the two varieties of this plant, *foliola ovato-oblonga* and *foliola lineari-lanceolata*, of which Roxburgh made two species, and distinguish between their gum-resins. Birdwood remarks that from his observations the gum-resin has been either stalactiform, like the runnings of a wax candle, or in small tears, and always so soft that when kept in a bottle, in a short time, it ran into an oleo-resinous mass, with the smell of frankincense, but more turpentiny. He adds that he often tried to get regular tears of olibanum from this plant, but never succeeded in getting anything else than soft, oleo-resinous " runnings " from it, which even after months' exposure on the trunk still remained quite soft. In Khandeish the olibanum produced by this plant is sold under the name of Dup-Salai (*i.e.* incense of Salai) in the village bazaars.

As regards the " Lubân Meyeti " above-mentioned, it has been proved to be the produce of a strongly marked species of *Boswellia*,

* Ann. Ch. Pharm. xxxv. p. 306.
† Kurbatow, Zeitsch. f. Chem. [2] vii. p. 201.
‡ Flor. Ind. ii. p. 383.
§ 'Asiatic Researches,' ix. p. 377, tab. 5.
‖ Cor. ii. tab. 207.

viz., the *B. Frereana,* Birdwood. Kempthorne thus describes it * :—" The tree is one of the most extraordinary plants I ever saw, quite a *lusus naturæ* of the vegetable world, for the trees actually grow out of the sides of the almost polished rocks. . . . The trees were about 40 feet high, the stem was about 2 feet in circumference, rising straight up, with a bend outwards of 6 or 7 inches. They are attached most firmly to the rocks by a thick oval mass of substance, about a foot or so in diameter, something resembling a mixture of lime and mortar. Branches spring out rather scantily at the top and extend a few feet down the stem ; the leaves are 5 inches or so long and 1½ broad, narrowing and rounding towards the point, but not serrated at the edges; the upper surface is of a rich dark shining green, while the lower is of a lighter hue ; they are thin and smooth, and crimped like that beautiful species of seaweed so often found on the coast of England. The tree has four layers of bark, the outer being coarse and loose, like that of the beech, while the next two are as it were glued to the trunk and delicately fine, resembling oiled paper or gold-beaters' skin, and of a bright amber colour ; this bark is perfectly transparent, and can be stripped off easily in large sheets; the natives use it for writing on. The inner bark of all is an inch or so in thickness, adhering closely to the stem ; it is tough, not unlike leather, striped red and white, and yields a strong aromatic perfume. The timber is white, soft, porous, and of little use except as firewood. A deep incision into the bark causes the odoriferous resin to exude in large quantities, which is of a milky white and of the consistency of honey ; but it soon hardens by exposure to the atmosphere. It is a remarkable fact that not a single frankincense-tree did I perceive growing upon any other rocks than those of almost pure limestone."

Dr. Birdwood concludes his description of this tree as follows † : " As I saw this plant in Playfair's garden at Aden in September last, in young leaf and covered with bloom, I was much struck by its elegant singularity. The long racemes of green star-like flowers, tipped with the red anthers of the stamens (like aigrettes of little stars of emerald set with minute rubies), droop gracefully over the clusters of glossy glaucous leaves ; and every part of the

* Trans. Bombay Geo. Society, xiii. 1857.
† Trans. Linn. Soc. xxvii. p. 148.

plant,—bark, leaves, and flowers,—gives out the most refreshing lemon-like fragrance."

The Lubân Meyeti is collected chiefly by the Abardagahala tribe of Soumalis. The season for piercing the trees from which it is produced is during the north-east monsoon in the months of July and August.

This oleo-resin is shipped from the ports east of Karam, 45° 41' E. long., to Egypt, Trieste, the Red Sea ports, and Bombay, a small portion only being consigned to the United Kingdom.

The external appearance of Lubân Meyeti of the finest description is on the whole very different from the various sorts of olibanum ; its bulk being not constituted of separate tears but of stalactitic masses.

It is moreover different inasmuch as it is an oleo-resin, not a gum-resin. At a meeting of the Pharmaceutical Society, Nov. 1876, attention was drawn by the Curator of the Museum to a fine specimen showing the peculiar papery bark on its under surface, and distinguished from other varieties of olibanum by a peculiar whitish efflorescence on its surface and stratified opaque white layers in its interior.

West African Frankincense.

The tree furnishing this product is the *Daniellia thurifera,* Bennett. It is plentiful in the peninsula of Sierra Leone and circumjacent regions. In Sierra Leone it is called the " Bungo tree." The mountainous districts to the westward of Freetown, and the wooded slopes in the neighbourhood of York, Lumley, and Gooderich, are the localities in which it principally abounds, although it has been observed on the banks of the Sherbro and other adjoining rivers.

The frankincense-tree grows to a large size, and may be distinguished without difficulty by the erect and stately trunk and beautiful foliage. When of advanced age its recognition is rendered still more certain by the peculiar grey or ash-like colour of the bark and massive divergent branches, which expand into a mass of foliage at an altitude of fifty or sixty feet from the ground to a considerable distance around. During the early years of growth, the young plant has the bark of a deep brown, which changes gradually in colour as it enlarges in magnitude. When of moderate size the entire circumference of the trunk is studded by a series of horizontal excrescences or oblong elevations in the cortical covering, of a pale or yellowish brown hue, appearing in dense parallel but disconnected strata, a quarter of an inch or more apart, and varying from one sixth to one inch in length and about one or two lines in breadth. Being of a lighter tint than the surrounding portions they are distinctly perceptible, and answer as a diagnostic peculiarity to identify the lesser shoots. As the tree approaches maturity, these elevated projections proportionately diminish, and the cortex, while partially retaining its smoothness, becomes traced by irregular patches of white or grey. In the course of time these patches enlarge to such an extent as to embrace in many instances the entire surface of the exterior. The inner cortical layers in plants of an immature development present a peculiar fibrous character, are delicately organized, and may be peeled off in smooth ribbon-like layers, which cannot be effected in the older specimens.

The leaves are bi-pinnate, of a pale green pervaded by a greyish tint that also characterizes the trunk. In the younger productions they are of larger development, but the pinnæ are less numerous than in those of a later growth.

The gum, when a natural exudation, mostly appears in a liquid state, of a white or pale straw-colour, in some seasons oozing so copiously from the branches that the ground and shrubs beneath are from successive excretions thickly covered with white spots. This effusion, however, does not occur so abundantly from the cortex, and when so produced appears in thin and shallow layers, that mark their course by whitish streaks which, after their exsiccation on the trunk, present all the aspects of a saline efflorescence.

The frankincense-tree is subject to the attack of a certain insect termed by the natives *Tumbo*, which deeply perforates the bark in various directions. Its progress is attended by long and sinuous passages, the woody débris from which is ejected externally by a circular orifice about an inch in diameter. In the course of a few days the gummy liquid issues largely from this aperture, blended with minute ligneous particles, which in their transit through these excavations acquire a ruddy or brown tint, by degrees accumulating in small masses and falling to the earth. In this state they become converted into dark brown fragments after a short interval, and are then gathered by the negro women and children who resort to the woods with the express purpose of collecting them. Another

mode of procuring the gum consists in stripping the dead or un-
sound bark from the wood, the more decayed portions of which are
commonly saturated with the gummy exudation, and are found
amalgamated with the woody fibre beneath in black crusts.

Two kinds of frankincense are found in the market of Freetown, both of which are evidently the product of the same tree.
The first can be partly recognized by the dark brown or black
shining irregular fragments, and from apparently having a larger
amount of gum blended with the woody fibre than the other,
which is met with in smaller and less compact pieces, more friable,
and of a lighter brown or yellow tint, being chiefly constituted of
white woody particles cemented into masses by the excreted gum.
Of these two varieties the latter is the least valued. The incense-
like fragrance of these woody resinous excretions renders them
available for a variety of uses, as a perfume and for fumigation.
The native females triturate the gum with lime manufactured from
sea-shells between two purposely adapted stones, and rub their
bodies with the fine powder so produced. The bark is endowed with
similar odoriferous properties in a lesser degree. The ignition of
the gum is rapid; a bright yellow flame, attended with a black
carbonaceous smoke, resulting. This is followed by the deposition
of a viscid oleaginous matter and the evolution of a remarkably
aromatic and semi-resinous odour, approximating to that produced
from the common pastilles, for which they would form a substitute
for use in the sick chamber or for perfuming the atmosphere of
apartments.

The gum is sold by the native traders under the designation
Bungo, the same name as applied to the tree. The above information is abstracted from a paper by Dr. Daniell contributed to
the ' Pharmaceutical Journal' [1] xiv. p. 400. A complete
botanical description of this tree, which is of the Order *Leguminosæ,* suborder *Cæsalpinieæ,* is given by Bennett at page 251 of the
same volume.

CHAPTER XIII.

LIGN-ALOES.—PATCHOULI.—VETIVER.—CAMEL GRASS.— HENNAH.

ORIENTAL LIGN-ALOES.

AN enquiry into the history of Lign-aloes shows that an odoriferous substance bearing the name of *Ahalim* or *Ahaloth* was known to the ancient Jews [*], which in the Greek was called ἀλόης [†], also at a later date " Agallochon " by the Greeks and Romans. The Arabs corrupted the term into Agha-lúkhi, Agulugin, and Yelunjooj, but subsequently adopted the terms Ood or Aúd, meaning " wood," and Aúd-hindi, " Indian wood," as technical names for Aloes-wood. In Sanskrit it is called Agaru, from which is derived the Western Hindi name Agar, Aggur, etc. In Persian, it is Owd-hindee; in Malay, Garu ; in Siamese, Nwahmi; in Chinese, Niaw-cha ; and in Portuguese, Poa d'Agila. It is known in English as Eagle-wood, Aloe-wood, Oriental Lign-aloes, and by the old commercial and pharmacological names Lignum Aquilæ, Agallochum, and Agallage.

Indian Lign-aloes or " Eagle-wood " is obtained from the *Aquilaria Agallocha* identified and described by Roxburgh [‡]. It is a native of the mountainous districts to the east and south-east of Silhet (especially on the Jaintiya Hills), the most easterly province of Bengal, in about lat. 24°–25° N. The order *Aquilarineæ* only comprises three (supposed) species of the genus *Aquilaria*,

---

[*] Ps. xlv. 8 ; Prov. vii. 17 ; Song of Sol. iv. 14.
[†] John xix. 39. ἀλό s, meaning " wood of " aloe.
[‡] Fl. Ind. ii. p. 422 ; Trans. Linn. Soc. xxi. p. 199, where it is figured ; and De Cand. Prodr. xiv. p. 601.

inhabitants of tropical South-west Asia, the Malay Peninsula and Archipelago, and Borneo. On the Jaintiya Hills the *A. Agallocha* attains a height of 100 feet, with a trunk about 12 feet in circumference. In Assam it is found even of still larger size. The tree is found both in sandy and clayey soils ; both in the plains and on the sides and tops of the hills. The trunk and branches are generally crooked. The wood is white, very light, soft and porous like deal, but inodorous, nearly tasteless, and quite useless, except that part containing the perfume.

The wood is chiefly cut at one period of the year, viz. the dry season ; the collection is described as a precarious and tedious business, as few trees contain the perfume, and such as do have it partially distributed in the trunk and branches. The people employed in this business proceed two or three days' journey among the hills, jungles, and mountains, and without discrimination cut down the trees as they are found, young, old, and withered, but the latter are generally preferred (the trees are known on the spot by the Bengal name *Tuggur*). They then on the spot search for the *Aggur* or perfumed part, which is done by chopping off the bark, and into the wood until they observe dark-coloured veins, yielding the perfume which guides them to the place containing the *aggur*, which generally extends but a short way through the centre of the trunk or branch. In this manner they search through the whole tree and bring away only such portions as contain the oil, or have the smell of it. Neither root, leaves, nor bark yield any oil. The formation of the *aggur* is really the result of disease, sometimes occurring where the tree has at some previous time been wounded or injured by a branch being broken off. To expedite this condition pieces of the wood are often buried in moist ground to decay, and afterwards dug up. Parts which have undergone this change become oily, heavy, and black. They occur in fragments of various sizes and shapes, and when cut out are tested by being thrown into water ; those parts which sink are the most valuable and are called *Gharki*. Any portions of unmellowed wood are carefully separated from it. Specimens which sink but partially are termed *Nimgharki* or *Samaleh-i-aala*, and those which float *Semleh* or dregs, and are the least esteemed. This fragrant substance is of various sorts, distinguished by the names Aúd Hindi Agar, which is the darkest ; the Samadúrí, named from the district in which it is produced, having a more unctuous appearance than the first named; the *Kumarí* and the *Mandalí,* both also named after districts; the Kumarí being of a lighter colour and the Mandalí (or Gomandalí) the most fragrant of all.

The names *Barí* and *Jabali* have also been used to identify certain pale sorts marked with black lines or streaks.

The oil or " Uttur," called also " Agar-atar " and " Agar-ká-itr," is a perfume much admired in India. The process of extraction consists in bruising the wood in a mortar, or rasping it, macerating it in water and then distilling it. It is sometimes adulterated by adding raspings of santal-wood to the still, and some fraudulent dealers sell lumps of the Agar after the oil has been distilled from it.

Dr. Dymock states * that the varieties of Agar found in the Bombay market are three :—the Siam or Mawurdhee, the Singapore, and the Gargulee, also a wood resembling aloes in appearance called *Tagger* (probably obtained from Zanzibar), which is odoriferous, and is used to adulterate the true *Agar.*

Some of the specimens found in commerce in India may be derived from the two other species of *Aquilaria,* as the *A. secundaria,* De C., a native of the Moluccas †, it only differs from the *A. Agallocha* of Roxburgh in that its leaves are gradually acuminated and not abruptly so ; also, although the *A. Malaccensis* is defined by Lamarck ‡ as distinct from the other two species of *Aquilaria,* Roxburgh considered it identical with his *A. Agallocha* ; it is a native of Malacca and called *Garo de Malacca,* also *Bois d'aigle.* Cavanilles describes and gives a figure of the Garo de Malacca in his seventh Dissertation, p. 377, t. 224, under the name *Aquilaria ovata,* which is continued by Willdenow in his edition of the ' Species Plantarum of Linnæus,' vol. ii. p. 629. His description differs very little from that of Lamarck, and his figures are considered by Roxburgh to agree with his *A. Agallocha* ; but whether the three species be distinct or not, the perfumed substance is yielded by all of them.

Gamble says § that " *Akyan* (the Burmese name for Agallocha)

. * Mat. Med. Ind. pp. 239–241.
† Prodr. ii. p. 559, and Rumph. Amb. ii. tab. 10.
‡ Encycl. i. p. 49.
§ ' List of Trees and Shrubs of the Darjeeling District,' 1878, and ' Manual of Indian Timbers,' 1881.

is the most important produce of the forests of South Tenasserim and the Mergui Archipelago." Another writer (Kurz) says "the wood is very light, yellowish white, coarse, fibrous, and scentless, but closely-grained and takes a pale brown polish," and that "the fragrant wood ' Ood ' is also largely used for making jewel-cases, and indeed precious stones are frequently set in it. It is also used for making ornaments and rosary beads." It is stated to be worth about £30 per cwt. for 1st quality (Sumatran) ; £20, 2nd quality (Malaccan); and £2 10s., 3rd quality (Malaccan and Indian). It should melt like wax when fresh and emit an agreeable odour.

It is used as incense. Reduced to powder, mixed with cedar dust and clay, it is manufactured into joss-sticks.

CHINESE LIGN-ALOES.

The finest Lign-aloes is produced by the *Aloexylon*, Loureiro * (*Cynometra agallocha*, Spreng. †). It is a native of Cochin-China, on the highest mountains, and of the Molucca Islands.

The genus *Aloexylon*, described by Loureiro and said to include but this one species, is not very well known, and is rather doubtfully referred to the natural order " Leguminosæ." It is described as being of about 60 feet in height, with erect branches, and simple, alternate, lanceolate, stalked, entire leaves and terminal flowers. The legume is described as woody, smooth, falcate, and 1-seeded. The account given by Loureiro of the nature and production of Lign-aloes by this tree in the 'Memorias de la Academia Real das Sciencias,' i. pp. 402–415, presents a striking similarity in many respects to the tree yielding Agar in the vicinity of Silhet, viz. :— smoothness and fibrous texture of the bark (of which paper is made in both countries), shape, texture, and appearance of the leaves, in the want of odour and taste in every part of the tree except the part yielding the drug itself ; in the wood being light, white, porous, etc. Loureiro's description being considered incomplete or inconclusive, his genus was set aside by some writers of botanical text-books, as by Bentham and Hooker in their ' Genera Plantarum.' Roxburgh placed but little confidence in Loureiro's description of the parts of fructification, as he

---

* Flor. Cochinchinensis, p. 267 ; De C. Prodr. ii. p. 518.
† Syst. ii. p. 327.

(Loureiro) acknowledges, in Willdenow's edition of his 'Flora Cochinchinensis,' to have only once seen a mutilated branch of the tree in flower, "which, by long carriage, had the petals, anthers, and stigma much bruised and torn." As not much credence can be placed in the natural character of a plant written under such circumstances, and as the natives of Cochin-China may have supplied him with the fruit of some other tree for that of his *Aloexylum*, Roxburgh, in his description of the genus *Aquilaria*, was inclined to think that the trees producing the aloes-wood of Cochin-China and the Agar from the vicinity of Silhet were the same. On the other hand, there is the opinion of the celebrated botanist De Candolle, who classed the *Aloexylon Agallochum* in Leguminosæ. In any case it is an acknowledged fact that this perfumed wood is found in the greatest perfection in the mountainous country to the east of the Gulf of Siam, including Camboja and Cochin-China between the 8th and 14th degrees of N. lat., and that the quality found near Silhet in Bengal is inferior to it.

The Malayan name specifically for the wood of *Aloexylon Agallochum* is *Kalambak*.

The perfume extracted from the swellings produced on the tree where branches have been broken off is known in Cochin-China under the name of *Tramtoc*. A tree producing Lign-aloes has also been found on the island of Hainan.

Large forests of trees yielding "Kalambak" are found in Campar, on the eastern side of Sumatra, and opposite to Malacca.

A wood having a strong resemblance in perfume to Kalambak or Agallocha (and according to Rumphius hardly distinguishable from it) is yielded by the *Excœcaria Agallocha*, L., a small tree belonging to the family of the *Euphorbiaceæ*, found along the coast of Burmah, from Chittagong to Tenasserim. Its name *Excœcaria* is due to the fact that it contains an acid lactescent sap, which the labourer has to beware of in cutting the wood, as if he happens to get a drop of it in the eyes it is apt to produce blindness. The colour of the wood is of a rusty brown. It is hard and brittle as glass, very bitter, very resinous, and very inflammable. In comparing this to Agallocha, Rumphius evidently referred to the perfume alone, as genuine fresh Agallocha is soft enough to be easily scratched by the thumb-nail, although hardening by age.

MEXICAN LIGN-ALOE.

This wood is derived from an entirely different plant to any which yield the Oriental Lign-aloes, and has not the slightest claim to be regarded as the Lign-aloes of the Bible.

The first importation of the wood into London, in 1869, was in the form of squared logs, consisting of a central portion of irregular outline, and of a pale ferruginous-brown, surrounded with wavy, darker, band-like markings, the contiguous outer portion being of a dull iron-grey. In Guibourt's Hist. des Drogues, iii. p. 491, it is referred to under the name of "Bois de Citron du Mexique," as being internally white, with very irregular, slightly brownish, longitudinal veins; very light and porous, and having a strong odour of citron.

It has been described as found in abundance in the Misteca, and the meridian of Matamoros, also on the mountains about the valley of Colima.

Although the essential oil of Mexican Lign-aloe has been a commercial article in Europe for many years, and was noticed in the columns of the 'Pharmaceutical Journal' in 1869 by Mr. Collins, the Curator at that time of the Society's Museum, nothing definite was published concerning its botanical source until 1884, when a description of the trees yielding the oil was published by Poisson in the 'Bulletin de l'Assoc. Franç. pour l'avancement des Sciences,' xiii. p. 305, pl. x. (Blois, 1884). The author of that article was led to enquire into the botanical source of the product through seeing specimens of the wood and oil at the Paris Exhibition of 1878, where they were exhibited by Ollivier and Rousseau, of Paris, who obtained specimens of the leaves, flowers, and fruit from their correspondent in Mexico, M. Delpech, in whose honour the tree has been named by M. Poisson. The complete botanical description of this tree, now known as *Bursera Delpechiana*, is reprinted from the French journal above-named, into the 'Pharmaceutical Journal' of London, 14th August, 1887, p. 132, with additional notes by Mr. Holmes, the present Curator of the Society's Museum.

It belongs to a set of species peculiar to Mexico, including *B. Aloexylon*, Engl., and *B. penicillata*, Engl. The tree is of medium height.

According to Delpech the trees are felled by the native Indians in a reckless manner, so that they have almost entirely disappeared from Cuantla Morelos, where they formerly abounded. He states that old trunks afford as much as 10 or 12 per cent. of oil by distillation with steam, costing 20 to 25 francs per kilogram; an inferior oil prepared by the natives being sold at a lower price. The structure of the wood presents the following characters:— " The fibres are of medium length, with the walls only slightly thickened; each is divided transversely by numerous thin walls constituting a kind of ligneous parenchyma, of which the whole wood is formed. On transverse section the fibres are seen to be all of equal thickness, so that it is not easy to distinguish the zones of growth of the wood. The vessels are of large size, with numerous transverse trabeculæ, which on longitudinal section are seen to give a moniliform appearance to the vessels; they are dotted all over, the dots being surrounded with areolæ. The medullary rays are thin, and have two to four courses of cells in thickness. It is chiefly in the fibres and medullary rays that the nearly solid odorous substance occurs. It is of a yellowish resinoid aspect under the microscope, and fills them either wholly or partially. All the fibres, however, do not contain it, and it is most abundant where the wood is streaked with dark veins. This matter is soluble in alcohol, so that the wood treated with spirit becomes transparent under the microscope."

In the early notice of this wood by Collins, above referred to, it is stated that the wood, "and *even the twigs*, are rich in essential oil;" but in Poisson's recent paper, here abstracted, he continues to say that "in the green and healthy state the wood presents the same appearance" (under the microscope) as above detailed, but "without any trace of oil, although at the same time the oil may be perceived in the fruits and bark by rubbing them." In Leon Marchand's memoir on the "Organization of the Burseriaceæ," a somewhat similar occurrence is mentioned. The resinous and perfumed matter of *Balsamodendron Myrrha, B. Africanum,* and *Protium obtusifolium* is localized in the pith of the young branches to some degree, but is abundant in the bark and pericarp of the fruits of these plants.

This oil of Lign-aloe has been examined by Verneuil and Poisson. Their experiments show that the wood cut into shavings readily yields the oil by distillation with steam, 7 to 9 per cent. being thus obtained. Being dried over calcium chloride it distils

U

almost entirely between 189° and 192°, a small quantity of a much less volatile and resinous body remaining in the still. These authorities describe the oil as "an oxygenated body having the formula $2(C_{10}H_8)5H_2O$, this formula answering to that of a hydrate of terebenthene or of an isomer. The oil slowly absorbs oxygen and becomes resinified. It does not combine with sodium bisulphite. The red-brown coloration which it takes under concentrated sulphuric acid is analogous to that which turpentine produces with the same acid." The odour of the oil is likened by Poisson to a mixture of lemon and jasmin, but by others it is thought to resemble bergamot.

The odoriferous constituent of oil of Mexican Lign-aloe was discovered by Semmler[*] and termed *Linalool*. This has recently been isolated by Messrs. Schimmel, who find that it is the principal constituent of the oil, and is the sole bearer of the most delicate odour of the latter which is so valued in perfumery. It is described as an almost colourless liquid, very soluble in alcohol, of sp. gr. 0·878 at 15° C.; boiling-point between 197° and 198° C.; optical rotation +2.

The acetic ester of linalool constitutes about 40 per cent. of oil of bergamot, so it is not surprising that the odour of Lign-aloe has been thought to resemble that of bergamot. Lign-aloe oil also contains a small quantity of Geraniol.

It is difficult to say whether other species of *Bursera* yield this oil or not. Poisson suggests that it is probably obtained also from *Bursera Aloexylon*, Engl. (*Elaphrium Aloexylon*, Schiede).

The new Mexican Pharmacopœia (1884), p. 75, also gives *Amyris linaloe*, La Llave, which is a synonym of *Bursera Aloexylon*, Engl., as the source of the oil. Schlechtendal, however, in 'Linnæa' (1843), xviii. p. 303, remarks that this species has a fennel-like odour.

Several other species of *Bursera* grow in the same district as *B. Delpechiana*, including *B. bicolor*, Engl., *B. Schiedeana*, Engl., and *B. jorullensis*, Engl., but nothing appears to be known about the oil of these trees. Schlechtendal mentions (*loc. cit.*) that *Elaphrium glabrifolium* (*Bursera penicillata*, Engl.) has a strong aromatic odour, and that *Amyris ventricosa* (*Bursera fagaroides*, Engl., var.) has an odour of caraways.

---

[*] Ber. Deutsch. chem. Ges. 1891, xxiv. p. 207.

The Mexican species of the genus appear to be very numerous, and require further examination as to their economic products.

GUIANA AND BRAZILIAN LIGN-ALOE.

An " oil of Lign-aloes " has also been produced in Cayenne from another plant belonging to the Burseriacæ, viz., the *Icica altissima*, Aublet* (*Amyris altissima*, Willd.†). The *Icica* is a genus of Amyridaceæ, or Burseriaceæ, found chiefly in the tropics of the Western hemisphere; only two or three of the twenty species occurring in the Eastern. In the forests of Guiana the *Icica altissima* attains a height of 100 feet. The odour of the wood is compared to that of rose and citron, and on this account it is used for the inside fittings of houses, book-cases, etc. There are two varieties of the tree, called in the French colony " Cèdre Blanc " and " Cèdre rouge." The wood is also known as " Bois de Rose femelle," also " Licari Kanali." An examination made of the oil by Morin in 1881‡ seems to indicate its identity with the main constituent of oil of bergamot, and its close affinity to oil of Mexican Lign-aloe above described. When freed from a small quantity of water that it usually contains, it has a density of 0·868 at 15° C., boils at 198° C. under a pressure of 775 mm., and is optically lævogyre. Its composition is $C_{10}H_{18}O$, and it is soluble in alcohol and ether. (This alcohol and its acetic ester form the principal constituents of oil of lavender.)

The balsam obtained from the trunks of many of the species of *Icica* is highly odoriferous, and is commonly used as a perfume in South America. That of *I. heptaphylla*, Aublet §, a native of the woods of Guiana, is called " l'arbre d'encens." It is the *Amyris ambrosiaca* of Willdenow‖. It is also called *Hyawa* in Guinea. The fruit of this tree contains four stones wrapped up in a viscid pulp which has a balsamic smell and taste, hardens into a grey resin, and is used to burn as a perfume. The whole tree is very sweet scented and exudes a very odorous, clear balsam from the wounded trunk or branches which is used in houses and churches as incense. This tree is very abundant on the banks of the Maroni River, and is also found plentifully along the Itoori-

---

* Guian. i. tab. 132.   † Spec. ii. p. 336.
‡ Comptes Rendus, xcii. p. 998.
§ Guian. i. p. 337, tab. 130.   ‖ Spec. ii. p. 335.

bisci Creek of the Essequibo River, in loose, sandy soil; its wood is little used, as it decays rapidly on exposure to the weather. The average height of the tree is about 50 feet. The Caribbee name for this is "Arouaou." Another wood of strong aromatic scent, resembling the "Hyawa," also found along the Itoori-bisci Creek, Essequibo River, and growing plentifully in the loose, sandy soil, is locally called *Oulu*; its average height is about 90 feet. This produces a gum resembling "Hyawa," but in much smaller quantities*. Its wood is the colour of pale cedar, and should be useful for drawers and shelves of wardrobes. The timber will square 16 to 18 inches.

The *Icica Icicariba*, D.C.†, is a native of Brazil, where it is called "Icicariba" and "Resina Icica." It produces a sweet-scented gum which is used as incense. The *Icica decandra*, Aublet‡, is a native of the woods of Guiana, where it is called *Chipa*. It is the *Amyris decandra* of Willdenow §, and is synonymous with *I. pentandra*, Aublet ‖. When the bark of this tree is wounded a whitish liquid flows out, which has a scent of citrons, when dry becoming a yellow transparent resin, which is found in small pieces under the bark. The resin is carried by the Caribbees to Cayenne, where it is employed in churches as incense.

The balsams obtained from some of the species of *Icica* remain fluid for a considerable time, but ultimately harden, and are then used for burning in the churches. Some of the trees are so highly charged with resin that the branches can be used as torches.

Other aromatic woods from British Guiana, derived from trees not botanically identified, are locally known as "Keritee" or "Kretti" from the Aroua-pia-kooroo Creek, Pomeroon River; a wood which is plentiful in some localities. It has a strong aromatic scent, is light, and in colour and appearance resembles satin-wood. Its average height is 80 feet, and its timber will square 20 inches. The "Yellow Cirouballi" or "Sirua-balli," also from the Aroua-pia-kooroo Creek of the Pomeroon River, is light, of bright yellow colour, and strong aromatic scent. It often grows to a very large size, averaging in height 60 feet, and prefers a sandy soil. Its bark is useful for tanning.

---

* Catalogue of Exhibits from British Guiana to Paris Exhibition, 1878.
† Prodr. ii. p. 77.   ‡ Guian. i. p. 346.   § Spec. ii. p. 335.
‖ Guian. i. p. 135.

Patchouli.

Pogostemon Patchouli, Pelletier-Sautelet, Pharm. Journ. [1] viii. p. 574, with figure.

Pogostemon Patchouli, var. *suavis,* W. J. Hooker, Hooker's Journ. of Bot. and Kew Mis. i. p. 328, tab. 11, and Hooker's Flor. of Brit. Ind. iv. p. 634 *.

A labiate plant cultivated at Silhet, Penang, the Straits Settlements, Java, Island of Bourbon, and Mauritius.

In appearance the plant much resembles a *Coleus.* It grows 2 or 3 feet high, and sometimes higher; the branches are obtusely 4-cornered. The leaves juicy, somewhat fleshy, and covered, especially on the inferior surface, with a soft pallid pubescence; they are opposite, petioled, broadly ovate, obtuse, at the base wedge-shaped and shortly attenuated; rather acutely, unequally duplicato-dentate, somewhat serrate, the teeth being obtusely serrated, green above, the under surface pallid, strong smelling, with innumerable small glands visible by the aid of a powerful lens. The leaves measure 2 to 4 inches. All the young parts are densely villous. The under surface of the leaf has a very thick rib and nerves, and largely reticulated veins. The large stems are round and woody, and when cut transversely show the pith surrounded by a thick layer of wood remarkable for the distinct medullary rays. The inodorous flowers are minutely described by Pelletier-Sautelet.

The generally accepted name, "*Pogostemon Patchouli,*" was given by Pelletier-Sautelet †. Bentham was of opinion that this plant was identical with, or not really specifically distinct from, his *Pogostemon intermedius* ‡, of Silhet, Penang, and the opposite shore of the Malay Peninsula, or from *P. parviflorus,* of Silhet, Assam, and Saharunpur, or even from *P. Heyneanum,* Benth., which Drury describes as "probably merely a variety with larger spikes, and more drooping in habit," and says that it

---

* Also Journal de Pharmacie, 1826, xii. p. 261. Pharm. Journ. [1] iv. p. 80; vi. p. 432; ix. p. 382; [3] iv. p. 362; and xi. pp. 409, 813.

† Mém. de la Soc. Roy. des Sciences d'Orléans, v. no. 6, 1845. Bentham in De C. Prodr. xii. p. 153.

‡ Wall. Cat. p. 2327.

is found wild in the Concans, also that it is probably Rheede's synonym " *Cottam* " *.

Apparently there are *several* varieties of Patchouli. The minute botanical description given by Pelletier-Sautelet agrees, as regards the leaf-structure and habit, with the plant now cultivated commercially, and so does the description given by Sir William Hooker, but both of these authorities figure and describe the *flower* of the plant. The plant as cultivated on Fisher's estate in province Wellesley *does not flower*, neither does the cultivated variety grown on another estate near Singapore.

Fig. 9.

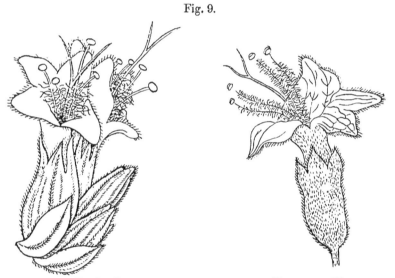

Pogostemon Patchouli.   Pogostemon Heyneanum.
Both highly magnified.

The Curator of the Government Museum, Perak, states, in a recent communication to the ' Journal of the Agricultural Society of India 't, that " Patchouli is a very shy flowerer, so much so that by the natives it is said never to flower, and Mr. Hardouin told me that though he had grown and bought it for the last thirty years, he had never seen or heard of such a thing as a flower." Thus, it is evident that under cultivation it does not flower, but is propagated by cuttings ; yet, in its natural state of growth, it

* Hort. Mal. x. tab. 77 ; and Wallich, Plant. As. Rar. i. t. 31 ; D.C. Prod. xii. p. 153 ; Wight, Icon. t. 1440.

† Reproduced in ' Kew Bulletin,' June 1889.

must flower and drop seed, or else, by its nature and manner of growth, become extinct. Fisher mentions a flowering variety (see under).

A plant introduced from Penang into the Botanic Garden at

Fig. 10.

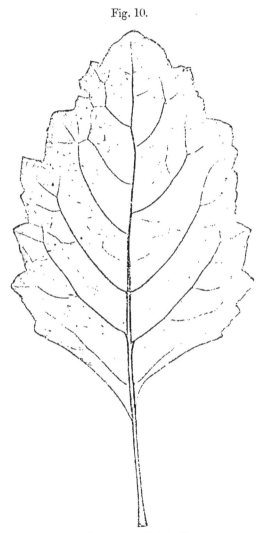

Pogostemon Patchouli.

Calcutta did not exhibit during ten years any disposition to blossom. Specimens have been known to flower in the stoves at

Kew and at Orleans, but specimens received from Louis Van Houtte, of Ghent, and carefully grown by myself in a moist stove in London, never attempted to flower, although they otherwise throve exceedingly and agreed in structure of leaf and stalk with the Kew plant, and with Hooker's description. An impression of a leaf is shown in fig. 10.

Professor Oliver, of Kew, expresses the opinion that it is doubtful whether this particular form, which is the economic plant of commerce, is indigenous to any part of India, and Mr. Thiselton Dyer thinks that it may ultimately prove to have originated in China *.

It is grown and much esteemed by the aboriginal tribes of Perak and Pallang, and is found at an altitude of nearly 5000 feet amongst the Sakais of the mountains, at the source of the Pallang River, far away from any Malayan villages, also among the same people in the Bernam, Batang-Padang, and Kinta Districts of Perak, and among the Semangs in Upper Perak and Selama.

The Sekais of Batang-Padang call the plant *Boon kalif*; and as this is not a corruption or derivation of the Malayan name, it indicates its being known to them prior to their coming in contact with the Malays, and points to the conclusion that it is indigenous. The Malay name of the plant at Perak is *Poko nilam*, and the leaves *Doun Nilam*, which slightly differs from the Malay name commonly applied to it in Province Wellesley. Mr. Fisher informs me that there are three sorts of the plant, viz. :—*Dhelum Utan* or *Delam outan*, or Wild Patchouli; *Dhelum Boonga*, a flowering variety; and *Dhelum Wangi*, which he cultivates, the leaves of which are called *Doun Dhelum Wangi*. The word *Doun* in Malay means "leaf;" *Dhelum* is either mis-spelt from the pronunciation, or is a Javanese word meaning the same thing, a "mattress," the Malay and Javanese word for which is *Tilâm*; *Wangi* means "fragrant" and "healthy." The three words jointly mean that the natives stuff their beds and pillows with the fragrant herb, and believe in its health-giving properties. The word *Utan* means "jungle," "forest," "wild," and "unculti-vated," consequently the *Tilâm* or *Dhelum outan* is the uncultivated plant. *Bonga* or *Boonga* in Malay means "flowering," and was also applied by Rumphius to the Cananga tree; he calls it "*Bonga Cananga*."

* Kew Bulletin, March 1888, p. 73.

Mr. Fisher kindly supplied to me the following details of the method of cultivating the plant and preparing the oil as practised on his estate.

The variety selected for cultivation is known locally as "*Dhelum Wangi,*" which was originally obtained from a small island south of Penang, called Rhio (probably one of the Dindings). The soil most suitable is a rather stiff clay, containing only a small percentage of silica. Land of this description is found near the coast (containing traces of marine deposits), and is planted in rows 4 or 5 feet apart. The plants are propagated by cuttings struck in the open air, which, until rooted, are sheltered from the sun by pieces of cocoa-nut shell. The harvest is made in dry weather, and when the sun has drawn up the dew from the leaves; the tops and green parts of the plant are taken off, rejecting all yellow and decayed leaves, and as much as possible the woody stems. The selected parts are then dried in the shade, under large sheds (as the sun would draw out the perfume), and to ensure evenness in drying they are spread on bamboo racks, allowing the air to penetrate from beneath. During this process they are frequently turned over, and when so far dried as to leave just sufficient moisture to permit of a slight fermentation they are piled in heaps and allowed to heat gently; after this they are again spread out and dried—but not to absolute dryness—and are immediately distilled. The addition of about 25 per cent. of the wild herb *Dhelum outan* is said to increase the fragrance of the distillate. The distillation is effected by passing steam, generated in a boiler apart, through the leaves in the stills. The pressure of steam is not allowed to rise above 20 lbs., the yield under these conditions being about $\frac{1}{4}$ oz. per lb. of leaves; by high-pressure steam the yield would be a little increased, but rank in quality. The stills are sometimes jacketed, and, by passing a separate current of steam into the jacket, condensation in the body of the still at the commencement of the operation is avoided. This oil, called "Singapore" oil, is sent to London in cases of 12 bottles, containing 22 ounces in each bottle, labelled with the manufacturer's name, and guaranteed by him to be pure.

The oil distilled from leaves that have been baled has rather an inferior odour, and has not the peculiar olive-brown tint of the Singapore or Penang oils.

It will be observed that Mr. Fisher adds to the still 25 per cent.

of uncultivated plant "outan." On my enquiring of Mr. Wray, of the Government Museum at Perak, a confirmation of the meaning of the Malay word "outan" as applied to this plant, he replied that it means "jungle, or forest" (consequently "wild or uncultivated"), but adds that at Perak "the plant is not known in a wild state." From this I infer that the natives so value the plant that they always cultivate it, even if it be indigenous, as we should treat any useful plant indigenous to England; the only alternative idea is, that it is not indigenous, but a possible introduction from China, and probably] from the neighbourhood of Canton. Dr. Wallich, in the ' Transactions of the Medical and Physical Society of Calcutta,' 1835, says :—" Baron Hügel informs me that he has found a plant growing wild at Canton, which closely resembles that from Penang, cultivated in this Garden " (referring to the Patchouli).

The information furnished by Mr. Wray respecting the cultivation of the plant at Perak has been published in the ' Journal of the Agricultural and Horticultural Society of India,' and reprinted in the ' Kew Bulletin ' of June 1889, as follows :—

" The cultivation of patchouli is carried on almost exclusively by the Chinese in the Straits Settlements. They do not grow it on a large scale, but a man will plant a patch of perhaps ½ an acre or an acre at a time. The land is trenched and thrown up into long beds either 4 feet or 18 inches wide. The former will take 2 rows of plants and the latter only 1. The plants are put 2 feet apart along the rows. The planting should be done in the wet season, and the cuttings, which are about a foot long, require careful shading with leaves until rooted, or they wither and die, the plant being delicate, and very susceptible to the heat of the sun.

" The first cutting of the crop is made in about 6 months after planting, by which time the plants will have reached a height of 2 or 3 feet, and two other cuttings are made from the same plants at intervals of about 6 months. At the end of this time the old roots are dug up, the land re-trenched and manured, and fresh cuttings planted. Both flat and hill lands are suitable to its cultivation, and it seems to flourish best under slight shade, but probably the production of oil is less in that grown under shade than in that grown out in the sun, though the yield of leaf would be greater. It is often planted on new land between coffee, nutmegs, and other permanent crops, and it pays all the expenses of clearing and planting, leaving the permanent crops as clear profit.

"Of natural enemies, one was described to me as a beetle; but, as the young leaves which it is said to attack are dwarfed and deformed rather than eaten, I am inclined to think it is a bug. The older leaves are very much attacked by some insects, probably caterpillars and grasshoppers.

"The plants are cut down near the ground when they have reached a sufficient size, one stalk only being left to each bush. The patchouli is then laid out in the sun to dry in the day-time, and put under cover at night, and on the approach of rain. The time required to dry it varies with the weather, taking from 4 days to a week. When thoroughly dry it is done up into bales and sold either to dealers in the leaves or to distillers. In this state it fetches about $8 per pikul of 133⅓ lbs. The dealers cut it up and separate a great quantity of the larger stalks, and, according to its freedom from these, it is classed as 1st, 2nd, and 3rd quality. The best consists of leaves only and is valued at $30 to $32 per pikul; but, owing to the labour involved, this quality hardly pays to prepare. The 2nd quality is composed of leaves and young shoots, with little of the heavier stalk, and ranges in price from $17 to $20 per pikul. The 3rd quality contains less leaf and more stalk, and fetches about $14 per pikul. The best quality of all would be produced by picking from the plants the leaves and tops of the young shoots, and drying these in the shade, but it is doubtful if it would pay. Prepared in this way, 30 lbs. of green leaves produce 10 lbs. of dried patchouli. The percentage of essential oil in shade-dried leaves is higher than in those which have been exposed for many hours to the full heat of a tropical sun, which in this latitude often goes over 120° F.

"In distilling the oil, the dried patchouli is put into a large copper cylinder fitted with a perforated false bottom, and mounted on trunnions. Through one of these steam enters from a boiler, and is conducted by a tube beneath the false bottom. The remaining trunnion is also hollow, and the steam, after passing through the leaves, passes out by it and into a worm immersed in a tub of water in the ordinary way. The pressure of steam employed is about 10 lbs. per square inch, but it varies with the size of the worm and the temperature of the water used to cool it.

" One pikul of the dried plant just as it is cut yields from 24 to 30 ounces of oil, and a sample free from the heavier stalks yields about double that amount. (Other records show the yield to vary between 1·5 and 4 per cent.) In an ordinary still (boiling the plant with water) not more than one half of the oil can be extracted, the temperature not being high enough to volatilize the whole of it *.

" The green leaves yield little oil, therefore it is necessary that they be dried before distillation.

"The oil is of two varieties, the one being sage-green, and the other the colour of medium-coloured sherry. It is thought that the green oil is produced from young leaves, and the golden-brown from old leaves ; but this is doubtful. Soil and shade may have more to do with the colour than the age of the leaves. Sometimes one colour is in greater demand than the other, but the prices are the same for both. The golden-brown oil has a sp. gr. of ·9580 at 85° F., and the green oil ·9578 at the same temperature.

" The spectrum exhibited by the golden-brown oil is not crossed by any absorption bands. The red, yellow, and green light, as far as the b line, are transmitted with full intensity; but the blue-green from b to F is much absorbed, and beyond the latter line all is complete darkness. The limits of this spectrum in wave-lengths are 7140 to 4165, the oil being contained in a tube ·6 inch in diameter, both daylight and lamplight being used with the same results.

" The green oil gives a spectrum of full intensity from the c line to midway between b and F lines, from which point it shades off gradually and disappears a little before the h line is reached. At the red end it extends beyond the c line, but with reduced intensity as far as to between the A and a lines. In wave-lengths the limits of this spectrum are 7390 to 4130 in daylight. Lamplight gave a greater extension towards the red end, but much less in the violet."

The following information respecting the cultivation of patchouli

* This suggests that a finer quality could be separated by first distilling with water at ordinary pressure, and then extracting a second quality,—the portion with a higher boiling-point,—by passing steam of a certain pressure through the mass, and by carefully regulating the steam a third quality could be obtained, and the plant totally exhausted of oil. Operating in this way on *leaves alone* the first fraction of low boiling-point might be of very superior quality.

at Penang is contained in a letter addressed by Mr. C. Curtis, Assistant Superintendent in the Forest Department of the Straits Settlements, to the Superintendent of the Royal Gardens, Kew, dated February 1888 :—" . . . . It may interest you to know the result of an experiment in cultivating and harvesting one-twentieth of an acre of patchouli in the Experimental Nursery, Penang. Cuttings were put in in January, and the last week in February one twentieth of an acre was planted three feet apart. The soil of the nursery is poor, and the only manure used was wood-ashes.

" On July 21st the whole was cut, and weighed in a green state 449 lbs. After being dried in a cool airy shed for 10 days the weight was 106 lbs. The leaves were then separated from the stems, and each weighed separately, the result being, 'good leaf,' 69 lbs. ; 'refuse,' 37 lbs. Samples were submitted to London Brokers. One valued it at 8*d.* to 10*d.* per lb., the other at 10*d.* to 11*d.*

" The same patch was cut again the first week in January of this year, and the yield and results were approximately the same as in July; so there is no doubt an acre will yield considerably over one ton of 'good leaf' per annum. No special skill is required in its cultivation. I am by no means sure that the system of cutting the whole patch at once is the best system possible. I think if only one side of the rows were cut first, and the remainder after they had commenced to grow again, there would be less exposure of the roots to the sun, and consequently less check to the growth.

" The principal point to be observed in drying is to dry slowly, and not to the point of crispness, otherwise the leaves get broken to powder, and are of less value " *.

Mr. Wray, in the report above referred to, says that large quantities of *Ocimum Basilicum*, L., var. *pilosum*, Benth., known by the Malay name, *Ruku*, are often used as an adulterant to the dried leaves ; and he was told by Hardouin (the principal distiller of patchouli oil in the States) that recently a Chinaman bought

---

* This observation of course applies to the case when the leaves are to be baled for export ; as when they are to be distilled on the spot they are not dried so much as when baled.

the whole of the *Ruku* growing wild in a cocoa-nut plantation in Province Wellesley, and 700 pikuls of this dried herb were collected and taken to Penang, to be used for the adulteration of patchouli. Therefore it is always preferable to a local distiller to buy a crop just as it is cut, as then it is easy to see if it is adulterated or not, but if the leaves are bought it is very tedious to detect the imposition. The *Ruku* leaves are rather whiter, and the stalks smaller and rounder. Seed-vessels are also often mixed with them. The smell of the two herbs is quite different, but if the sample has been baled for some time this would be imperceptible, except as communicating a twang to the general odour of the sample.

Another adulteration is the *Urena lobata,* L., var. *sinuata.* It is called *Perpulut* by the Malays. It is not cultivated, but it is a very common weed all over the Straits Settlements, and is to be had in any quantity for the trouble of collecting it. It is found in considerable quantity in cocoa-nut plantations and waste places near the coast. The leaves when dried are much like those of the herb it is used to adulterate, but, unlike it, are scentless. People who gather it obtain $3 per pikul, dried, for mixing with the patchouli.

A very complete examination of the leaves of patchouli as found in commerce has been made by Dr. Heinrich Paschkis, giving microscopic representations of sections of the true plant, also of the leaves used as adulterants *. He says :—" Even in a superficial examination of the dried leaves considerable differences are observed in the leaf-balls taken from the same sample. Some are of a light wood-brown colour, others dark red-brown, and others greenish coloured ; some have a sparse, others an abundant and even velvety pubescence. These differences become still more clearly apparent when the balls are soaked for a time in water, and the leaves then carefully spread out and smoothed.

" The *true leaf* is broad, ovate, coarsely crenate, dentate, 10 centimètres long, diminishing at the base into a long petiole, light brown, moderately thin, not very abundantly hairy on both sides, with one principal nerve and the secondary nerves forming curves running towards the margin. The microscopic examination reveals in the epidermis of the upper and under sides, deeply indented, mostly elongated flat cells, 0·063 millim. long, and 0·030 millim.

* Zeitschrift of the Austrian Pharm. Assoc. Republished in Pharm. Journ. [3] xi. p. 813.

broad ; among them, in greater number below and fewer above, are stomata, with a single contiguous cell. The epidermal cells of the upper side are *coarsely papillose,* here and there brownish-

Fig. 11.

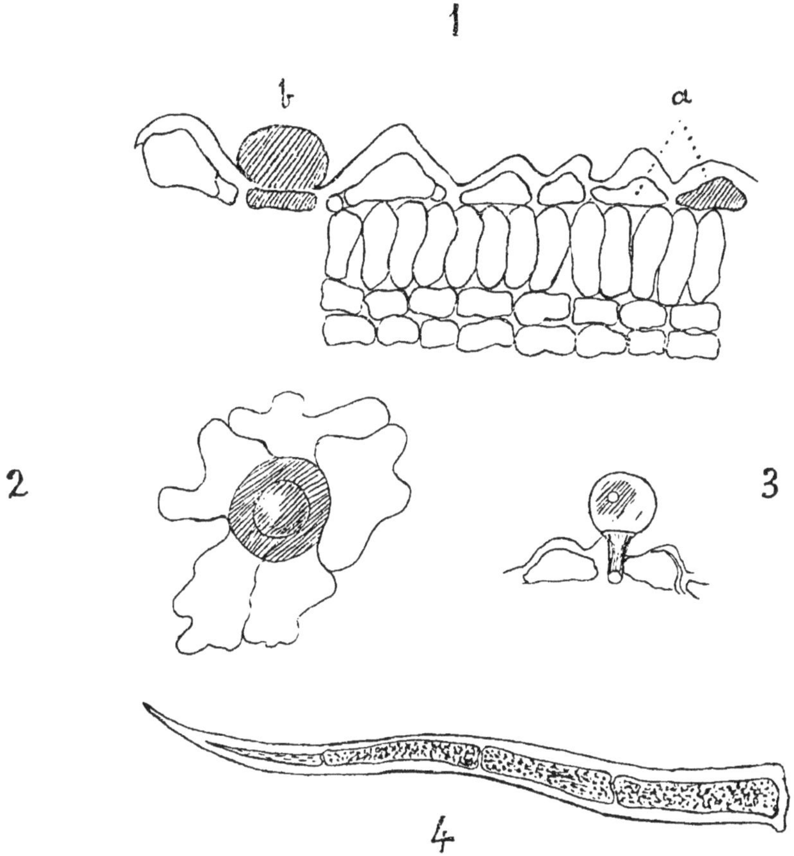

1. Portion of a section of leaf.
    a. Papillose epidermis cells.
    b. Large gland.
2. Large gland from the surface.
3. Small gland.
4. Hair.

coloured. The cuticle is thick. The mesophyll consists of a row of palisade cells, below which is a layer of longitudinally elongated cells. Here and there are rosettes of oxalate of lime. The hairs are *simple throughout,* several-celled (up to 6) ; the cuticle of the hairs is warty, which is especially very clearly perceptible in the younger hairs. On the upper and under sides are numerous glands, one kind bearing a small head upon a short stalk, the other (large glands 0·048 millim. thick) imbedded deeply in the epidermis and stalkless. In the former are drops of a greenish-yellow volatile oil, which occurs also in the hairs, as well as in the cells of the epidermis and the mesophyll. In the mesophyll, in the wall of the epidermis cells, and in the hairs, is a tannin substance, giving a green colour with iron."

Seven forms of leaves used to mix with the true leaves are described and figured by Dr. Paschkis. Observed microscopically, all of them differ from the true leaf in the characteristic structure of the hairs, cells, glands, etc.; and in the general form of the leaf six of them differ entirely in shape from the true leaf ; they also differ in their system of venation, which is radiate from the base of the leaf, and not branching from one main central nerve. Amongst the leaves used as adulterants, the one most nearly approaching in shape and microscopic structure is the *Plectranthus fruticosus* (frequently grown in England for window decoration) but it is easily recognizable. Some of the other sophisticated leaves have been identified, such as *Lavatera olbia* and *Pavonia Weldenii,* besides the *Urena lobata,* var. *sinuata,* and *Ocimum Basilicum,* var. *pilosum,* already referred to. The mixing of these leaves is in many cases carried to such an extent as to constitute 80 per cent. of the drug as found in bales. Dr. Paschkis states in his pamphlet that of the many samples examined by him only one was found to consist of the true leaf alone.

In some of the bales examined in London a quantity of earth and mud has sometimes been found attached to the leaves, and of course adding to the weight.

Under the name of *Puchá Pát,* the leaves are found in every Bazaar throughout Hindustan, and the Arabs use it in great quantities for stuffing mattresses and pillows.

An examination of oil of patchouli was made by Dr. Gladstone in 1864[*], on a sample which he belived to be quite genuine, obtained from Dr. Piesse ; also on a sample obtained from India. " Both specimens were brownish yellow and slightly viscid. They commenced to boil at 257° C., at which temperature nearly all

* Journ. Chem. Soc. [2] iii.

distilled over [*], and was found to be a hydrocarbon analogous to that from cubebs; but towards the end the thermometer rose much higher and the distillate became of a deep blue colour, owing to the presence of an intensely blue matter termed 'azulene' or 'cærulein,' which is also found in the oils of *Calamus aromaticus, Achillea Millefolium, Matricaria Chamomilla, Artemisia Absinthum,* and in a small quantity in the oils of Bergamot and Ceylon Lemon-grass." The analysis of this remarkable fluid shows its formula to be $C_{16}H_{13}O$. Its boiling-point is 576° F., and its sp. gr. ·910. There are but few liquids which give a coloured vapour when boiled, but azulene is one of them. Its vapour is blue. It is soluble in alcohol, fatty and volatile oils, and many other liquids, to which it imparts its colour, but not in water. It is very permanent, and bears a temperature of 700° to 800° F. in a sealed tube without alteration, and none but the strongest acids aided by heat will break up its constitution. It is most intensely blue, appearing almost black when in a concentrated state. It is not decolorized by sulphurous acid, sulphuretted hydrogen, or bromine-water. It does not attach itself to animal charcoal, nor does it dye wool, cotton, or silk. It has been found to exist to the extent of 6 per cent. in pure oil of patchouli.

Oil of patchouli contains a camphor called "Patchouli camphor." It crystallizes in regular hexagonal prisms, melts at 59° C., and boils at 296° C. Its composition is $C_{15}H_{26}O$. It was investigated by Gal [†] and by Montgolfier [‡]. It is considered to be an isomer of camphor of cubebs and of concrete oil of cedar. The conditions most favourable to the formation or deposition of this body are little known, but it has been remarked that it forms or separates more rapidly in samples of oil which have been desiccated by chloride of calcium. This camphor not having any commercial value, its formation is undesirable, but as it results from a simple molecular change, it may be difficult to prevent it; however, it is possible that the presence of a small quantity of water in the oil may at least retard it. A sample of oil filled up to the cork,

---

[*] This seems strange, considering the way in which it can be fractionated when distilling directly from the leaf, and leads to the conclusion that although his samples may not have been adulterated, they were not intact, and that the finer and most volatile portion may have been abstracted from them.

[†] Bull. Soc. Chim. 1869, p. 304.

[‡] Comptes Rendus, 1877, p. 88.

tightly corked, and kept in a cool dark place did not deposit a trace of it in 17 years. This was pure Singapore oil. By the action of hydrochloric acid or acetic anhydride this camphor is decomposed into water and *patchoulene*, $C_{15}H_{24}$, which boils at 252° –255° C.*

Commercial oil of patchouli is often adulterated to the extent of 60 per cent. with cheaper oils, generally those of cedar and cubebs. It is remarkable that these have been selected, as the camphor of patchouli is isomeric with that of cubebs and with the concrete oil of cedar †.

Gladstone observed ‡ the rotatory power of " Penang " oil of patchouli (determined for a column of liquid 10 inches long) to be −120°; the same for Cedar-wood oil +3 ; the hydrocarbon of patchouli oil, patchoulene, −90°, and oil of cubebs +55°.

The same authority gives the sp. gr. of three patchouli oils as follows :—" Indian," ·9554; " Penang," ·9592; " French," 1·0119 (!), all taken at 60° F.; and for their hydrocarbons :—

" Indian."	Sp. gr. at 20° C. ·9211.	Boiling-point 254° C.	
" Penang."	,, ,, ·9278.	,, 257° C.	
" French."	,, ,, ·9255.	,, 260° C.	

A flowering variety is known to grow on one of the islands near Sourabaya, south-east of Sumatra ; its leaf is odorous, though not so broadly ovate as the cultivated plant, and with shorter petioles ; this is grown simply for its flowers, which are sold in large quantities for a medicinal purpose in the various markets of Java, and fetch a high price. All labiate plants, especially the Coleus (which the Patchouli seems to belong to, or be nearly akin to) and the Mints, are apt to take a character and habit not true to the original plant when transplanted to a climate or soil other than is natural to them ; and under such conditions the development of the odorous principle is as much changed as is the development of medicinal properties in many drug-yielding plants.

A very similar plant to the cultivated one, and of very much the same odour, grows in the lofty range of hills northwards of Gowahatti in Assam (Hooker's Journ. of Botany, i. p. 22), and,

---

* Bull. Soc. Chim. xxviii. p. 414.
† Comptes Rendus, 8 Jan. 1877.
‡ Journ. Chem. Soc. xvii. p. 3.

according to Mr. Thiselton Dyer ('Kew Bulletin,' March 1888, p. 74), there appears to be evidence of the existence of a plant with a patchouli odour, native to Khasia and Assam, which is widely different from any form of *Pogostemon Patchouli*, although it may have the true odour and be of commercial use in India. It is mentioned in the 'Flora of British India,' iv. p. 624, as "a doubtful *Plectranthus* (*Plectranthus Patchouli*, Clarke)." He adds :—" *There is no inherent scientific improbability in widely different plants elaborating the same essential oil.*"

Other varieties of *Plectranthus* are much esteemed in India for their perfume, as the *P. aromaticus*, Roxb. (Flor. Ind. ii. p. 466), synonymous with *Coleus aromaticus* *, figured by Rumphius †, and in the 'Botanical Register,' tab. 1520. It is mentioned by Royle in his 'Illustrations of the Botany of the Himalayas,' i. p. 303. The vernacular name in India for this plant is Páthor chúr—consequently considered distinct from the patchouli, which is equally widely known as Puchá-Pát.

Amongst other plants to which the patchouli plant is described as having an affinity is, according to Dr. Wallich ‡, the *Marrubium odoratissimum Betonicæ folio* of Burmann's 'Thesaurus Zeylanicus.'

Its odour has likewise been compared with that of *Gendarussa vulgaris*, Nees. This is described by Wallich, Plant. As. Rar. iii. p. 104; DeCandolle, Prod. xi. p. 410; Wight, Icones, tab. 468; Rheede, Hortus Malabaricus, ix. tab. 42 ; and is also figured in the Botanical Register, tab. 635. It is the *Gandharusa* of Rumphius, Amboinensis, iv. p. 70, tab. 28, and the *Justicia Gendarussa*, Linn. Sp. Pl. ed. Willd. i. p. 87, under which name it is described by Dr. Roxburgh, Flor. Ind. i. p. 129, as a handsome shrub indigenous on the Malay Islands, common in gardens in India, growing readily from cuttings, and flowering during the wet season. The bark of the young parts is generally dark purple and very smooth, but in some varieties green. The branches are numerous and straggling except when cut back. The leaves are opposite, short-petioled and lanceolate, frequently a little scolloped, smooth ; the nerve and veins dark purple, from 3 to 6 inches long, and from half an inch to an inch broad ; spikes terminal, erect, verticilled.

---

* Flor. Brit. Ind. iv. p. 625.
† Amb. v. t. 102. f. 3.
‡ Transactions of the Med. and Phys. Soc. of Calcutta.

Anthers double, the lower ones have a spur-like process projecting downwards and outwards.

Origanum Indicum is only another name for *Pogostemon Heyne-anum* of Benth., figured in Wallich's Plant. As. Rar. i. p. 31, Rheede's Hortus Malabaricus, x. tab. 77, and Wight's Icones, tab. 1440. This plant, as above observed, is probably a mere variety of the true Patchouli, with longer spikes.

INDIAN GRASS-OILS.

Of the genus of Grasses belonging to the tribe *Andropogonæ* about 25 species are met with in India; of these, four or five are of commercial interest as yielding the oils known as " Grass-oils."

The greatest confusion has existed in the identification of the plants yielding the essential oils from this genus, and much uncertainty yet appears to exist in Europe in the assignment of each oil to its proper botanical source; that is to say, in the identification of nearly allied plants which afford distinct oils known commercially under various names in London, Paris, and the East. The trade names in London of the four principal oils, having such a variety of equivalents in the vernacular of Egypt, Turkey, and India, and the plants which these oils are derived from being known in the various provinces of India under such a number of local dialects, it is not surprising that errors creep into the literature of a subject so difficult as that of the identification of the plants which yield the four oils known on the London Market as " Citronella," Lemon-grass, Ginger-grass, and Vetiver. Had I not known personally one of the largest growers and distillers in the East, who was as well acquainted with the Malay and Indian dialects as he was with the cultivation of the plants, I might have been led by text-books to believe in the existence of a great number of plants yielding various oils under many names. The European and Vernacular names are very numerous, but the oils are four (unless rectified or adulterated oils be counted), and the plants yielding them are four (unless a sub-genus, or varieties somewhat modified by cultivation be counted).

As regards the Indian botanical nomenclature, there is great difficulty experienced by Europeans in ascertaining from the

natives the correct vernacular names of plants, and even greater
difficulty in expressing by any combination of the Roman
characters, or by accentuation, the guttural pronunciation,
peculiar aspiration, etc., of Arabic, Sanscrit, Malay, or of the
languages and dialects of the East—possibly they might be more
easily rendered in German than in English, as the German
language has sounds more approximate thereto.

A museum specimen of essential oil should be distilled by the
exhibitor himself, as all Oriental oils are adulterated; it should be
accompanied by a dried specimen of the plant taken when in
flower, a sample of the root, and a drawing of the living plant,
also a description of the aspect of the place where found and its
exact local name written in Oriental character, then—in London,
we know it. However, to summarize on the evidence at present
available, the four commercial oils derived from the four plants
are Ginger-grass, Citronella, Lemon-grass, and Vetiver. The
three first are already described.

Vetiver or Cus-Cus.—This is the root of the *Andropogon muricatus*, Retz., syn. *A. squarrosus*, Linn., *Vetiveria odorata*, Virey,
Anatherum muricatum, Retz., *Raphis muricatus*, Nees, *Phularis
zizanoides*, Linn. There is a verse in the Sanskrit language
composed of nine words arranged in two lines purporting to be
the[*] nine names under which the plant was known; doubtless they
were poetical names, as they are not to be found in the extensive
list of local names recently enumerated by Watts [†]. The roots
are universally known in Bengal as Chás or Khás-Khás, and in
Bombay Khása-Khása. It is a perennial tufted grass, very conspicuous, tall, and erect. It is very common on every part of the
coast of Coromandel, Mysore, also in Bengal and Burma, where
it meets with a low, moist, rich soil, especially on the banks of
water-courses. It covers large tracts of waste land in Cuttack.
It inhabits the plains of the Punjab and North-west Provinces,
and ascends into Kumaon 1000 or 2000 feet in altitude [‡]. It is
also found in Réunion, Mauritius and the Philippine Islands. With
the exception of lemon-grass it is the only species of the grasses
under discussion occurring in the New World, being abundant in

[*] 'Asiatic Researches,' iv. p. 306.
[†] Dict. Economic products of India, 1889.
[‡] Duthie's 'Grasses of the North-west Provinces,' 1883.

the Antilles, Porto Rico, Jamaica, Brazil, etc. It was observed by
Virey * that the word *Ver* in the Hindu language means "a long
creeping root." The roots of this grass closely resemble in
appearance the roots of the "Chien-dent-à-balai" (*A. Ischæmum*,
Linn.), roots which are used for making carpet-brooms—being
long, thin, and creeping, with a bark of a pale yellowish brown or
light tawny colour. The roots extend in a fibrous tangled mass.
In the 'Gazetteer of the Central Provinces' this grass is described
as " a nuisance to the agriculturalists, as it grows on the rich soil
and is very difficult to eradicate ; " but the 'Oudh Gazetteer,' iii.
p. 176, says "it is generally strictly preserved, as it takes time to
spread, and proprietors are averse to its being dug up for Khás."
This seems to indicate a different value being put on it in the
different localities within the wide range of its growth. This
plant is alluded to on some copper-plate inscriptions discovered
near Etawah, south-west of Agra, dated A.D. 1103 and 1174, as
being one of the articles of commerce on which the Kings of
Kananj levied taxes †. The leaves are inodorous. The roots
have a strong peculiar odour, somewhat like myrrh combined
with that of some flower of the mignonette type. This odour
partly disappears when the root is dried, but immediately manifests itself on the application of moisture, and is retained so
tenaciously as to be perceptible even after the root has been
scalded or insufficiently distilled. The root contains a resin of
a deep-brown colour having an acid taste and an odour like
myrrh, a colouring-matter partly soluble in water, a free acid,
a salt of lime, a considerable quantity of oxide of iron ‡, and a
powerful volatile oil which is rather difficult to thoroughly
extract in the ordinary way by reason of its high boiling-point
and its association with the resin. This difficulty may be overcome by placing the root in a steam-jacketed still with just
sufficient water to drench it, and allowing it to stand for a short
time so that the water may penetrate into the tissues. Then
by admitting steam of about 10 lbs. pressure into the jacket,
the light oil (for there is a light oil of a lower boiling-point)
will come over and may be collected separately, and a current
of steam of 10 lbs. gradually raised to 15 lbs. pressure afterwards

* Journal de Pharmacie, xiii. p. 499.
† Proc. Asiatic Soc. Bengal, Aug. 1873, p. 161.
‡ Vauquelin, 'Annales de Chimie,' lxxii. p. 302.

admitted into the still by a pipe at the bottom can be blown through the mass until oil ceases to drop into the receiver.

Dr. Piesse in his work on Perfumes states the yield to be 10 ozs. per cwt., but according to Watts * the yield of 100 lbs. of root is only 2 ozs.; other observers have found it to vary between 0·2 and 3·5 per cent.† The crude heavy oil is very viscid, consisting mainly of a liquid boiling at 280°–283° C. It does not appear that the light oil of vetiver is met with in England as an article of commerce, but it is worthy of attention on account of its delicate fragrance. It seems to blend well with the odours of orris root and cassie flowers (*Acacia Farnesiana*). The uses of Vetiver in England are confined to the distillation of the oil, which commands a very high price. The oil enters into the composition of many favourite perfumes, as "Mousseline des Indes," Maréchal, " Bouquet du Roi," etc., and it is known that in India the roots are woven into fans, screens to cool the atmosphere, ornamental baskets, etc. Dr. Irvine, in his " Medical Topography of Ajmere," mentions the oil in the preparation of sherbet. In India it enters into the composition of several cooling medicines. An aromatic bath is prepared by adding to a tub of water the following substances :—roots of *A. muricatus, Pavonia odorata,* santal-wood, and a fragrant wood called " Padma Kastha " ‡.

Camel-grass.—This aromatic grass seems to be very little known in England by name, and its essential oil does not appear to be known at all. Botanically it is the *Andropogon Laniyerum* of Desfontaines. It is identical with *Fœnum Camelorum* and *Juncus odoratus.* It has been termed *Cymbopogon Laniger,* and it partly agrees with Roxburgh's description of *A. Iwarancusa.* It has been long known to pharmacists in the East as *Herba Schœnanthus.* It is figured by Pomet as " Squenanthe " §. In Bengal it is known as " Ibharankusha," in the North-western Provinces (amongst other names) as " Ganguli-ban." The name in Bombay and Arabia (for the culms of the plant, with or without a portion of the root) is " Izkhir"; this name, as given

---

* Watts' Dict. Chemistry, 1868, v. p. 999.

† " The plant is very common in the Mascarenes ; it often borders the fields of sugar-cane, and does not require the least care for its cultivation. Of this essence 400 or 500 grammes are furnished by 50 kilos of roots . . ."—*Gardener's Chronicle,* 16 July, 1892.

‡ Hindu Mat. Med., p. 271. § Hist. des Drogues, p. 173.

in the best lexicons, is derived from the same Arabic root which furnishes the derivative Zákhira, a common term in India for stored-up forage, etc. The name *Fœnum Camelorum* signifies its use as a forage for Camels. It is a native of Arabia, growing plentifully in the Desert and in the hot, arid regions of Algeria. The Arabians call it "Helsi Meccavi" and "Idhir Mecchi." It is said that in the deserts between Syria and Egypt it is the only grass eaten by camels. This plant has a wide distribution but is not cultivated. It is found growing on the lower Himalayan tracts and in Thibet at an altitude of 11,000 feet, extending through the plains of the North-west Provinces to Sind. Roxburgh says it grows in large tufts, each tuft composed of a number of plants adhering together by the roots. This description corresponds with Pomet's figure alluded to above. It is common about Kurrachee, and is used as a perfume by the natives. Lemery, commenting on Pomet, says that this *Fœnum Camelorum* is a kind of fragrant rush or grass growing plentifully in Arabia Felix at the foot of Mount Libanus, where it serves for fodder and litter for the camels. The stalk is about a foot high, divided into several hard stems, of the size, figure, and colour of barley-straw, being much smaller towards the top. The leaves are about half a foot long, narrow, rough, pointed, of a pale-green colour. The flowers growing on the top are arranged in double order, small, hairy, of a carnation colour, ... all the plant, and particularly the flower, is of a strong smell and bitter taste. This plant is also figured in Plukenett's 'Phytographia,' 1691, tab. 109. fig. 1.

The yield of oil from the fresh plant is said to be 1 per cent. A sample of the oil, distilled by Dr. Dymock of Bombay, has been examined by Messrs. Schimmel with the following result :— Its odour recalls that of Elemi oil; its sp. gr. is 0·915; its optical rotation +34° 38'. It boils between 170° and 250° C., and contains Phellandrene.

The *Andropogon laniger* has recently been discovered in British Baluchistan by J. H. Lace, Deputy Commissioner of Forests of India. In his valuable paper on the "Vegetation of the Hurnai Railway Route," recently read before the Linnean Society of London,[*] he mentions having found it covering large tracts of land on the lower hills.

* Journ. Linn. Soc. xxviii. p. 293.

There are writers who frivolously waste time by worrying themselves and the world with their ideas about the identity of the "Sweet Cane" of Scripture; some fancying it to be one plant, some another; they refer back to Dioscorides, even to Jeremiah. The Ancients mixed up many plants under one poetical name and led us Moderns into much useless confusion and dispute (instance Spikenard); their writings, in languages not over rich in botanical terms, are misty, abrupt in expression, and have been mauled in translation and re-translation. Whether the "Sweet Cane" was the *Fœnum Camelorum* or not, now matters little.

HENNAH.

This plant has been cultivated for ages in India, Egypt, North Africa, Syria, and the Levant. In the "Song of Solomon," written about 1000 B.C., it is referred to under the Hebrew word *Copher* and has been wrongly translated into English as *Camphire*, a word which was used in old English to signify Camphor (the product of an entirely different tree). Hennah is the "Cyprus of Egypt" referred to by Pliny. It appears to be a native of Arabia and to have been distributed by the Arabs into Turkey (Asiatic and European), Egypt, and along the coast of the Mediterranean. Its modern Arabic name is Thamar-ul-hinnâ'i (the hinna shrub), the leaves are also called hinnâ. Dr. Dymock states the vernacular name in Bengalee to be Mehedi, in the Bombay and Hindee dialects Mehndee, and in Tamil, Marutouri and Aivanam. Arabic and Persian works give Arkán and Fakúliyún as the Greek names. The plant is figured in Van Rheede's 'Hortus Malabaricus,' i. tab. 40, under the name of Mail-anschi. Botanically it is known as the *Lawsonia inermis* (Linnæus). According to Dr. Roxburgh (Flora Indica, ii. p. 446) it is called in the Telinga dialect Gounta, and it is indigenous on the Coast of Coromandel, where he found it in the state of a large shrub, though it is naturally a small ramous tree. It flowers and seeds most part of the year. The flowers are small, greenish yellow, and remarkably fragrant whether fresh or dry, being particularly grateful at a distance. It is much used for hedges, growing readily from cuttings. Fertile seeds are not often met with. He considers the species called *spinosa* to be nothing more than the same plant growing on a dry sterile soil; it is then very thorny, the branchlets being thin,

short, and rigid, with sharp thorny points. The fresh leaves beaten
up with catechu dyes the nails and skin of a reddish-orange
colour which is much admired by the women all over India.
The freshly-made paste is laid on at bed-time and removed in the
morning; the colour remains till the nails or epidermis is
renewed or removed. The leaves yield in decoction a porter-
coloured or deep orange-coloured liquor which acids destroy
although alkalies and infusions of astringent vegetables deepen
it. Although this decoction dyes the fingers, it does not communicate any colour to cloth variously prepared, or yield scarcely
any precipitate.

A large business is done in the leaves of this plant. The stems
are cut several times a year and stripped of their leaves, which
are dried in the sun and reduced to powder. The plants, which
are cut down almost to the ground, throw out fresh shoots and
suckers, which are cut with a sickle like the first. Several crops
are thus yielded during many successive years.

To obtain the flowers, the shrub is not pruned but allowed to
attain the height of 2 or 3 mètres, when it flowers in the second
year after planting. Delchevalerie, writing in the 'Belgique
Horticole,' says the culture is easy and might probably be carried
on successfully in Italy and the south of France. He considers
the *inermis*, which has larger leaves, to be a different variety
from the *spinosa*. Specimens of Hennah from Senegal, under
the vernacular name of " Foudeen," were exhibited at the Paris
Exhibition, 1878. Samples of the essential oil from Lucknow,
under the local name of *Mehndee-Ka-utter*, were exhibited at the
London Exhibition, 1862; the word *Utter* in Hindee being
equivalent to otto and applying to all fragrant essential oils.

The Hennah has been naturalized in the West Indies, where
it is called the "*Jamaica Mignonette.*" The powerful fragrance
of its flowers has, however, more the resemblance to a combination
of mignonette and rose.

CHAPTER XIV.

SANTAL.—CEDAR.

SANTAL.

DR. BERTHOLD SEEMANN, the botanist, in calling attention to the commercial importance of Santal wood (sometimes called " Sandal," " Yellow Sandal," and " Santal Citrin ") remarked that " the trade in this fragrant wood has been going on since the dawn of history and will probably not cease until the connection between santal trees and idolators, existing from time immemorial, shall have been broken up, by either the one or the other becoming as extinct a race as the Archæopteryx or the Dodo." The religious sentiment of millions of human beings is still intimately associated with this wood. Some of the most ancient records inform us of the prominent part played by the wood in India; and since the introduction of Buddhism into China, that country, destitute of santal trees, has become the principal market for this important production. A piece of wood the diameter of 4 to 6 inches is considered as the most acceptable offering a person can make to the idols of the temple. Large pieces are presented by the rich on particular occasions.

Santal wood is the product of several species of the genus *Santalum*, of the natural order *Santalaceæ*. The genus is composed of about 20 species, spread over Asia, Australia, and Polynesia, and in habit may be compared with the Myrtles. It is possible that other species now said to exist in Madagascar and New Guinea may be recognized by botanists. The East-Indian santal wood and probably also the so-called Macassar santal wood are furnished by *Santalum album*, Linn.* It is

---

* Sp. Pl. ed. Willd. i. p. 691.

botanically described by Roxburgh (Flor. Ind. i. p. 462); and by
Rumphius (Amb. ii. p. 42, t. 11). In the Asiatic Researches (iv.
p. 253) it is named *Chandana*, the Sanskrit names being given as
Gandhasára, Malayaja, and *Bhadrasrí*, and the Persian (by Roxburgh) as *Sundul-sufed*. In Bengali it is called *Chandan*, and
in Hindi *Safed-Chandan*.

Santalum album is a native of the mountainous parts of India,
but is found more especially in the Mysore, Malabar, and Coim-
batore, extending northwards into Canara, and has been found
in the thickets of Midnapúr. It likewise grows on the Coromandel Coast, in Madura and Assam, and is frequently cultivated
as a garden plant. The same tree, or a variety, is met with in
several islands of the Eastern Archipelago, viz., Eastern Java,
the Santal Wood Islands, Sumba, and Timor. It grows freely
in hedges and gardens, and in a rich soil attains a large size;
but in such localities the timber is of little value and has scarcely
any perfume. Soil and elevation have great influence on the
amount of oil produced. It thrives up to an elevation of 4000
feet, and yields the largest quantity of oil when grown in dry,
sunny, rocky, mountainous districts, or soils of volcanic origin,
although it does not reach so great a height. It is rarely found
in forests. In the Madras Presidency and the Mysore it is now
grown in Government plantations from seeds; but it also springs
from roots which have been left in the ground.

The height of the tree is only from 20 to 30 feet. The bark
is greyish brown and somewhat scabrous, with longitudinal
fissures. The branches are numerous, opposite, slender, and
much divided, rising in every direction and forming nearly a
spherical head. The young twigs are round and smooth.

The leaves are lanceolate-obtuse, opposite, entire, and smooth,
their under surface glaucous; their length is from 1½ to 3 inches,
on stalks of scarcely one fourth the length of the leaves. The
numerous small inodorous flowers are of straw-colour when they
first expand, but change to a deep rusty purple, as do all the
exterior parts of the growing plant, even when bruised. The
succulent fruit is black when ripe and of the size of a cherry.
The seed is solitary. The tree is an evergreen and produces
flowers and ripe seed nearly all the year, chiefly, however, from
March to July.

In a young state the plant appears to be parasitic. W. B.

Hemsley says *:—" Dr. King, Superintendent of the Calcutta Botanic Gardens, assures us from his own observations that the *S. album* is sometimes parasitic on the roots of other plants when young. This assertion is borne out by the frequent failure of cultivators to raise seedlings of this plant when its seeds are sown alone, also by the fact that its usual habitat in a wild state is in hedges and thickets. But that it is not always so is evident from the fact that germinating seeds sent from Ceylon to Mauritius in 1877 have since grown, and fruited twice a year. On the other hand, as we learn from an article in the current volume of the 'Indian Forester,' decaying vegetable matter is necessary for the successful raising of seedling santal trees. The editor of the serial named inserts the following note (p. 205) on this subject :—' Colonel Doveton, Conservator of Forests, found santal seedlings growing as a root-parasite on the wild Date-palm (*Phœnix sylvestris*), and such seedlings were more vigorous than others rooted independently in the soil.' " In regular plantations the seeds are sown two or three in one hole, with a chili seed. It is certain that other genera of this order are parasitic, and, indeed, other species of this genus, for Nadeaud has stated (' Enumération des plantes indigènes de l'Ile de Tahiti ') that the Tahitian santal wood (*Santalum insulare*, Betero), is parasitic on the roots of other trees, generally on *Commersonia echinata* or *Alphitonia ziziphoides*. Also Scott showed in 1871† that *Santalum* was parasitic, its roots becoming attached to those of many other plants by tuber-like processes.

When santal-wood trees have reached perfection, which they do in from 20 to 30 years, having then a diameter of from 9 to 12 inches just above the root and a height of about 25 feet, they are either cut down, at the latter part of the year, or dug up; if the former, the roots are generally dug up afterwards. If a tree be allowed to exceed these dimensions it is generally found rotten at the core, which is the most valuable part. The tree being felled, worthless branches are removed, also the bark, and the trunk is either buried for 6 or 8 weeks, or left lying on the ground, in order that the white ants may eat off the inodorous sap-wood. It is then taken up and sent to the Depôt, where it is cut into billets of from 2 to 4 feet long, carefully trimmed,

* Challenger Reports, Botany, vol. i. 1885.
† Journ. of Agricult. and Horticult. Soc. of India, ii. part i. p. 287.

and sorted according to quality. In cutting down the trees, the earth is removed from about the root, so that the collectors may cut as low as possible. The billet taken from the trunk immediately above the root is called the " root billet " and is of superior quality, and by its comparative scarceness realizes the highest price.

Concerning the sorting, Buchanan says * :—"The deeper the colour the higher the perfume, hence the merchants sometimes divide santal wood into red, yellow, and white ; but these are only different shades of the same colour, and do not arise from any difference in species." He distinctly implies that the three kinds are derived from the heart-wood. The following statement by Udoy Chand Dutt † is to the same effect, when writing of the white and yellow santal wood :—" These varieties are founded on the difference in depth of colour of the heart-wood." From this it appears that the statement which has been made by some writers, that white santal wood is the sap-wood of *S. album*, is, at least, somewhat misleading. The sap-wood is nevertheless found in commerce, for Balfour describes it as " coated with ' thick, compact bark, has a grey and brownish epidermis, it is nearly inodorous and has a slightly bitter taste." Kirkby, who has studied the santal-woods, says‡ :—" As the white is doubtless a lighter shade of the yellow, so is the red (apart from that obtained from *Pterocarpus santalinus*, which is truly red and inodorous) a deeper shade. The red kind was not recognized in Sanskrit medical works, the only kinds mentioned being *Strikhanda* or the white wood and *Pitachandana* the yellow wood."

The trees being now carefully protected by the Mysore Government, and only cut down when they reach maturity, it might not be thought that the wood much varied in quality, but it varies considerably; the pieces which are straight and have the most heart-wood fetch the highest price. The chips and fragments removed in the process of trimming the billets and squaring the ends are also sold.

It is very difficult to estimate the value of santal wood by reason of the very variable amount of oil contained in each billet, and parcels purchased by distillers often give a very disappointing

* ' A Journey from Madras through the countries of the Mysore, Canara, and Malabar,' ii. p. 133.

† The Mat. Med. of the Hindus, 1877, p. 133.

‡ Pharm. Journ. [3] xvi. p. 858.

yield. In common wood and chips it may be as low as 1½ per cent. and in some fine wood as much as 5 per cent. The amount extracted greatly depends upon the fine state of division to which the wood is brought before distillation. This is effected by first incising or chipping the logs by powerful machinery and then disintegrating them with special tearing or rasping apparatus or with mill-stones.

The wood is all sold by weight at the annual Government auctions, native merchants congregating from all parts of India to make purchases. The Lots being classed according to quality are catalogued under the following denominations :—

1st class billets.	
2nd „ „	Selected logs, only obtainable in small quantities.
3rd „ „	
4th „ „	
·5th „ „	Logs. The superior santal-wood of commerce.
Large Roots.	Roots in large pieces.
Small Roots.	„ small „
Jug Pokal	Ordinary commercial, consisting of small logs and large branches.
Bagar Adad	Similar to the former but somewhat smaller.
Jyen Bagar......	Inferior woods.
Jyen Chilta......	Common wood.
Milva Chilta	Chips in large pieces.
Milva Chilta split.	„ split „
Hathri Chilta	„ „ „
Adzed chips	„ good „
White chips	„ pale inferior.
Jyen chips	„ most inferior.
Sawdust.	

All logs bear the stamp of the particular class to which they belong.

The wood is imported into Europe from Tellichery and Bombay in logs from 2 to 4 feet long, and from 3 to 8 (rarely 14) inches in diameter. It is very homogeneous, rather hard, and ponderous, although it does not sink in water. It is somewhat hard to cut transversely but it splits comparatively easily. In colour it is yellow, fawn-coloured, or reddish-brown, being darker in the centre than at the periphery, and is marked with darker concentric zones or annual rings. In the inner part of the wood the zones are sometimes very wide, measuring sometimes as much as seven millimetres; possibly, therefore, they do not correspond to one year's growth*, but to that

---

* C. A. de Bary, 'Vergleichende Anatomie der Vegetationsorgane,' 1877, p. 519.

of a longer period. The transverse section, examined by means of a lens, displays the numerous narrow medullary rays and wide vessels, partly empty, partly loaded with yellow resin. It has a strong, persistent, agreeable and characteristic odour, especially when freshly cut. Perfumes are difficult to compare, but, by some persons, that of santal wood is thought to resemble a mixture of musk and rose. The taste is aromatic. " Three sorts are recognized in the commercial houses of China, viz. ' South Sea Island,' ' Timor,' and ' Malabar '; the last fetches from three to four times as high a price as either of the others " *. " Malabar" santal is included amongst parcels commercially called " Bombay Santal " †. Under the microscope the wood is seen to consist of tracheides interspersed with solitary pitted vessels and traversed by narrow medullary rays. Among the tracheides are parenchymatous cells of about the same diameter as the tracheides ; it is in these the essential oil is contained ; they are isolated, but two, or perhaps three, are found in close proximity. In transverse section, single large crystals of calcium oxalate are seen. These are contained in angular cells arranged in vertical rows. The medullary rays consist of two rows of thickened, pitted cells, and contain resin. In the sap-wood, which is scentless, both vessels and medullary rays are less distinct ‡.

The Mysore Government have establishments for distilling the oil, which is also sold at the annual auctions along with the wood, and chiefly bought up for exportation to China and Arabia The roots yield the largest quantity and the finest quality of oil.

Dr. Bidie describes the method of distillation in India as follows :—" The body of the still is a large globular clay pot of about $2\frac{1}{2}$ feet deep by about 6 feet circumference at its widest part. No capital is used to this still, but when in use, the mouth is closed with a clay lid having a small hole in its centre, through which a bent copper tube about $5\frac{1}{2}$ feet long is passed for the escape of the vapour. The lower end of the tube is conveyed inside a copper receiver, placed in a large porous vessel containing cold water. When preparing the santal for distillation, the white or sap-wood is rejected and the heart-wood is cut into small chips, of which

---

* Flückiger and Hanbury, ' Pharmacographia,' 1874, p. 543.

† Durand, " Etude sur les Santalacées," Thèses de la Faculté de Médecine de Paris, 1874.

‡ The microscopic sections of various santal woods are delineated by Petersen in Pharm. Journ. [3] xvi. p. 758 ; and by Kirkby, ibid. pp. 859 and 1065.

about 50 lbs. are put into the still. As much water is then added as will just cover the chips, and distillation is carried on slowly for ten days and nights, by which time the whole of the oil is extracted. As the water from time to time gets low in the still, fresh supplies are added from the heated contents of the refrigerator. The quantity of oil yielded by wood of good quality is (by this process) at the rate of 10 ozs. per maund of 25 lbs. or 2·5 per cent. It is transparent and of a pale yellow colour, and has a resinous taste and sweet peculiar smell. The sp. gr. is about 0·980 " *.

The new British Pharmacopœia requires the sp. gr. to be 0·96, a figure which by large distillers is considered too low, although a sample examined from the India Museum had a sp. gr. at 65° F. of 0·9901; and a perfectly pure specimen distilled in Mysore specially for the Pharmaceutical Society was found to have a sp. gr. at 16° C. of 0·9896. These great densities are attributed to the crude method of distilling over a bare fire. Oils distilled in Germany from the best materials have been found to mark a sp. gr. at 15° C. of 0·970 to 0·978. A larger yield is obtained in Germany by reason of more effective appliances being used to thoroughly disintegrate the wood and distil out the oil.

According to Dr. Dymock, as much as 12,000 lbs. of this "Indian" oil are imported into Bombay from the Malabar coast, being worth 8½ rupees per lb., and used chiefly for perfumery. It is imported into this country to the extent of not more than 1500 lbs., and often reaches England in a discoloured state, adulterated with fixed oil and containing traces of water. It is rejected by wholesale druggists for medicinal purposes, and is entirely used in perfumery.

Although the English distilled oil realizes a higher price than either Indian or German oil, it is believed that the Continental distillers are far in advance of the British in their superiority both of plant and of the knowledge necessary for the production of this and other essential oils, two well-known German firms probably distilling three-quarters of the whole quantity used in Europe.

The Indian oil may be adulterated before shipment with various fixed oils, such as those of sesame and cotton-seed, and even with santal-tree-seed oil, which is used as lamp oil in Mysore. A

---

* 'Pharmacopœia of India,' p. 461.

volatile oil which becomes an easy adulterant is cedar-wood oil, as various oxidizing agents have very much the same action on both oils, and reduction of sp. gr. caused by admixture of cedar-wood oil is not a sufficiently reliable indication by reason of the high figures of some of the Indian distillates.

Pure German distilled oil of Indian santal wood, sp. gr. 0·9797, is perfectly soluble in its own volume of methylated alcohol of sp. gr. 0·839. Cedar-wood oil forms a white cloudy mixture in its own volume of the same solvent; but one volume of an equal mixture of santal-wood and cedar oil under the same conditions dissolves perfectly. (See Addenda, p. 374.)

With alcohol of sp. gr. 0·920, the same Indian santal oil also dissolves in an equal volume; one c. cm. of "Macassar" santal oil of sp. gr. 0·9738 requiring 1·3 c. cm. of the solvent. One c. cm. of cedar oil requires 5 c. cm. of alcohol of same strength for solution. From these facts, observed by Mr. Holmes *, it appears that whilst oil of cedar may be recognized by its insolubility in an equal volume of alcohol of sp. gr. 0·920, its admixture with santal-wood oil to the extent of 10 per cent. cannot be easily detected.

The admixture of " West Indian " and " West Australian " santal oil can be detected by observing the optical rotation, and the addition of "South Australian" santal oil would increase the sp. gr. (these figures are hereafter given).

Pure Indian oil of santal contains a body called *Santalal*, $C_{15}H_{24}O$, boiling at 300° C., together with a smaller amount of *Santalol*, $C_{15}H_{26}O$, which boils at 310°, and is converted by distillation with phosphorous pentoxide into *Santalene*, $C_{15}H_{24}$, boiling at 260°; *Santalal*, on the other hand, when subjected to similar treatment, yields a hydrocarbon $C_{15}H_{22}$ which is probably identical with *Cedrene* (?) †. *Santalyl acetate*, $C_{15}H_{25} . C_2H_3O_2$, is prepared by heating *santalol* to 150° with glacial acetic acid; this liquid boils at 298° and has a fruity odour.

The *Santalum myrtifolium*, Roxb., differs in one or two particulars from *S. album*, and Roxburgh considered the differences to be of sufficient importance to warrant him in giving it a specific name ‡. It has lanceolate, waved leaves about 2 inches long and ¾ inch broad. It is a native of the Circar mountains, where it is

---

* Pharm. Journ. [3] xvi. p. 822.
⊥ Bull. Soc. Chim. xxxvii. p. 303.
‡ Flor. Ind. i. p. 464, and Corom. Pl. i. no. 2.

but of small size. Its wood is much less odorous than *S. album*, and yields an oil of poorer quality. Birds greedily eat the berries, which are smooth, black, juicy, and about the size of a large pea; by this means it is propagated extensively. At present it is considered to be a variety of *S. album*, Linn. (var. *β. myrtifolium*, D.C.). Bentley and Trimen have included it in their description of *S. album* *.

Macassar santal wood is brought from Timor and Sumba, the " Santal Wood Islands" off the north-west coast of Australia, to the market in Macassar, and is thence consigned to Holland. The tree yielding it is probably the *Santalum album*, L.

The " Macassar " santal wood is lighter than water, cuts with difficulty transversely, but cleaves easily. It is of pale fawn-colour to yellowish brown, with darker concentric zones spreading over the whole. Its odour is similar to the East-Indian but not so powerful, and its microscopical structure is similar except that the distance apart of the medullary rays and diameter of the vessels are less.

By distillation, from 2 to 5 per cent. of oil is obtainable, very similar to the East Indian oil.

An examination of various santal woods was made by Kirkby in 1886, and illustrations of his microscopic analysis published †. He distinguishes between the Macassar and Indian woods by the following chemical reaction :—" When a section of the former is treated with the microscopist's ordinary solution of iodine and potassium iodide it becomes coloured an intense black . . .; the Indian oil of santal abstracts iodine from its aqueous solution, but no change of colour takes place, . . ." He considers that the oil " contains something in solution which has a remarkable attraction for iodine when compared with the Indian santal-wood oil, which may be accounted for by differences of climate, soil, or elevation." However this may be, there is very little difference between the two oils as regards their sp. gr. and optical activity; the figures of which, as determined by Schimmel, are as follows :—

	Sp. gr. at 15° C.		Rotation	
East Indian.........	0·970 to 0·978.		Rotation	−17° 20'.
Macassar............	,,	0·976.	,,	−18° 40'.
West Australian...	,,	0·953.	,,	+ 5° 20'.
West Indian	,,	0·965.	,,	+26° 10'.

* Medicinal Plants, iv.
† Pharm. Journ. [3] xvi. p. 859.

The " *West Australian* " *santal* is derived from *Fusanus spicatus*,
R. Br. (*S. spicatum*, D.C., and *S. Cignorum*, Miq.), from Swan River.
A specimen of a " West Australian " santal wood was exhibited in
Paris in 1878, and was said to be derived from *S. latifolium*. The
Australian santal woods arriving in England from Adelaide and
Freemantle are less fragrant than the Indian, and are not much
valued in Europe; their principal market is Singapore. The other
Australian santal woods are *Fusanus persicarius*, F. Muel., in
West Australia; *S. lanceolatum*, R. Br., in North Australia, New
South Wales, and Queensland; and *Fusanus acuminatus* of South
Australia.

The *Santalum Preissii* of the interior of South Australia yields
a wood which is said to be quite different from that of *S. Cignorum*,
Miq. (*Fusanus spicatus*, R. Br.), of West Australia. The tree
is locally called *Quandong*. The wood is dark brown in colour,
with unusually close tenacious texture, and is extraordinarily
hard and heavy. Messrs. Schimmel state * that 75 kilos of the
wood directly imported by them from South Australia yielded
the comparatively large quantity of 3 kilos 800 grams of essential oil, which is quite equal to 5 per cent., and is thus one of
the richest santal woods for oil. In many respects this oil is
characteristic and interesting. It is viscid, of a cherry-red
colour, and heavier than water; at 15° C. its sp. gr. being 1·022.
It possesses the property of solidifying at medium temperatures,
and separating a crystalline body which forms in colourless prisms
melting at 104°–105°, so that in the process of distillation the
cooling must be effected very carefully, otherwise the condensing
tubes become blocked. This phenomenon occurs especially in
the medium fractions of the oil. When in raspings the wood has
an agreeable balsamic odour, with a suggestion of rose oil that
unfortunately is not perceptible in the normal oil. By separating
the oil into a number of fractions by steam distillation, the Rose
odour could be recognized distinctly in some of the middle
fractions.

African santal-wood oil is another novelty recently acquired by
the same firm. A parcel of 17,000 kilos was received by them
from Tamatave (Madagascar). The wood is brownish red in
colour, and uncommonly hard and close. Distilled with water

---

* Bericht, April 1891, p. 63.

it yielded 3 per cent. of a ruby-red oil having the consistence of East Indian oil. Its sp. gr. at 15° C. was found to be 0·969. Its odour is poor, resembling that of "West Indian" oil. This wood is not botanically identified, but it appears * that "in the northern parts of Madagascar a wood with properties similar to santal wood is known under the Sakalava name of *Hasoranto*."

The "*West Indian*" *santal wood* of commerce comes from Puerto Cabello in Venezuela. The tree producing it does not appear to have been identified. In appearance it differs totally from the Indian santal and the Macassar, and probably does not belong at all to the Santalaceæ. It has been examined by Kirkby, in the paper above referred to, also by Andreas Petersen of Copenhagen †, who says that "a transverse section exhibits very distinctly a well-marked irregular limit between the brown heart-wood and the yellowish sap-wood. It is very hard, tough, and ponderous, and sinks in water, to which it yields a faintly yellow matter. It is difficult to cut and split. Medullary rays or annual rings cannot be discovered even by means of a lens, whereas long radiate rows of vessels loaded with resin make their appearance in the heart-wood if examined by a lens. On a radiate section the vessels may also be seen to contain a bright glittering resin. The odour is weak but agreeable. The yield of oil is about $2\frac{1}{2}$ per cent." Its sp. gr. is stated above.

Fiji santal wood is yielded by *Santalum Yasi*, Seem. It is now rather rare through the trees having been mostly cut down by traders. The microscopic structure of this wood bears a close resemblance to that of *S. album*, but is less rich in vessels, and its parenchymatous cells as seen in the transverse section are rather more regularly arranged in rows than in the latter, but the differences are not very conspicuous (*Petersen*). A sample of this wood obtained from the Colonial Exhibition and distilled by Mr. Umney gave the unusually large yield of $6\frac{1}{4}$ per cent. of oil, the sp. gr. of which was determined by McEwan at 16° C. as 0·9768. This oil was of a pale straw-colour and soluble in less than its own volume of rectified spirit; thus a mixture of equal parts of spirit and oil takes up another part of oil. Its rotatory power, observed by Dr. Symes, was −25°·50 ‡.

* Kew Bulletin, May 1888, p. 135.
† Pharm. Journ. [3] xvi. p. 757.
‡ Ibid. xviii. p. 661.

Sandwich Islands santal wood was furnished for a number of years, until the reckless cutting down of the young trees nearly destroyed it, by *Santalum Freycinetianum*, Gaud., and its varieties *ellipticum*, *paniculatum*, and *pyrularium*, Gray, natives of these islands (where they were called *Lau-ala*); only a few isolated specimens being left of the magnificent groves that formerly covered parts of the Islands of Hawaii, Maui, Oahu, and Kauai.

The supply of *Santalum Homei* from the island of Eromanga seems to be equally worked out. In Tahiti the *S. insulare*, Betero, formerly grew and has probably met with the same fate; it was known by the native name "*Eai*."

In *New Caledonia* the wood was yielded by *S. austro-caledonicum*, Vieill., known to the natives as *Tibean*. The tree in its natural state having been nearly exhausted, it is now cultivated, and small parcels are sent from thence to France.

In *New Zealand* the wood is said to be yielded by *S. Cunninghami*, Hook.; it is locally known as *Mairi*.

In the Percy Isles, Repulse Bay, Cape Upstart, and the Palm Islands, an inferior kind has been met with, it is the wood of *Exocarpus latifolia*, R. Br.*. Balfour also states that a white santal wood called *Lava* or *Lawa* is imported from Zanzibar into Bombay. "In the Kew Museum there is a specimen of wood labelled 'Santal vert' (*Croton* sp.), exported from Madagascar and Zanzibar into India, where it is said to be used for burning the bodies of Hindus." Also a specimen from J. Heathcote from Professor MacOwan, received 6 Feb. 1886, labelled "Wood like Santal wood (*Croton* sp.)." It is ground and mixed with water, and used by the natives at Inhambane to anoint themselves. "These latter are not properly santal woods. They are mentioned as indicating the possible source of what is called santal wood at Madagascar" †.

The wood of *Plumieria alba*, belonging to the *Apocynaceæ*, is sometimes substituted for santal wood.

Mexican santal-wood bark.—This name is applied in Mexico to what is considered to be the bark of a species of *Myroxylon* or *Myrospermum* (leguminous plants), indigenous to Mexico and some of the Central American Republics. It is used as incense in the

---

* Balfour, Cyclo. of India, 1873, v.

† Kew Bulletin, May 1888.

churches.   As described by Dr. Stieren *, this bark occurs in irregular, more or less smooth, or unevenly corrugated pieces, of a light whitish cinnamon colour, with dark, hard epidermis, and of an agreeable custard-like smell, and aromatic, slightly acid, balsamic, bitterish taste.   A small quantity coarsely powdered and sprinkled over burning coals emitted a balsamic, mixed aromatic odour.   A thin cross-section manifested, at about 75 diameters linear, in the microscope, oil-cells interstriated with apparently semi-viscid resinous matter.   By exhaustion with alcohol, and slow evaporation of the extract to the consistence of syrup, a clear, rich brown, sweet-scented balsam, not unlike Peruvian balsam in appearance, was obtained, amounting to 15 per cent. of the bark. Experiments proved that the odorous principles rest in an oily substance, cinnamic acid and its combinations and resinous matter.

A yellow wood which is sometimes used as a substitute for santal in religious ceremonies, is that of the *Xymenia Americana*, found in the Circar mountains, the Andaman Isles, Malacca, Ceylon, and distributed in tropical Africa and America.   It is a large ramous thorny shrub, bearing white fragrant flowers.   It is described by Roxburgh, Flor. Ind. ii. p. 252, and in Hooker's 'Flora of British India,' i. p. 574.   This wood may be powdered and mixed with other ingredients into the form of "joss-sticks" for fumigations, in the same way as santal wood is in China. Such sticks are sometimes so made as to smoulder continuously for a fortnight (and are actually used as time-pieces).

Possibly some of the santal woods of the Middle Ages were not exclusively furnished by species of *Santalum*.   The *Epicharis* (*Dyoxylum*) *Loureirii*, Pierre, and *Epicharis Bailloni*, Pierre, trees belonging to the order *Meliaceæ*, and growing in Yunnan and Cochin China, are mentioned by Baillon † as sources of santal wood.

The *Algum* trees "out of Lebanon" of 2 Chron. ii. 8, and ix. 11, and *Almug* trees "from Ophir" of 1 Kings, x. 11, 12 (both references being about B.C. 500), have been generally identified with the true Santal wood (*S. album*) because one of its Sanskrit names is evidently the same word as the Hebrew *algum* or *almug*.

* Pharm. Journ. [3] xv. p. 680.
† Traité de Botanique Méd. 1884, p. 974.

But considering the use to which Algum or Almug wood was put by Solomon, for flooring and pillars, and to make musical instruments, it was probably not santal wood, but cedar, or some hard close-grained wood like *shishem* or *sissoo* (*Dalbergia* sp.), well known as "Bombay Blackwood," or else Red Sanders wood (of which most of the musical instruments in India are now made). Nevertheless santal wood is used in India for the pillars and doors of temples. The famous gates of the temple Somnath, carried off to Afghanistan by Mahmud of Ghazni, A.D. 1025, and restored to India by Lord Ellenborough in 1842, were found on examination not to be, as was generally said, of santal wood but of *Cedrus Deodar* (Indian Cedar). They are still lying in the old palace in the fort of Agra.

The wood known as Red Sanders Wood, sometimes wrongly called "Sandal Wood," contains no oil and is quite odourless. It is obtained from *Pterocarpus santalinus*, a leguminous tree found in Ceylon, and is a native of Southern India. The wood is imported in heavy dark-red billets and in chips. It is employed in pharmacy for colouring tinctures and is feebly astringent. Its red colouring-matter is termed *Santalin*, $C_{15}H_{14}O_5$, which is insoluble in water, turpentine, and fixed oils, but soluble in alcohol, ether, and acetic acid. In most essential oils it is nearly if not quite insoluble, but owing to its free solubility in rectified spirit it is useful to detect the presence of spirit in many essential oils. (A crystal of aniline red, magenta, can be used in the same way— a drop of oil let fall on to it and gently pressed with a paper-knife, the colour is not miscible unless spirit is contained in the oil as an adulterant. An exception to this test is oil of cloves, which dissolves the colour, and possibly other oils containing eugenol may dissolve it.)

CEDAR.

The bulk of the oil of cedar of commerce is economically produced by distilling the sawdust and waste wood of the lead-pencil factories. In some factories in Germany this refuse accumulates to such an extent that it is sold at a very low price to get rid of it, as it would otherwise be used only as fuel. The wood is called "Red Cedar" or "pencil cedar" and is yielded by the *Juniperus Virginiana*, Linn., a coniferous tree native of the greater part of the United States.

J. Virginiana is the largest, the widest spread, and the most useful of the American Junipers. It is the only conifer (and one of the very few trees) which is found East as well as West, and certainly the only one which at the same time extends through so many degrees of latitude. It is well known from the St. Lawrence to the Cedar Keys of Florida, and from the Atlantic to the Rocky Mountains and even to the Pacific Coast of British Columbia; on the Upper Missouri (Cedar Island) it attains large dimensions. It is commonly of a pyramidal form with shreddy bark and red aromatic heart-wood; soft and easily splitting, but extremely durable, in fact almost imperishable. The branchlets are slender and 4-angled. The leaves of the young plants and of vigorous shoots are acicular, subulate and spreading; but on the older trees they are nearly all very minute, scale-like, and closely imbricate. From the great disparity in the proportion of scale-like leaves and subulate leaves in different individuals, as well as the more or less distinct habit, it is difficult to find two trees exactly alike, even in a large plantation. The prevailing hue is dark sombre green; but in the variety *glauca* (syn. *alba argentea*) the foliage is of a silvery glaucous tinge, and this variety has a more compact conical habit.

The *J. V. humilis* is a dwarf spreading form of reddish tinge, and the *J. V. pendula,* of which there are two or three varieties, has long, slender, pendulous branches. There are also variegated varieties, *aura* and *alba,* in the ordinary form.

The yield of essential oil from *Juniperus Virginiana* has been estimated at as much as $3\frac{1}{2}$ per cent. It distils over as a soft semi-fluid mass, consisting of a liquid hydrocarbon *Cedrene,* $C_{15}H_{24}$, and an oxygenated solid camphor or stereoptene, $C_{15}H_{26}O$. To separate the camphor, the crude oil is distilled, the distillate is pressed between linen to free it from the greater portion of the liquid cedrene which adheres to it, and then crystallized from alcohol of ordinary strength, which retains the rest of the cedrene in solution. Cedar camphor thus purified is a silky crystalline mass of great beauty and lustre and of aromatic odour. It melts at 74° C. and boils at 282° C. without alteration. It is very sparingly soluble in water, but freely so in alcohol, from whence it crystallizes in needles of a silky lustre. It gives by analysis 81 per cent. of carbon and 11·8 of hydrogen, agreeing with the preceding formula; hence it is isomeric with the camphor of cubebs, also with camphor of patchouli; probably for this reason and being of an odour which is not antagonistic, oil of cedar is selected as a convenient adulterant to oil of patchouli.

With pentachloride of phosphorus, camphor of cedar yields an aromatic substance which has not yet been fully investigated. By distillation with phosphoric anhydride it is resolved into water and cedrene $C_{15}H_{24}$, identical with the liquid portion squeezed out of the original crude product, the sp. gr. of which, at 15° C., is 0·984, and boiling-point 248° C.[*] Bertagnini found [†] that oil of cedar combines with the acid sulphites of the alkali metals.

In the American pencil factories some oil is obtained by collecting the condensed vapours of the drying-chambers, but it does not realize the price of the ordinary distilled oil.

Oil of " Jamaica Cedar " or " Honduras Cedar " is distilled from the *Cedrella odorata*, Linn. [‡], a native of the Caribbee Islands and Barbadoes. The bark of the tree is rough, marked with longitudinal fissures; this, as well as the berries and leaves, has a smell like Asafœtida when fresh. The timber, however, has a very pleasant odour of cedar, whence the name of " Cedar " is commonly applied to it in the British West India Islands, although it belongs to a totally different natural order to the tree yielding the pencil cedar. It is the cedar of which cigar boxes are mostly made, but other species of *Cedrella* are also used, as can be observed by the differences apparent in the woods of cigar boxes arriving from different localities of cigar manufacture. All these woods are not yet botanically identified, but they are evidently nearly allied and are probably all derived from species of *Cedrella*.

The *C. odorata* forms an immense tree, with a trunk sometimes 6 feet in diameter, and furnishes one of the most useful woods in Jamaica. There are canoes in the West Indies 40 feet long formed out of these trunks, a purpose for which it is extremely well adapted; the wood being soft is easily hollowed out, and being light will carry a great weight on the water. It is also used for the wainscoting of rooms and to make chests, because of

---

[*] Walter, Ann. Chim. Phys. [3] i. p. 498.

[†] Compt. Rend. xxxv. p. 800.

[‡] Sloane's 'Voyage to Madeira, Barbadoes, Jamaica,' etc., 1767, ii. tab. 220. fig. 2; 'Browne's Civil and Nat. Hist. of Jamaica,' 1798, p. 159, t. 10. f. 1; Lamarck, 'Illust. des genres,' t. 137.

its fragrance and because vermin do not so easily breed in it as in many other sorts of wood. It is the "Sweet-scented Bastard Cedar." The "Bastard Cedar" grown at Kew under the name *Cedrella odorata* is the *Cedrella velutina*, D.C. *, possibly identical with *C. villosa*, Roxb. †, a native of the East Indies at Tipperah. It attains a height of 50 feet.

Cedrella angustifolia, D.C. ‡, the narrow-leaved Bastard Cedar, is a native of New Spain and attains a height of 50 feet.

Cedrella Braziliensis, St. Hilaire §, forms a large pyramidal tree in the province of Minas Geraes in Brazil; it grows to a height of 40 feet, and is called the "Brazilian Bastard Cedar." There is a variety of it called *australis* according to St. Hilaire.

Cedrella Toona, Roxb. ‖, is a native of the East Indies, where it is called *Toon*, and by Europeans "Toon Bastard Cedar." Its size is enormous. Hooker mentions ¶ having measured one which was thirty feet in girth at five feet above the ground. This was in East Nepal.

Cedrella alternifolia **, native of Campechy, is a little-known species and probably belongs to a distinct genus.

Juniperus Bermudiana, L., and *J. Barbadensis*, L., are also found in Jamaica and the oil distilled. The method adopted in Jamaica for obtaining the oil is to chip the logs up with an adze and pack the pieces into an iron cylinder, through which the steam generated in a separate boiler is passed. This oil is light brown in colour and very aromatic.

The *Cedar of Lebanon* is the *Cedrus Libani*, a majestic tree found on Mounts Lebanon, Taurus, and Aman. This tree was introduced into England nearly two centuries ago, and there are now many hundreds of fine specimens in various parts of this country. It yields a brownish-yellow oil of pleasant odour, the yield being 2·9 per cent. Its sp. gr. is 0·985. Optical rotation −10° 48′ in 100 mm. tube (Schimmel).

The *Deodar* or *Indian Cedar* is *Cedrus Deodara*, a pyramidal tree when young, with dense, slender, drooping branches, thickly clothed with glaucous green leaves. There are two or three striking varieties :—*C. D. robusta, C. D. crassifolia,* and *C. D.*

* Prodr. i. p. 625.   † Hort. Beng. p. 18.
‡ Prodr. i. p. 624.   § Floræ Braziliæ meridionalis, ii. p. 86, t. 101.
‖ Cor. iii. t. 238.   ¶ Himalayan Journ. i. p. 183.
** Steudel, 'Nomenclator botanicus,' p. 170.

viridis. The species is a native of the mountains of North India, where it forms vast forests up to an elevation of 12,000 feet. It attains a height of 100 to 150 feet and a girth of 20 to 30 feet.

The cedars of California are the *Libocedrus decurrens, Cupressus fragrans, C. Lawsoniana,* and *C. macrocarpa.* The *L. decurrens,* called " White Cedar," is a noble evergreen tree, thriving in sandy soils, and attaining a height of 140 feet and a diameter of 5 to 7 feet. The generic name signifies " incense cedar " on account of the fragrant odour it emits when burned. It is found on the Sierra Nevada Mountains. The *Cupressus fragrans* is synonymous with *C. aromatica.*

The price-lists of American wholesale dealers also quote an " *Oil of Florida Cedar* " at one-third per lb. that of American cedar *.

Oil of White Cedar distilled from *Cupressus thuyoides* is also sold as oil of cedar in America.

---

* American Druggist, Aug. 1886, p. 159.

CHAPTER XV.

CAMPHOR.

The word Camphor is derived from the Arabic *Káfúr*, which, in turn, was derived from the Sanskrit *Kapúra* signifying a pure, white substance. The old English name for Camphor was spelt *Camphire*, and this word has evidently been used by mistake by the translators of the Bible in rendering into English the Hebrew word *Kopher*, which twice occurs in the "Song of Solomon" (i. 14 and iv. 13), and refers to the plant Hennah. The translators have apparently confounded it with *Káfúr*.

Camphor, as it was first known, was obtained from the land known as Kaisûr, the present Sumatra, consequently was the substance we now know as Borneo Camphor, Malay Camphor, Baros Camphor, or *Borneol*, a substance which exists ready-formed in the pith-cavities of the trunk of the *Dryobanalops Camphora*, Colebrook (*D. aromatica*, Gaertner), a magnificent forest tree growing in abundance in Borneo *, and on the West and North-west coast of Sumatra, in forests generally less than 1000 feet above the level of the sea. It is found between Ayer Bangis and Singkel, and as far north as Bacongan and Barus. It is also found extensively on the small British island of Labuan.

A few years ago a vast forest of these trees was found in Jahore, in the protected State of Perak (Straits Settlements), and is conserved by the Government. The tree is there known as *Kayo Kapur* and the camphor as *Kapur baroos*. It is said that all this camphor goes to China, where it is sold for more than its weight in silver (which is not surprising, as five dollars per ounce have been paid for specimens in Borneo). In Jahore the tree frequently attains a height of 150 feet, with a diameter of 6 or 7 feet at 5 feet from the ground.

* As. Res. xii. p. 537.

This tree is also described by Macdonald in the fourth volume of the ' Asiatic Researches.' He says it is found growing in a rich red loam, tending to a blackish clay mixed with a crumbling stone of the colour of marl, and that it grows principally on the West side of Sumatra, from the Equator to nearly 3 degrees North. The tree is straight, extraordinarily tall, and has a gigantic crown which often overtops the other forest trees by many feet. The trunks of these immense trees are sometimes seven feet in diameter, rising straight up to a height of 100 feet without a single branch.

This camphor can only be obtained by the destruction of the entire tree. The trees do not all contain camphor. Many of them contain an oil which is supposed to be the first stage of the formation of the drug, and this would develop into camphor were the tree left unmolested. Both oil and camphor, when found, occur in the heart of the tree, not occupying the whole length of the pith-cavity, but often in spaces of a foot or a foot and a half in length at intervals. The method of extracting the oil is simply by making a deep incision with a Malay axe about 14 or 18 feet from the ground till near the heart of the tree, when a narrower and deeper incision is made, and the oil, if any in the tree, gushes out and is received in bamboos or other utensils. In this manner a party proceeds through the woods, wounding the camphor trees till they attain their object. From a tree containing both oil and camphor, two gallons of the former and three pounds of the latter may sometimes be obtained, but hundreds of trees may be mutilated before camphor is discovered, as the natives have no certain means of ascertaining which trees contain it either in the solid or the liquid state. When camphor is found the tree is felled and cut into junks a few feet long ; these are then split and the camphor is removed from the heart of the tree, where it sometimes occupies a space of the thickness of a man's arm. The quantity varies considerably, from a few ounces up to 15 lbs., and rarely as much as 20 lbs. are obtained. Some trees when felled are not found to contain any at all. The rarity of this description of camphor commands for it a very high price—from 35 to 70 shillings a pound, according to quality. It very rarely arrives on the London market at all, but has some-times been received under the name of Native Camphor. In Sumatra it is used to some extent as incense and for embalming the bodies of the dead. The rest is exported to China, Japan, and other places in Eastern Asia for similar purposes. It is heavier than the "Laurel Camphor" and sinks in water. It has the odour of common camphor, mixed with something of an odour of patchouli. It is less volatile than laurel camphor and rather harder. The best quality occurs in the form of flat colourless crystals, the largest of which rarely exceeds half an inch across. An inferior quality is coarsely pulverulent and of a grey colour. Its chemical composition is $C_{10}H_{18}O$, that of laurel camphor being $C_{10}H_{16}O$. By the Chinese it is called Ping-peün.

COMMON CAMPHOR.

The camphor of European commerce is derived from *Cinnamomum Camphora*, Nees & Ebermaier (*Laurus Camphora*, Linn., *Camphora officinarum*, C. Bahn).

The date at which the Chinese discovered the production of camphor from *Laurus Camphora* is unknown. The tree is distributed throughout the eastern provinces of Central China, on the island of Hainan, and very extensively in Formosa. It also occurs as a forest tree on the islands of Kiushiu and Shikoku of South Japan, its growth being much more vigorous there than in the more northern localities. This description is called "Laurel Camphor" or "Common Camphor"; it is the ordinary camphor of European commerce, and is produced almost exclusively from the Camphor Laurels of Formosa and Japan.

The large and increasing quantities of this drug consumed in all civilized countries make the question of its continuous production and regular supply a matter of considerable importance. It is a well known fact that the sublimation of the crude camphor from the wood is conducted in a primitive, careless way which causes great waste. The Camphor Laurels of Formosa are gradually being destroyed under the careless system employed by the Chinese gatherers. In fact they have been entirely exterminated along the sea-board, and the wood is now obtained from the forests along the frontier between the settlements of the Chinese and the inland mountainous regions still occupied by the aboriginal population. The camphor-gatherers are thus continually exposed to the assaults of the natives, which interrupt the profitable prosecution of this industry. No attempts are made to cultivate laurels to take the place of those destroyed, and a sufficient quantity of the drug is only obtained by constant encroachments upon the territory of the Formosans, so destroying the trees still further into the interior at every new move.

The trees are felled and the small branches chopped up; these, with the chips and twigs, are alone used, the heavy wood being abandoned. A long trough made from a hollow tree, and coated with clay, is placed over 8 or 10 hearth fires, and is half filled with water. Boards perforated with holes are put across the trough, and above each hole is a jar filled with chips of the wood, with earthenware pots inverted above them, the joints being made tight by hemp and clay. The water in the trough is heated to boiling, and the steam passing through the holes saturates the chips, causing the camphor to sublime and condense in crystals in the inverted pots above. The camphor thus obtained is sent from the interior of the island to Tasmin, the principal port, packed in baskets covered with cloths and large leaves. On arrival it is re-packed in tubs or lead-lined cases for export by Chinese vessels to Hong Kong, Shanghai, or Canton; the loss by evaporation while in transit from the place of its production being very large. A yellow oil exudes from the packages of this crude camphor, which is collected and locally known as " oil of camphor." The Formosa camphor sometimes goes by the name of " Chinese Camphor " and it sometimes arrives in India in a semi-fluid state, owing to the addition of water before shipment.

The Japan camphor used to be extracted by boiling the wood with water in an iron kettle and condensing the vapour in an earthenware dome closed at the top with rice-straw. The modern practice is to distil the wood with water in an iron retort fitted with a wooden dome, from which the vapours are led through a bamboo tube to the cooling apparatus. This consists of a wooden box containing seven transverse compartments, and is enclosed in a second box through which water is allowed to flow; the vapours are conducted through all the compartments in succession by means of holes placed alternately at either end of the dividing walls.

The Japan camphor arrives dry; it is lighter in colour than the Formosan and somewhat pinkish. It arrives in double tubs, one within the other, without metal lining; hence it is sometimes called " tub-camphor."

According to a paper read by Professor Maisch at the meeting of the Philadelphia College of Pharmacy in October 1890 *, the camphor-tree is being cultivated successfully in Florida. It seems to flourish in almost any soil and the tree grows rapidly. It is believed that in ten years there will be more camphor-trees than orange-trees in Florida, and that the camphor industry will prove to be more profitable than that of sugar. The camphor obtained from Florida trees approaches more nearly to that of Japan than to Chinese camphor, since the odour of safrol is distinctly recognizable.

Attention to this industry has since been called by the American Pharmaceutical Association, and further statements in reference to this subject were made †. "Messrs. Beach and Son for some time have been experimenting at Palatka, Florida, and from them is gathered some interesting information :—For the growth of the camphor-tree the preparation of the soil is very easy and simple, the tree growing with very little care after first starting. All timber should be cut and piled, not burned, and left to rot on the ground. The soil is thoroughly grubbed where the tree is to stand. The tree is planted and carefully staked to prevent the wind from swaying it when it is first set. For three or four years the ground should be carefully worked around the trees and kept clean, the growth of timber being kept down and grubbed out, but after this time the tree will take care of itself. The first trimming of the camphor-tree is to a sufficient height for the body of the tree; after that thin out the top and shear off the outside of the top, as the young wood is the most productive of camphor." The yield of camphor obtained in this experimental plantation is said to be as much as four per cent., i. e. 1 lb. of crude camphor from 25 lbs. of boughs and leaves. A still of the very simplest description is used; the condenser is a straight pipe, running horizontally through a trough of water; the camphor sublimes on the inner surface of the condenser and is afterwards got out by standing the pipe on end and tapping the sides. A little water is put into the still with the wood and leaves to prevent burning.

Common camphor as it arrives in England contains from 2 to 10 per cent. of impurities, consisting of gypsum, common salt, sulphur, and fragments of vegetable refuse.

---

* Am. Journ. Pharm., Nov. 1890, p. 565.
† Pharmaceutical Era, June 1, 1891.

The European process of refining camphor was long kept a secret, and until towards the end of the seventeenth century the entire camphor used in Europe had to be sent to Holland to be refined. A monopoly was also held for some time in Venice, but at the present day camphor refining is largely accomplished in England, Holland, Hamburg, Paris, New York, and Philadelphia. The method formerly adopted in India was so arranged as to get as much interstitial water as possible into the " camphorcake." The method may be still in use in India; in any case, the apparatus consisted of a tinned cylindrical copper drum, one end of which was removable; into this was put about 14 parts of crude camphor and 2½ parts of water. The cover was then luted with clay, and the drum being placed upon a small furnace made of clay was also luted to the top of the furnace. In Bombay four such furnaces were built together, so that the tops formed a square platform. The sublimation was completed in about three hours. During the process the drums were constantly irrigated with cold water. The same practice was followed at Delhi and at a few other cities in India, and may be still in use.

Camphor sublimed in this way is not stored, but disposed of at once before it has had time to lose weight by drying. It is sold at the same price as the crude article, the refiner's profit being derived from the introduction of water.

The process adopted in Philadelphia has been described as follows * :—The subliming-chamber is a cylindrical iron vessel 20 feet long and 4 feet in diameter, provided with the necessary openings for filling and for the escape of the volatilized camphor. This vessel is so set that it is not touched at any point by the direct fire, and to this end it rests throughout its entire length upon an arch of fire-brick, and the flame is kept under complete control by means of dampers. The condensing-chamber is 30 feet long, 16 feet wide, and 11 feet high, the floor, the sides, and the arched roof being constructed of enamelled bricks set in Portland cement. The object of the inventor of this apparatus was to obtain the sublimate in the form of a finely pulverulent snowy mass, and this was obtained by adding about one tenth per cent. of water to the crude material before sublimation. After an operation was finished the apparatus was allowed to remain

* Oil and Drug News, March 7, 1882.

undisturbed overnight, to become sufficiently cool, and the next day the sublimed camphor was removed and subjected in moulds to a pressure of 2500 lbs. per square inch in an hydraulic press. The finished product was obtained in small cakes highly compressed weighing one ounce. This description, forming part of a paper read before the German Technical Society of Philadelphia, adds that " the refining of camphor until within the past few years was conducted in quite as primitive a manner as the preparation of the crude product," and as a description of the European method is then given, it is to be presumed that the Philadelphia method is the superior one.

The European method is as follows :—The crude camphor is broken up and mixed with about three per cent. of quick-lime and the same quantity of animal charcoal, both in powder. One or two per cent. of iron filings is added, and the mixture, after being thoroughly blended, is introduced through a funnel into a series of globular glass flasks with slightly flattened bottoms and wide necks. When the flasks are about half full, the necks are carefully freed from particles that might have attached themselves, and they are then sunk in a sand-bath, 50 or 100 together, and heat cautiously applied. The heat is suddenly raised from 120° to 190° C., and kept at that point for half an hour, so as to expel the water. The temperature is then raised to 204° C., and maintained at that point for 24 hours. When the crude camphor has melted, the sand is removed from the upper half of each of the flasks and a paper cork placed in the neck. This allows of a lower temperature in the exposed part, and the vapour of camphor not being permitted to escape condenses on the upper part of the flask as a pure cake, leaving all impurities at the bottom. Air, if freely admitted, would render the camphor opaque, but that is prevented by placing a glass bell-jar over the neck of each flask just as the vapour begins to be given off. The whole process lasts about 48 hours, and when completed the flasks are removed from the sand-bath and cold water sprinkled on them. They are thus broken, and a large cake of refined camphor 10 or 12 inches in diameter and 3 inches thick (weighing 9 to 12 lbs.) is removed from each flask. The quick-lime retains the resin or empyreumatic oil ; the iron fixes any sulphur that may be present, and the charcoal removes colouring-matter. Sand is sometimes mixed

z 2

with the crude camphor to allow of a more uniform escape of vapour, and thus save "bumping" or the sudden evolution of confined volumes of vapour. During the operation the temperature must be maintained uniformly at the point of volatilization. The process requires great care, for in addition to the very inflammable nature of the vapour, if too much heat be applied, the sublimate would re-melt and fall back again to the bottom of the flask; and if the heat be not great enough the camphor sublimes in loose flakes instead of a compact cake. After breaking the flasks, the glass fragments have to be carefully separated by hand from the adhering camphor.

Another process consists in subliming the camphor mixed with the other ingredients in any convenient vessel furnished with a large and well-cooled receiver and re-melting the product in close vessels under pressure, cooling the liquid mass as rapidly as possible. In this case the operation cannot be so conveniently watched, involving a difficulty in regulating the heat, and the condenser must be made of some material not subject to rust in order that the product may not be contaminated.

Camphor forms a white, tough, semi-crystalline solid mass which can only be powdered when moistened with alcohol or some other solvent. It dissolves in 1300 parts of water at $20°$ C., and at $12°$ C. in 0.8 part of alcohol of sp. gr. 0.806. It is readily soluble in ether, acetone, chloroform, benzene, and other hydrocarbons, also in glacial acetic acid and in carbon disulphide. It melts at $175°$ C., and boils at $204°$ C., but volatilizes very rapidly at the ordinary temperature, and sublimes when kept in close vessels in lustrous hexagonal crystals which frequently form splendid stars. A fragment of pure camphor placed on a heated spoon or in a warm situation will wholly disappear, and if the sample be quite pure the evolved fumes are fragrant and quite free from acid or terebinthinate odour. As a test distinguishing natural camphor from artificial camphor,—ammonia gives but a slight precipitate in an alcoholic solution of natural camphor, and this precipitate is dissolved on shaking the mixture; a similar solution of artificial camphor under the like treatment gives a flocculent precipitate which remains undissolved. Another distinguishing test is the application of polarized light :—If small fragments of natural and artificial camphor be placed separately on glass slides, and a drop of alcohol added to each, they dissolve and speedily re-crystallize.

If the crystallization of the natural camphor is watched by means of the microscope and polarized light, a most beautiful display of *coloured* crystals is seen, while with the artificial product nothing of the kind is witnessed *.

Pinene hydrochloride or *Pinyl chloride*, $C_{10}H_{17}Cl$, was discovered in 1803 and described as *artificial camphor* †. This hydrochloride of oil of turpentine, or terpene monohydrochloride as it was subsequently called, is obtained by passing dry hydrochloric acid into French oil of turpentine ‡ diluted with carbon disulphide § or benzene ‖ ; according to Wallach these diluents are superfluous, it being only necessary to avoid the presence of any trace of water and to prevent the temperature rising. Pinyl chloride is a crystalline mass which appears and smells like camphor; it is deposited from alcoholic solution in feathery crystals which possess the unpleasant property of welding to a viscous mass, which adheres firmly to all objects with which it comes in contact. It melts at $125°$ and boils at $210°$.

Camphene may be readily obtained by heating equal parts of pinyl chloride and anhydrous sodium acetate to $200°$ for three or four hours with twice the weight of glacial acetic acid or by simply heating a mixture of pinyl bromide and glacial acetic acid for some time in a flask connected with an inverted condenser ¶. It forms a crystalline mass resembling paraffin, of a smell which is considered to resemble a mixture of turpentine and camphor.

Camphoric acid, $C_{10}H_{16}O_4$. According to Wreden **, 150 grams of camphor and 2 litres of nitric acid (sp. gr. 1·27) are brought into flasks of 4 litres capacity, a conducting-tube for the nitrous fumes being fastened in the neck by means of plaster-of-Paris. The mixture is heated on a briskly-boiling water-bath until the vapours are only slightly coloured, the operation lasting about 50 hours. The product is then converted into the sodium salt and this once re-crystallized. About 725 to 800 grams of pure camphoric acid are obtained from 1500 grams of camphor. It is

* Silliman's Journal, May 1851.
† Trommsdorff's 'Journal of Pharmacy,' xi. 2. p. 132.
‡ Pogg. Ann. xxii. p. 89.
§ Ann. Chim. Phys. [3] xl. p. 5.
‖ Ber. Deutsch. chem. Ges. xii. p. 1131.
¶ Bull. Soc. Chim. ccxxxix. p. 6.
** Ann. Chem. Pharm. clxiii. p. 323.

readily soluble in alcohol; 100 parts of water dissolve 0·625 at 12° and 10 parts at the boiling-point. It crystallizes in small plates melting at 178°.

Camphorophone, $C_9H_{14}O$, is formed when camphor is heated to 100° with sulphuric acid [*]. It is an aromatic liquid, boiling at about 210°, and is considered identical with *Isophrone*—obtained, together with other products, by the dry distillation of cane- or grape-sugar with lime [†]. The Phorone which is formed by heating acetone with lime or sodium also appears to be camphorophone [‡].

Bornyl acetate is formed when Borneol is heated to 150° with acetic anhydride. It is a thick pleasant-smelling liquid which boils at 227° and crystallizes on standing for some time in masses which melt at 24°.

Bornyl formate and Bornyl valerate have been prepared, and according to Haller [§] the borneol separated from these ethers is identical with Ngai camphor.

Borneo camphor (Borneol, or Bornyl alcohol) can be artificially prepared by heating an alcoholic solution of ordinary camphor with sodium [||], the following method being employed:—"Fifty grammes of ordinary camphor are dissolved in 500 cb. cms. of alcohol of 96 per cent. in a capacious flask, fitted with a wide reversed condenser, through which 60 grms. of sodium are gradually added in small pieces. The operation must last about an hour and the rise of temperature not be prevented by cooling; it is even advisable to accelerate the completion of the reaction by finally adding about 50 grms. of water, the mixture being well agitated during this process. The product is then poured into 3 or 4 litres of cold water, the separated borneol collected on a filter cloth, well washed, and crystallized from petroleum ether after drying [¶].

According to Berthelot [**], Borneo camphor can be prepared artificially by heating common camphor with alcoholic potash; its formation being attended either with evolution of oxygen or with simultaneous production of camphic acid. The action takes place

* Jahresb. Chem., 1857, p. 483; Ann. Chem. Pharm. cxxiii. p. 298.

† Ann. Chem. Pharm. xv. p. 278, and ibid. c. p. 353 & clxii. p. 303.

‡ Ibid. cx. p. 32; cxii. p. 309; & cxi. p. 279.

§ Compt. Rend. ciii. p. 151.

|| Ber. Deutsch. chem. Ges. xvi. p. 2930; ibid. xviii. Ref. p. 335; ibid. xvii. p. 1036

¶ Wallach, Ann. Chem. Pharm. ccxxx. p. 225.

** Ann. Ch. Phys. [3] lvi. p. 78.

slowly at 100° C., more quickly at higher temperatures in sealed tubes.

Although the Chinese make large use of "ordinary camphor" produced in Japan and Formosa, they attach a much higher value to the camphor obtained in Sumatra and Borneo from the *Dryobanalops aromatica*, Gærtner (*D. camphora*, Colebrooke).

Attention was drawn by Daniel Hanbury * to a third kind of camphor, standing intermediate in value between the two above-mentioned descriptions, that is to say :—

When the Formosan camphor was worth $25 per pecul (133⅓ lbs.),

the Japan	,,	,,	30	,,
the Ngai	,,	,,	250	,,
the Malay, 1st quality		,,	2000	,,
,, 2nd ,,		,,	1000	,,

This Ngai camphor is closely allied to the rare Malay camphor, being identical in chemical composition and very similar in odour, and yet now and from time immemorial the Chinese appreciate the Malay camphor enormously higher than the Ngai which grows in their own country.

The Ngai camphor is produced by the *Blumea balsamifera*, D.C. (*Coniza balsamifera*, Linn.), a very large, herbaceous, or bushy member of the Compositæ. It is figured in Oliver's 'Icones Plantarum' for April 1891, tab. 1957.

This species is a native of India from the Himalayas southwards ; it is common throughout the Eastern Himalaya at altitudes ranging from 1000 to 4000 feet. It occurs in the Khásia Hills, in Chittagong and Pegu, and extends to Singapore and the Indian Archipelago. It is a wonderfully common weed in Burmah and Assam, and is distributed throughout the Eastern peninsula to China. It occurs at Hainan and Kwangtung and on the coast of Formosa. The crude product is known to the Chinese as Ngaï-fên. It is refined at Canton, and is then known as Ngaï-p-ien, or refined camphor, about 10,000 lbs. of which are annually exported from thence. It is also exported from the port of Hoihow in Hainan to the extent of about 15,000 lbs. annually.

The abundance of the plant in Burmah is mentioned by Mr. Thiselton Dyer in the ' Journal of the Linnean Society ' (Bot.), xx.

* Pharm. Journ. [3] iv. p. 709.

p. 414; and Hanbury, who has described this camphor ('Science Papers,' p. 394), mentions that a crude form of the drug is prepared in Burmah.

The properties of Ngai camphor are referred to in 'Neues Repert. für Pharmacie,' xxiii. p. 325, and the observations of Plowman on this subject * are hereafter abstracted.

In the crude state this camphor appears in the form of greyish-white crystalline grains intermixed with fragments of vegetable tissue. By sublimation it crystallizes in the same brilliant distinct form as Borneo camphor; it has the same chemical composition, odour, and density as the latter, being rather heavier than water and not volatilizing quite so quickly as common camphor. Ngai camphor does not appear to come into the European markets at all. In China it is used medicinally and in the preparation of some of the finer kinds of " Chinese ink," to which it imparts the peculiar perfume thought by some to resemble patchouli and ambergris. It is probable that camphors are obtained from two or three species of *Blumea* in Burmah, as there is a variety called *Bang Phien,* also used in the manufacture of scented ink, which forms crystalline masses saturated with a greenish oil, and has a still more powerful odour than ordinary Ngai; or possibly a Ngai may be obtained from some other plant which is not a *Blumea* at all, the native character for the word Ngai being used to designate several plants including both Labiatæ and Compositæ †.

The chemistry of this camphor has been studied by Plowman ‡, who found it to be isomeric with Borneo camphor, that is to say, possessing the same elements in the same centesimal proportions, the same number of atoms in the molecule and the same chemical constitution, but much more volatile, perceptibly different in odour, and of somewhat greater hardness and brittleness. Under the microscope the Borneo and Ngai camphor crystals were found to resemble each other in a remarkable degree, so as to render it impossible to give any characters by which they could be distinguished one from the other. They were principally of the pyramidal form with a varying number of sides, generally truncated, but sometimes perfect, then appearing as the halves of octahedra, while a few seemed to belong to the doubly oblique prismatic system. The same observer remarks that the sublimate of laurel

* Pharm. Journ. [3] iv. pp. 710, 712.
† Ibid. iv. p. 709.　　　　　　　　　　　‡ Ibid. p. 710.

camphor, when viewed microscopically, was found to consist of masses of six-sided tabular crystals with a few six-sided prisms scattered amongst them. Their volatility was so marked that even while under the microscope they lost the sharpness of their angles and soon degenerated into ill-defined masses. He found the sp. gr. of the three sorts of camphor (after exposure *in vacuo* to remove interstitial and adherent air) to be as follows :—

Laurel camphor	. . .	0·995
Ngai	,,	. . .	1·02
Borneo	,,	. . .	1·011

(Unless deprived of adherent air both the Borneo and Ngai camphors will swim on water and turn on it in the same way as common camphor.)

The melting-point of each was found by introducing a small quantity into a thin narrow tube, sealing one extremity and immersing in melted paraffin, then gradually heating till the little column of camphor became transparent and noting the temperature of the surrounding paraffin ; by so doing the mean of the melting- and solidifying-points was found to be 177° C. for the laurel camphor, 204° C. for the Ngai, and 206° C. for the Borneo.

The physical characters of the three varieties of camphor were subsequently studied by Professor Flückiger *. He observed that crystals of common camphor appear in forms belonging to the hexagonal system, as previously shown by Des Cloiseaux †, and they exhibit in polarized light brilliant colours like other crystals not belonging to the cubic system. This is easily demonstrated by examining a clear splinter of ordinary camphor under the polarizing microscope. or by melting a little camphor between two slips of glass and examining it in the same way. According to Des Cloiseaux ‡ the crystals of Borneo camphor derived from *Dryobanalops* belong to the cubic system, an observation confirmed by Flückiger, who adds that they display no action on polarized light when examined in the way above described ; regular cubes and all allied forms being devoid of that optical power. Crystals of Ngai camphor examined by means of the polarizing microscope proved likewise to belong to the cubic system.

---

* Pharm. Journ. [3] iv. p. 829.

† " Etudes du Camphre ordinaire," Comptes Rendus, 1859, p. 1064.

‡ Ibid. 1870, p. 1209, and Pogg. Annalen, 1870, p. 302.

An alcoholic solution of common camphor turns the plane of polarization to the right hand, a similar solution of Borneo camphor also possesses a dextrogyre power but little inferior in degree, but on examining a solution of Ngai camphor in the same way Professor Flückiger found a marked difference : it deviates the ray of polarized light to the *left* hand, being as much lævogyre as the solution of Borneo camphor is dextrogyre.

In 1856 Jeanjean * examined the product of the fermentation of the sugar contained in Madder, *Rubia tinctorum*, Linn. From the fusel-oil which distilled over with the alcohol produced by this fermentation, he separated out a camphor having the composition of Borneo camphor, but possessing exactly the same lævo-rotatory power as Ngai camphor. That chemist examined the nature of the crystals of Rubia camphor, and by boiling them for a short time with nitric acid he obtained the compound $C_{10}H_{16}O$, agreeing in composition with Laurel camphor, but differing from it by its alcoholic solution deviating the ray of polarized light as much to the left- as that of Laurel camphor does to the right-hand.

Operating in the same way on Ngai camphor, Flückiger obtained crystals which were entirely devoid of the peculiar odour of Ngai and rather reminding of ordinary camphor. Its alcoholic solution was lævogyre, and the polarizing microscope proved the crystals did not belong to the same system as their mother substance ; they displayed brilliant colours, showing that there was no longer any question of the cubic system.

It has been observed by Chautard † that oil of Feverfew (*Chrysanthemum Parthenium*, Pers.), on cooling, or by treating it with nitric acid, yields a lævogyre camphor, $C_{10}H_{16}O$, which is in all probability identical with Rubia camphor.

Jeanjean found ‡ that when Rubia camphor is distilled with phosphoric anhydride or chloride of zinc, it yields a hydrocarbon resembling oil of lemon or Bergamot.

The exhaustive researches of Professor Flückiger conclusively prove that the camphor $C_{10}H_{16}O$ obtained by oxidizing the Malay camphor is absolutely identical with ordinary camphor ; the crystals after washing and drying, being examined under the polarizing-microscope, proved most brilliantly not to belong to the cubic

* Comptes Rendus, xlii. p. 857, and Ann. Chem. Pharm. ci. p. 94.
† Journ. de Pharm. xliv. p. 22.
‡ Ann. Chem. Pharm. ci. p. 94.

system but to the hexagonal, and, as he says, the microscope thus enables us to demonstrate very manifestly the transformation of the optically indifferent crystals of *Dryobanalops* camphor into Laurel camphor, even in the smallest fragment.

Oil of camphor is a bye-product in the preparation of camphor in Japan. When the branches of *Laurus Camphora* are distilled with water, the oil volatilizes over and is condensed with the camphor. It is collected separately, and a further quantity drains off from the packages in which the crude camphor is put up.

A Paper on the subject of camphor oil was communicated to the British Pharmaceutical Conference by Mr. John Moss, in which analytical results of a number of samples were given, showing the oil to be of a very variable nature. A Paper was also communicated to the Chemical Society by Mr. Yoshida, a Japanese chemist*, who showed that the oil consists of two hydrocarbons boiling at 150° C. and 172–173° C., about 25 per cent. of camphor and an oxygenated oil, *camphorogenol*, which, through the influence of heat and oxidation, changes into camphor, thus accounting for the deposition of camphor in the oil through age.

A Paper describing observations on numerous samples of this oil was read at the Meeting of the British Pharmaceutical Conference in 1887 by Mr. P. MacEwan, from which it may be gathered that great range of quality was apparent in the samples—some being almost colourless, others very dark and varying greatly in their physical characters. In the discussion on Mr. Moss's paper referred to, it transpired, on the evidence of Mr. Yoshida, that the dark coloration of some samples was the result of heat, or partial carbonization during distillation, they being in fact " residues."

It appears that at present (also in 1887) immense quantities of crude camphor-oil are imported into Europe, principally to Germany. It is then refined, the camphor being separated by a freezing process, and the lighter oil (sp. gr. 0·910) distilled off and sold for making varnishes, drying paints, mixing with soaps, as a protective for leather against insects, and other purposes. The heavier oil (sp. gr. 0·970, boiling between 240°–300° C.) is said to much resemble the natural oil of sassafras, and is also used as a perfume for soap. After safrol has been separated from it a certain quantity of eugenol (identical with the heavy portion of oil of cloves) is obtainable. Heavy camphor-oil is of a pale green colour

* Journ. Chem. Soc., Oct. 1885, p. 779.

and, although volatile, of oily consistence. It inflames with great difficulty and, like all essential oils, has a powerful antiseptic and disinfectant action. Its peculiar power of dissolving resins of all kinds, as well as india-rubber, gives it the power of making varnishes smooth and flexible without seriously injuring their drying qualities. It is also excellently suited for covering the strong smell of mineral oil, and on account of its very low price is useful for perfuming any preparations which have an unpleasant smell.

Oil of Borneo Camphor or *Borneene* is a liquid hydrocarbon isomeric with oil of turpentine, secreted by the *Dryobanalops Camphora*; it is obtained by tapping or felling the trees. It is recorded that from a tree felled in the Island of Lebua about 5 gallons of oil were removed from a cavity scooped in the trunk *. This oil holds in solution a resin and a solid camphor, Borneol. By fractional distillation the oil can be separated into two portions, one lighter than the other, but both of like composition.

Borneene can be obtained by distilling the wood of *Dryobanalops Camphora* with water. The distillate can be fractionated into two oils, one boiling between 180° and 190° C., the other at 260°. A resin can also be separated which melts at a temperature a little above 100° C.

Borneene is formed when Borneol is gently heated with phosphoric anhydride. It can also be obtained from essential oil of valerian by submitting that oil to fractional distillation and heating the first portions of the distillate with potassium hydrate, which takes up valerol while Borneene passes as a distillate.

Planks are sawn from the finest parts of the wood and used for making chests, which are useful by reason of their strong odour for preserving furs and clothes from moth.

* F. & H., Hist. des Drogues, ii. p. 260.

CHAPTER XVI.

CAJEPUT.—LAVENDER.—ROSEMARY.

CAJEPUT.

OIL of Cajeput is distilled from the leaves of several species of *Melaleuca*, myrtaceous shrubs or trees abundant in the Indian Archipelago, the Malay peninsula, Northern Australia, Queensland, and New South Wales. The bulk of the oil commercially dealt in is shipped from Batavia and Singapore and yielded by the *Melaleuca minor*, Smith *, and was designated by Rumphius, who passed fifty years in the Dutch East-Indian possessions, as *Arbor alba minor*, to distinguish it from other closely allied trees which are also called in the East Indies *Kaya-pootie* (white wood) †. Its other native names are *Daun-Kilsjil* and *Caju-Kilan*. It is described by Roxburgh ‡ under the name *Melaleuca Cajuputi*.

Melaleuca minor forms a tolerably erect tree, but crooked and slender; the bark is very light or whitish ash colour and smooth, the exterior bark peeling off from time to time in thin flakes like that of the birch-tree, and the interior part separable into numerous laminæ like the leaves of a book. The branches are scattered, with the slender twigs often drooping as completely as in the willow, round and smooth and, when young, silky. The leaves are alternate, projecting in every direction, but most frequently vertical, short-stalked, narrow-lanceolate; while young, silky, when full grown, smooth, deep green, from 3 to 5 inches long and from half to three-quarters of an inch broad; very aromatic when bruised. The white globular flowers are borne on terminal spikes; while in

---

* Rees' Cyclo. xxiii. p. 2; DeC. Prodr. iii. p. 212: Mueller in Benth. Flora Australiensis, iii. p. 142: Bentley & Trimen, Med. Plants, t. 108.

† Rumph. Amb. ii. p. 72, t. 16.

‡ Flor. Ind. iii. p. 394.

flower there is a scaly conic bud at the apex, which soon advances into a leafy branchlet.

This tree was included by Linnæus in his species *Melaleuca Leucadendron*, a taller tree with a thick, spongy, black bark, which detaches itself in flakes ; its branches, however, are white; no doubt the derivation of the word *Melaleuca* is to be attributed to this strange appearance, μέλας meaning black and λευκὸς white. The form and texture of the leaves of *M. Leucadendron* are very variable, as is also the colour of its flowers. It does not yield so much essential oil as the *M. minor* ; but both oils are very similar in character and hardly to be distinguished.

The distillation is conducted in a primitive way and the oil put up in beer bottles.

The oil has a very powerful aroma, reminding of a mixture of camphor, cardamoms, and turpentine. The sp. gr. of oil of cajeput at 15°·5 C. has been found by West *, on examination of 14 samples obtained from different sources, to be 0·922 to 0·924. In taking the sp. gr. of this oil, it is important to take accurate note of the temperature, in consequence of the high expansion-equivalent of the oil, which has been found by Cripps † to be ·0009 for each degree Centigrade between 13° C. and 23° C. The boiling-point of the 14 samples above mentioned was found to be 174° to 174°·5 C., and copper was detected in all of them. It is generally supposed that the pale greenish tinge in oil of cajuput is necessarily the result of copper, acquired by contamination with the metal of the still, but it has been ascertained that recently-distilled oil has a natural and beautiful green tint.

Professor Tichomirow ‡ examined oil of cajuput spectroscopically and satisfied himself that the green colour of a sample that had been demonstrated to be free from copper was due to the presence of chlorophyllan, or oxidized chlorophyll. As the green colour due to chlorophyllan may in old oils change to brown, it is probable that oil is kept for a time in copper vessels that it may become impregnated with the metal, with the intention of causing it to retain its beautiful natural colour. Copper may be detected by agitating the oil with dilute hydrochloric acid, pouring the liquid into a platinum capsule and inserting a slip of zinc ; if copper be

---

present, it immediately forms a deposit on platinum. The liquid can then be decanted, the copper dissolved and identified by reagents.

Rectified oil is colourless, but acquires the green colour after being a short time in contact with metallic copper.

All genuine cajeput oil of the Moluccas, Malacca, and India are optically lævogyre; several of the Australian cajeput oils are dextrogyre in their action on polarized light.

Schmidl found * that oil of cajeput consists mainly of di-hydrate of cajuputene, $C_{10}H_{16} + 2H_2O$, which can be extracted from the crude oil by fractional distillation between 174° and 178° C.; smaller fractions, perhaps products of decomposition, are obtained from 178° to 250°, at which temperature only a small residue of carbonaceous matter is left in the still, mixed with metallic copper. On treating this residue with ether, a green solution is obtained which, when evaporated, leaves a green resin soluble in the fraction which boils between 174° and 178° and capable of restoring the original colour.

When the di-hydrate is heated to the boiling-point and sulphuric acid added by degrees, it becomes coloured, and on fractional distillation there passes over, between 170° and 175°, the monohydrate, $C_{10}H_{16} + H_2O$.

When the di-hydrate is cohobated with phosphoric anhydride for half an hour and distilled, there passes over at 160°–165° the hydrocarbon cajuputene, $C_{10}H_{16}$, which is a colourless liquid possessing the odour of hyacinths; it is insoluble in alcohol, but soluble in ether and in oil of turpentine; its sp. gr. at 15° C. is 0·850. It is permanent in the air. With gaseous hydrochloric acid it forms a beautiful violet liquid. On continuing the distillation two isomeric hydrocarbons pass over, iso-cajuputene at 176°–178° and para-cajuputene at 310°–316°.

Chloride of cajuputene, $C_{10}H_{16}Cl_2$, is produced by the action of nascent chlorine on the di-hydrate (the rectified oil of cajuput). When the portion of the oil distilling between 174°–178° is mixed with very dilute nitric acid, and hydrochloric acid gas is passed into the liquid, a violent action takes place in a few minutes, chlorine and nitrous gas being evolved, and if the passage of the hydrochloric acid gas is continued, chloride of cajuputene ulti-

---

* Trans. Roy. Soc. Ed. xxii. (6) p. 360, and Journ. Chem. Soc. xiv. p. 63.

mately sinks to the bottom as a limpid brown oil which may be
freed from adhering nitric and nitrous acid by distillation over
strong potash-lye. It has a pungent odour and may be kept without alteration for any length of time, but is decomposed by distillation.

According to the researches of Voiry *, a sample of green oil of
cajeput examined by him presented considerable analogy to oil of
Eucalyptus globulus; on fractional distillation two thirds passed
over between 175° and 180° C. Below that temperature were
obtained butyric, valerianic, and benzylic aldehydes, also a lævogyre terpene, $C_{10}H_{16}$, that formed a crystalline monohydrochlorate.
After 180° the distillation was continued at reduced pressure, and
there was separated, amongst other bodies, a small fraction of
terpineol, $C_{10}H_{17}$. OH, identical with borneol, in the solid form.
This body has since been obtained by Messrs. Schimmel (April,
1892) from the fraction of cajeput oil boiling between 215° and
220° C., by means of cooling in a refrigerating mixture of solid
carbonic acid and ether. The terpineol thus obtained crystallizes
at ordinary temperatures, its melting-point after several crystallizations being from 33° to 34° C.

Melaleuca viridifolia †. Synonymous with *M. leucodendron
angustifolia*, Lin. fil.‡; *Metrosideros quinquenervia*, Cav. §; *Metrosideros coriacea*, Poir.‖; *Metrosideros albida*, Siebl.¶ A native
of New South Wales and New Caledonia; it attains a height of
20 feet and bears pale greenish flowers. The oil distilled from its
leaves in New Caledonia is locally called *Niauli*. The physical
properties of this oil do not appear to have been examined.

The following Australian oils of cajeput are mentioned in
Maiden's ' Useful Native Plants of Australia,' and samples of them
were exhibited at the London Exhibition of 1862, but they do not
appear to have any commercial demand in England :—

> *Melaleuca decussata*, R. Brown ; dark yellow; resembles cajeput
> in taste and odour ; yield, about ⅔ per cent.; sp. gr. 0·938 ;
> boiling-point 185°–209°. This shrub is only 3 to 6 feet
> in height **; it is found in Victoria and South Australia;

---

* Journ. Pharm. Chim. Aug. 1888, p. 149.
† Gærtn. Fruct. i. p. 173, t. 35.   ‡ Suppl. 342.
§ Icon. iv. t. 33.   ‖ Suppl. iii. p. 365.
¶ Pl. exsic. Nov. Holl. p. 349.   ** Ait. Hort. Kew. p. 415.

flowers lilac. A coloured figure is given in the ' Botanical Magazine,' t. 2268.

M. ericifolia, Smith ; bright yellow ; resembles cajeput in taste and odour ; yield, 5 ozs. of oil from 1000 lbs. of leaves ; sp. gr. 0·899–0·902 ; boiling-point 149°–184°. Rotation +26° (*Gladstone*) ; shrub of 4 to 6 feet in height ; native of New South Wales and found in all parts of the Colony except West Australia ; flowers pale yellow *.

M. genistifolia, Smith † ; yellowish green ; mild odour and taste ; 100 lbs. of leaves are said to yield 10 ozs. 2 dr. of oil. Native of New South Wales, where it is called by the English " White tea tree " ; occurs also in North Australia. This shrub varies in height from 6 to 20 feet ; its flowers have reddish petals and yellow stamens.

M. linarifolia, Smith ‡ ; a bright yellow limpid oil with pleasant cajeput-like odour and agreeable mace-like taste, with an after-taste reminding of mint ; sp. gr. 0·903 ; boiling-point 175°–187°. Yield, according to Bosisto, 28 ozs. per 100 lbs. of leaves ; rotation +11° (*Gladstone*). This tree is synonymous with *Metrosideros hyssopifolia*, Cavanilles §. It is a native of New South Wales and Queensland ; it attains a height of 20 or 30 feet ; flowers cream-coloured ‖.

M. squarrosa, Smith ¶ ; green, in very small quantity, and (according to Bosisto) of unpleasant taste. This tree attains a height of from 20 to 40 feet : it is found in South Australia, Victoria, New S. Wales, and Tasmania. Flowers yellowish. It is figured in Curtis' ' Botanical Magazine,' t. 1935.

M. Wilsonii, F. von Mueller ; a bright yellow oil very similar to cajeput ; sp. gr. 0·925 ; 100 lbs. of fresh leaves yield 4 ozs. of oil. Native of Victoria and South Australia.

M. uncinata, R. Brown ** ; a green oil ; sp. gr. 0·925 ; optical rotation +1° 40'. The principal part distils between 175° and 180° C. The lowest boiling fraction possesses a decided Spike or Rosemary odour. The second fraction has a pure

---

* Smith, Exot. Bot. i. t. 34. † Ibid. i. t. 55.
‡ Ibid. t. 56. § Icones Plantarum, iv. t. 336. f. 1.
‖ See F. Mueller in Bentham's ' Flora Australiensis,' iii. p. 140.
¶ Linn. Trans. vi. p. 300. ** Ait. Hort. Kew. iv. p. 414.

2 A

odour of Cineol, which is the most important constituent of the oil. The highest boiling portion probably consists of Terpineol *. The shrub is found in South and West Australia, New Zealand, and Queensland; in height it is only about 4 to 6 feet; its flowers are yellowish.

M. leucadendron var. *lancifolia*. Like common cajeput oil it consists mainly of Cineol; has a sp. gr. of 0·955 and rotates polarized light 3° 38' to the left †.

M. acuminata. Colourless oil, of an odour slightly resembling that of Juniper berries; sp. gr. 0·892; optical rotation −15° 20'; contains a considerable quantity of Cineol ‡.

Lavender §.

The general aspect of the various species which compose this genus of labiate plants, although presenting very characteristic differences, merges gradually from one species to another; all are, in their native habitat, small ligneous undershrubs, from 1 to 2 feet in height, with a thin bark, which detaches itself in scales; the leaves are linear, persistent (until the stems become woody), and covered with numerous hairs, which give the plant a hoary appearance. The most commonly known species are *Lavandula vera, L. spica*, and *L. Stœchas*.

Commercially the *L. vera* is the most valuable by reason of the superior delicacy of its perfume; it is found on the sterile hills and stony declivities at the foot of the Alps of Provence, the lower Alps of Dauphiné and Cevannes (growing in some places at an altitude of 4500 feet above the sea-level), also northwards, in exposed situations, as far as Monton, near Lyons, but not beyond the 46th degree of latitude; in Piedmont as far as Tarantaise, and in Switzerland, in Lower Vallais, near Nyon, in the Canton of Vaud, and at Vuilly. It has been gathered between Nice and Cosni, in the neighbourhood of Limoné, on the elevated slopes of the mountains of Western Liguria, and in Etruria on hills near the sea.

---

* Schimmel, April 1892.   † Ibid.   ‡ Ibid.

§ Abstracted and revised from the Author's original articles in ' The Chemist and Druggist,' Feb. 28 and Mar. 21, 1891.

The *L. spica*, which is the only species besides *L. vera* hardy in this country, was formerly considered only a variety of *L. vera*. It is distinguished by its lower habit; much whiter colour; the leaves more congested at the base of the branches, more persistent; the spikes denser and shorter; the floral leaves lanceolate or linear; and the presence of linear and subulate bracts. *L. spica* yields by distillation an oil termed " oil of spike," or, to distinguish it from oil of *L. Stœchas*, " true oil of spike." It is darker in colour than the oil of *L. vera*, and much less grateful in odour, reminding of turpentine and rancid cocoa-nut oil. It is used by painters on porcelain and in the manufacture of varnishes; it is often largely admixed with oil of turpentine.

L. Stœchas (Στιχὰς) was discovered prior to the year 50 A.D. in the Stœchades Islands (now the Islands of Hyères, south of Toulon), hence the name. At present it is found wild in the South of Europe and North of Africa, also at Teneriffe. Its leaves are oblong linear, about half an inch long in the wild state and fully an inch long when cultivated, with revolute edges, and clothed with hoary tomentum on both surfaces; the spike is tetra-gonal, compact, with a tuft of purple leaves at the top; the calyces are ovate and slightly shorter than the tube of the corolla. The whole plant has a strong aromatic and agreeable flavour. There is a variety of this species (*L. macrostachya*), native of Corsica, Sicily, and Naples, which has broader leaves and thicker octagonal spikes.

L. Stœchas is known in Spain as " Romero Santo " (Holy Rose-mary). Its essential oil (also that of *L. dentata*) is there extracted for household use, by suspending the fresh flowering stalks, flowers downwards, in closed bottles and exposing them for some time in the sun's rays ; a mixture of water and essential oil collects at the bottom, which is used as a hæmostatic and for cleansing wounds. The sp. gr. of Spanish oil of *L. Stœchas* is 0·942 at 15° C. It boils between 180° and 245° C. The odour of this oil is not at all sug-gestive of that of lavender, but resembles more that of rosemary, possessing also the camphoraceous odour of that oil. In India this oil is much prized as an expectorant and antispasmodic.

The other species which are distinctly characterized are *L. pe-dunculata*, *L. viridis*, *L. dentata*, *L. heterophylla*, *L. Pyrenaica*, *L. pinnata*, *L. coronopifolia*, *L. abrotanoides*, *L. Lawii*, and *L. multifida*. The *L. multifida* is synonymous with *L. Burmannii*.

In Spain the therapeutic properties of *L. dentata* are alleged to be even more marked than in the oils of any of the other species of lavender. In odour this oil strongly suggests rosemary and camphor. Its sp. gr. is 0·926 at 15° C. It distils almost completely between 170° and 200° C.

Although *L. Stœchas* was well known to the ancients, no allusion unquestionably referring to *L. vera* has been found in the writings of classical authors; the earliest mention of it being in the 12th century, in the writings of St. Hildegard. It was known to the Welsh physicians in the 13th century under the name of *Llafant* or *Llafantly*.

The best variety of *L. vera* (and there are several, although unnamed), improved by cultivation in England, presents the appearance of an evergreen undershrub of about 2 feet in height, with greyish-green linear leaves, rolled under at the edges when young; the branches are erect and give a bushy appearance to the plant; the flowers are borne on a spike composed of 6 to 10 verticillasters, more widely separated towards the base of the spike. In young plants 2 or 4 sub-spikes will branch alternately in pairs from the main spike; this indicates great vigour in the plant, and the flowering tendency is then so great that if these spikes are nipped off others immediately throw out. This strength rarely occurs after the second year of flowering.

The floral leaves are rhomboidal, acuminate, and membranous, the upper ones being shorter than the calyces; the bracts are obovate; the calyces bluish, nearly cylindrical, contracted towards the mouth, and ribbed with many veins. The corolla is of a pale bluish violet, of a deeper tint on the inner surface than the outer, tubular, 2-lipped, the upper lip with 2 and the lower with 3 lobes. Both the corolla and calyx are covered with stellate hairs, amongst which are imbedded shining oil-glands, to which the fragrance of the plant is due. These glands are fractured by very slight pressure, and then the fragrance is perceptible, but not otherwise, unless it be from remnants of oil from glands previously fractured by handling the flower or friction of one spike against another.

The *L. vera* was identified in 1541 and introduced into England in 1568, flourishing remarkably well under cultivation and yielding an oil far superior in delicacy of fragrance to that obtained from the wild plant, or from the same plant cultivated in any other country. In a favourable locality a single plant will form a bush

Fig. 12.

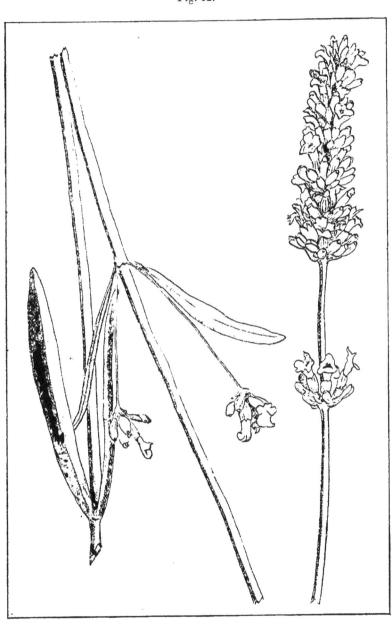

L. vera, 29th July.   (Natural size.)

5 feet in diameter if not crowded by other plants and throw up spikes nearly 5 feet high.

When it is remembered that north of the 50th degree of latitude the vine yields little but garlands of leaves, and that we should attempt in vain to cultivate the olive north of the 44th degree, it may seem strange that *Lavandula vera*, which is a native of about the same climate as these, should resist unprotected the vigorous frosts of this country; even at Upsala, lat. 59° 51′ N., in the Botanic Garden, it merely requires the shelter of a few branches to

Fig. 13.

L. vera.   ½ natural size.   Appearance in middle of March.

protect it in the winter; but this hardiness may be accounted for by several physiological reasons.   Like all fruticose labiates which have a hard compact tissue, and contain much oily matter, the lavender absorbs less moisture than herbs which are soft and spongy, and as it always prefers a dry, calcareous, even stony soil, the northern cultivators find that by selecting such localities, the tissues of the plant take up so little water that our ordinary frosts do not injure it.   In a northern climate the length of the days in summer, and the natural dryness of the air, compensates in some measure the reduction of temperature, and the plant is matured only to the extent sufficient for the purpose for which it is grown. Perhaps the suspension of vital action during winter, which must be more complete in northern latitudes, as our frosts are more severe, tends to preserve certain plants natives of the South, for it is observed that all plants are more sensitive to cold when vegetation is active than when it is at rest. The vine is an instance of this. On the other hand, when the plant is cultivated further south than its natural boundary, the same causes seem to exert their influence. Lavender is cultivated on the mountains of Yémen in Arabia; the humidity, increasing inversely to the latitude, compensates the exhaling force of the sun's rays, and the elevation of the locality the effects of the heat. Thus is confirmed, both in North and South, the law of vegetable physiology observed by De Candolle in the temperate climates of France and published in his ' Essai de Géographie Botanique,' that " plants can best resist the effects of cold in a dry atmosphere, and the effects of heat in a humid atmosphere." A mild, damp winter, like the one of 1889–90, does more harm than a seasonable frost, as the plants are apt to make green shoots prematurely, and the late frosts nip off these tender portions, each of which would otherwise have produced a flower-spike; but the frosts of 1890-91 and of 1891–92 were so exceptionally severe that English growers lost many plants.

The stems and branches of lavender being ligneous and strong, are able to resist the force of the wind, and the plant thrives best in a perfectly open locality, where the air circulates freely; the oil and resin which it contains enables it to resist the parching action of the wind and sun; thus on the most arid and sterile ground on the mountain-sides in the South, especially in Spain, plants of this genus flourish with more vigour in the season when most other vegetation is scorched up by the ardent rays of the sun. The *L. vera* seems to have a predilection for such spots. Certainly the plants then assume a more stunted appearance than in richer soil, but at the same time the perfume is stronger and sweeter; the calyces become charged with oil-glands and yield a greater abundance of volatile oil. In a very moist soil the water penetrates too much into the tissues, detaches the bark, the plant blackens at the root, and a white fungus attaches to the main stem and lower branches; it becomes feeble, diseased, and dies. A rich soil furnishes too much nutriment, the plant grows very straggling and herbaceous, becomes overcharged with water relatively to its assimilating and elaboratory power, especially if growing in a cold climate, and the equilibrium of the chemical proportion necessary for the formation of natural juices becomes deranged at the expense of the quantity and quality of the volatile oil produced.

These facts, long ago pointed out by Linnæus, have been verified in England ; some years ago a disease manifested itself in most of the plantations, and not being understood by the growers, was not remedied (in fact is not generally understood and remedied at the present time), the acreage under cultivation decreased, and partly owing to this and a scarcity occasioned by a failure in the crop, the price of the oil rapidly rose from 50s. to 200s. per lb. ; consequently, with the continually increasing demand and the continued rise in price, manufacturers of lavender-water and compound perfumes in which oil of lavender is a necessary ingredient commenced to buy the French oil, and venders of the English oil commenced to adulterate largely the English with the French oil. By degrees the French oil became almost entirely substituted, and at present it is difficult to purchase true English lavender-water of a quality equal to that vended twenty years ago, except at a few first-class houses. The exorbitant profits demanded by chemists and druggists and the incomprehensible will of the public to buy anything *cheap*, however bad, have encouraged a marvellous increase in the figures of the imports of French oil.

In 1880, when the price had reached 125s. per lb., it was pointed out by an eminent London Firm of druggists that unless the cultivation in England were extended, the price would become prohibitive ; inferior oils would be introduced into the market, and so destroy the popularity of this beautiful perfume. The price, still rising, did, in fact, induce this importation, and to this day the bulk of chemists and perfumers continue to use foreign oils at from 10s. to 18s. per lb. notwithstanding the fall in the price of English to 60s. (and it has been much lower). The constant demand, however, in America (where people still exist who will have things good) will yet support the price of English oil, which is the finest produced in any country.

Attempts were made by a French manufacturing perfumer to establish a plantation in the South of France of plants taken from parent stems grown in England. The result was, that the young plants deteriorated to their original condition—even in their native habitat. The character of a plant and the character of its produce depend on even more than a similarity of soil and geographical position : it is asserted that a good judge can distinguish between oils produced in two adjacent fields. The difference in odour is very apparent between the oils produced in Hertfordshire, Surrey, and Sussex.

The principal lavender-plantations of England are in the districts of Mitcham, Carshalton, and Beddington in Surrey, Hitchin in Hertfordshire, and Canterbury in Kent. The Surrey and Hertfordshire plantations are situated on the outcrop of the Chalk which surrounds the London basin. The most suitable conditions of soil are found to be light brown loam over chalk, the depth of the loam being very slight, varying from 6 to 20 inches. Of course on slopes there is hardly any soil at all in some parts, and in some of the hollows it is as deep as 4 feet. There is often a thin seam of Coombe rock, of a soft, dry, pulverulent nature, light brown in colour, between the loam and the chalk; this is very beneficial to the plant. In selecting ground, a site should be chosen which slopes rather to the south or south-west. A wood or copse on the south-west boundary is of some advantage to ward off or break the force of gales which may occur in July; but in the autumn and winter months, the plants having been clipped, present little resistance to the wind. A July gale may do some damage; the tall spikes wave like corn in the wind, but with such weight and momentum as to cause the woody branches to sometimes split at the junction with the stem,—yet tall trees in the immediate vicinity are objectionable, keeping off too much light and air. Hedges are useful to keep off dogs, but they should be cropped very close and low. Walls of any sort are very objectionable, as the wind is thrown back from them,—more dangerous in effect than a direct wind, which drives through the plants.

In cleaning the ground preparatory to planting, all weeds should be carefully rooted out, stacked in small heaps, and burned; the ashes being afterwards distributed over the ground. A soil of the nature described is generally full of large flints ; those on or near the surface should be taken out, as being quite unnecessary and only likely to break the tools. These flints can be used as foundations to paths or sold ; they have a marketable value for road-mending. It is advisable to prepare the soil previous to planting by trenching in a quantity of short straw and refuse from stables, dung mixed with straw, but not very much dung. The ashes of wood and cinder-ashes can be used with advantage. The ground should be allowed to lie fallow until the spring, when all weeds should be again cleared and the whole ploughed over.

In May, according to the weather, the young plants can be dibbled into their places, in rows north to south, 4 feet apart and 6 feet between the rows. Some growers plant much closer, hedge fashion, but it is a false economy; the flower-bearing capacity of the plant is decreased, absence of light by interweaving of the branches stunts the growth and causes young spikes to decay prematurely; also, if the rows are less than 5 feet apart, it becomes impossible to weed clean between them; and in harvest still more difficult to walk between them to cut the flower and deposit it until enough is cut to carry up to the still.

A stock cannot be obtained from seed, as in this country the plant does not mature its seed, and of course foreign seed would not produce the right variety. Neither should the roots of old plants be divided, as such a process engenders a fungoid disease in the root, and such plants die; but cuttings from established plants will freely strike between May and October; they should be of young growth and taken at the joint with a heel. Young shoots strike more readily than woody branches and produce more compact plants. They can be put 3 or 4 inches apart, shaded from the sun, and watered. They can be transplanted the following spring to their proper place. Mild, moist weather should be selected for putting them out. The process of roughly taking cuttings or clippings of branches and striking them thickly planted together in rows or trenches, as adopted by some growers, is about the worst method of propagation. Plants propagated in this way have to be dragged apart, tearing the bark and injuring the delicate roots, so inducing the fungoid disease above-mentioned on the wounded organs. The effect of this disease is that in July, when the plant is making its greatest demand on the root, the young spikes, which are just beginning to show colour, droop and wither. Plants presenting these symptoms should be at once rooted out and burned, and their places left vacant until the next May.

After planting, the weeds should be carefully kept under, and, if cut down when young, they perish on the spot; but any large weeds should be lifted bodily out and placed in a heap on any vacant ground to rot or to be burned. As the roots of the lavender often spread near the surface it is dangerous to hoe deeply. The weeds, as a rule, can be got out with a hoe only 1 inch deep in the blade; the blade should be 9 or 10 inches long, in order to reach under the branches without dragging them.

The presence of the blue corn-flower and the poppy indicate a dry soil suitable to lavender; these and such weeds as groundsel, thistles, and chickweed, which seed in abundance, should be eradicated whilst quite young, otherwise the work will be endless and the plants choked.

The ground-ivy is a fearful enemy, the disagreeable odour of its foliage being so strong that a few pieces of the weed accidentally put into the still would spoil the entire charge of flower. Weeds cut down and buried before they seed mostly decay; but couch-grass, the creeping-thistle, and some others are not destroyed in that way, and the only plan is to burn them.

The seeds of some weeds (such as charlock) will retain vitality for centuries, and germinate on the first convenient opportunity. Some weeds, such as dock and dandelion, seem to be strengthened by having their tops chopped off, and shoot out again with increased vigour.

When lands are ploughed up once a year these difficulties with weeds are not so great, but lavender is expected to last 4 or 5 years at least, so the land cannot be ploughed over.

For the convenience of gathering in the harvest and removing weeds, it is well to make paths, about 4 feet wide, intersecting the rows, about 50 yards apart, and leading up to the stillery. Thus if rows are planted north and south, the paths should run east and west and lead to a main path.

The young plants should be prevented as far as possible from flowering the first year, by clipping them with shears (sheep-shears); this throws the strength into the lateral shoots, and makes the plants bushy and compact.

Some growers clear out all the plants after five years, and substitute a crop of potatoes or other vegetables. Such crops are not so remunerative, and the elements drawn from the soil can be replaced without changing the crop. If, for instance, a beginner start a plantation of five acres, he may find it difficult to procure more than enough plants for one acre at first; he then, by propagation, gradually fills up the other four acres, and each acre will contain plants of a different age; there will then be four acres in flowering condition and one acre of young cuttings. In the autumn of the fifth year, acre No. 1 can be cleaned, the old plants stacked in heaps and burned, and the ashes distributed over the land. This acre can then be ploughed, manured, and cross-ploughed, and left fallow till the next May, when it can be stocked with young plants out of No. 5, and there will still be enough young plants to plant two or three more acres, and so on.

The harvest depends upon the season : white frosts in May retard the growth (especially of plants near overhanging trees), but black frosts do not injure them so much. As a general rule the harvest may commence the first week of August if the weather be dry. The cutting commences as early as possible in the morning, before the dew is off. Flowers so cut seem to yield more oil than those cut in the heat of the day ; but the necessity of getting the whole crop cut quickly, when once ripe and ready, requires cutting to go on all day long, unless many men be employed. In wet weather it is better not to cut at all. If the days are bright and hot during June and July, the yield will be better in quality and quantity than if wet and dull. There are sometimes seasons of unusual heat and drought ; I have then noticed a smaller quantity yielded, but an infinitely superior quality. Such may be termed " comet " years, as in wine ; and the oil of such years is sold on *date* accordingly, at a higher price than the ordinary. There are also years of unusual rainfall, such as 1887 ; the quality of the oil of such years is poor, and plants are apt to become diseased through excessive moisture.

In taking the crop, the entire length of flower-stalk should be cut away from the plant, but as little of the stalk as possible should be put into the still. It pays the time and labour of the distiller to cut away a great deal of this length of stalk, which would otherwise only fill up the still with useless material, and the still will hold quite four times as much flower when this is done. The flower should be put into the still as soon as possible to prevent fermentation and loss of oil by evaporation ; and until it is so put in, it should not be allowed to lie in the sun.

Different materials require different methods of distillation, partly by reason of difficulty or facility which the oil finds in escaping from the glands or ducts which contain it ; partly by reason of the higher or lower boiling-point of the oil, also taking into consideration whether it can deteriorate by excessive heat or by contact with water.

As regards the flower under consideration, it has been proved that actual contact with water in the body of the still yields the finest product.

The body of the still should be very shallow compared with its width, to allow of the rapid passing over of the oil as soon as it escapes from the oil-glands of the flower, otherwise the first particles volatilized in the lower part of the still would condense in the upper part of the charge and remain there until volatilized a second time by the rising heat. The less the oil is subjected to the action of heat, the finer its aroma will be. With a broad shallow still, the area of the furnace must be proportionately large, *i. e.* if distillation by naked fire be the mode adopted, and by some distillers it is preferred to steam-jacketed stills. In order to prevent any chance of the flower scorching, also for the convenience of lifting the charge in and out, the flower should be packed in a basket made of sheet-copper, perforated with holes of about ⅛ of an inch diameter to allow free circulation of water. The perforations must be both at sides and bottom, and the basket furnished with a perforated copper lid and lugs on each side to allow of its being lifted in and out of the still by overhead tackle, such as a Tangye chain-pulley or a swing-crane. The copper basket should fit to within an inch of sides and bottom of still, and be provided with four copper legs to rest on. Sufficient water is let into the still to just reach the top of the basket. The head is then luted on and secured by clamps. The condenser consists of a coil of pure tin-pipe, and is connected with the head by a curved pipe secured at each end by a brass union, the pipe being passed through the unions and ground to a flush surface.

If well constructed, the vapours will begin to condense in about three-quarters of an hour from the time of lighting the furnace, and the condenser is then copiously supplied with a flow of cold water. The bulk of the oil and the finest part of it will condense during the first half-hour. The oil which condenses afterwards is inferior, and should be collected in a separate receiver.

Before loading in a fresh charge, the dirty water in the still should be siphoned out; the fire need not be drawn if fresh water be run in at the same time as the dirty water is coming out,—then, the fresh charge being packed, is lifted in, and the operation goes on as before, and can proceed continuously until the whole crop be distilled. On large plantations it may be more convenient to employ several small stills than one large one.

The process of distillation by driving dry steam through the flower is quite unsuitable to this flower as it scorches the oil.

Freshly distilled oil has a herby odour, but if put away in a cool place in the dark and the bottles be left *uncorked* and loosely covered with cotton wool to keep out dust, these odours will pass off in about three months. The herby odour is partly due to particles of water of vegetation being held in mechanical suspension or in solution; this may be removed by drying the oil with calcic chloride. The fine mellowness of matured oil is the result of chemical change *.

Spirit of wine should not be added to *new* oil; it would cause a slight etherification of very rancid odour.

Stocks of oil should always be kept in one uniform low temperature, in a dark place. The newly distilled oil is almost colourless, or very pale straw-colour when viewed in bulk. It improves or mellows by keeping, and is then catalogued as " matured " oil, with date affixed. This improvement is distinctly noticeable during the first five years; after which, it has a tendency to deteriorate by oxidation and resinification; this may be prevented by admixture with 20 per cent. of grape-spirit of 60 o.p.

Redistillation, or rectification, is said to improve the quality; but this is *very* questionable. The less it is submitted to the action of heat the better. However, if necessary to rectify it, the first step would be to wash it thoroughly by agitation with an equal volume of cold water to which a little carbonate of magnesia had been mixed; allow it to rest in a cool place to separate, then draw off the supernatant oil and distil it with an equal bulk of fresh water, the heat being applied by a steam-jacket, and the distillate collected in two portions, the first four-fifths being retained and the remainder kept apart as inferior. It may also be rectified by Dragendorff's process. (See Rosemary, p. 373.)

Pure English oil of lavender consists of a mixture in variable proportions of an oxygenated oil and a hydrocarbon $C_{10}H_{16}$, and

---

* It has recently been discovered by Raoul Pictet that the mellow flavour and fine bouquet observable in old brandy can be rapidly produced in young brandy by subjecting it to the action of intense cold (produced by the sudden evaporation of methyl-chloride).

various writers have affirmed that it contains in solution a stereoptene identical with camphor, but whether recently distilled oil contains it, is doubtful. Its sp. gr. is variable between 0·876 and 0·880 at 15° C. It is completely soluble in three volumes of alcohol of 0·894.

The principal adulterants are French oils of *L. vera* and *L. spica*, oil of turpentine, and alcohol.

The presence of French oils is recognized by the keen sense of smell possessed by accustomed buyers. The sp. gr. of pure French oils of *L. vera* varies from 0·885 to 0·887. The commercial French oils are very frequently adulterated with turpentine, which has a sp. gr. ranging from 0·856 to 0·870.

Oils containing turpentine and pure French oils of sp. gr. over 0·890 are less soluble in a given volume of alcohol of a given strength than are pure English oil and pure French oils of sp. gr. 0·885 to 0·889.

Dragendorff found * that 1 c. c. of oil of lavender (prepared by himself) dissolved in 2·3 c. c. of 65 per cent. alcohol (according to Tralle's alcoholometer, which gives the percentage volume for the temperature of 60° F.), the observation being made at a temperature between 15° and 20° C. He also found that on experimenting with ten different kinds of oil of turpentine, most of them required 3·5 to 3·75 of 92 per cent. alcohol to dissolve 1 c. c., and that 1 c. c. of a mixture composed of 10 parts oil of turpentine and 90 parts oil of lavender did not form quite a clear mixture with 5 c. c. of 65 per cent. alcohol.

On these figures it would be natural to expect that some reliance could be placed on a test which has been frequently given, viz. the solubility of 1 part of lavender oil in 3 parts of 70 volumes per cent. alcohol at 16° C.; but some recent investigations tend to show that pure oils may be brought under suspicion by this test, and that actual adulteration with 20 per cent. of turpentine is not detected by it †. It would therefore be well if distillers who have yet on hand some stock of their own make (in which of course they can rely for purity) to repeat the experiments.

The optical rotation of pure Mitcham oil has been recorded as —4° 2' in a column of 50 mm. and that of French oils —4° 15' to

* Neues Repert. für Pharm. xxii. i.: and Pharm. Journ. [3] vi. p. 541.
† Schimmel & Co., 'Bericht,' Oct. 1890.

—9° 20' in a column of 100 mm., but the figures are too variable
to serve as a test. The rotatory powers of turpentines also vary
according to the source from which they are derived.

The presence of alcohol may be detected by agitating the oil with
a few small pieces of calcic chloride; these remain unaltered in a
pure oil, but dissolve in one containing alcohol, and the resulting
solution separates, forming a distinct stratum at the bottom of the
vessel. When only a very little alcohol is present, the pieces
merely change their form and exhibit the action of the solvent on
their angles or edges, which become more or less obtuse or
rounded.

Santaline, the colouring-matter of Red Sanders wood, is nearly
insoluble in pure oil of lavender, and exerts no marked influence
on it, but it is freely soluble in oil adulterated with rectified spirit.
Aniline-red can be used in the same way; a drop of oil let fall on
a crystal of "magenta" and gently pressed will not dissolve the
colour unless alcohol be present.

By Davy's molybdic-acid test for alcohol in fluids generally,
it is asserted that an admixture of one per cent. of alcohol in
essential oils can be detected. One part of molybdic acid is
dissolved with gentle heat in 10 parts by weight of pure concentrated sulphuric acid, and kept in a well-stoppered bottle to
prevent absorption of moisture. A little of the oil under examination is agitated with a small quantity of distilled water in a
pipette, the small orifice being closed. When the oil and water
have separated, a few drops of the watery portion is then allowed
to run out and come in contact with 3 or 4 drops of the molybdic-
acid solution placed in a white porcelain capsule. If the oil has
been adulterated with alcohol, an intense azure-blue coloration
will develop in a few moments *.

The revenue of profit per acre depends much on the season and
age of the plants, but an average yield of 25 lbs. weight of oil per
acre may be reckoned if worked carefully, and the weather be
favourable. I have, in fact, gathered 35 lbs. per acre in a good
year. Much depends on the energy and personal superintendence
of the grower and care in the distillation. Land cultivated in a
slovenly manner rarely yields much profit in any kind of farming,
and if the distillation of this flower be left to the care of inex-

* Pharm. Journ. [3] viii. p. 201.

perienced or untrustworthy people, the oil may be either spoiled or "manipulated."

A scientific examination of perfectly pure French oil of lavender has recently been executed in the laboratory of those indefatigable investigators Messrs. Schimmel & Co. of Leipzig, the results of which are of such practical importance in the evaluation of the oil and of such interest generally, that it may here be allowable to quote at some length from the details published up to the present (April, 1892) by that firm :—Lavender oil contains only minute traces of low-boiling constituents. Cineol, which occurs, as is well known, in the *Lavandula spica* in considerable quantity, could not be discovered, nor could camphor, which, according to old authorities, is said to be a most important constituent of lavender oil.

The principal constituent of the oil is an alcohol, $C_{10}H_{18}O$, and its acetic ester. This alcohol boils between 197° and 199° C. ; sp. gr. 0·869 at 20° C. ; optical rotation −10° 35′ in 100 mm. tube. Refraction equivalent for sodium light, 1·464.

If this alcohol be heated with acid sulphate of potassium, or with dilute sulphuric acid, or other dehydrating agents, a mixture of hydrocarbons is obtained, amongst which Dipentine, characterized by the tetrabromide melting at 125° C., and Terpinen, melting-point of the nitrite 155° C., could be detected. By oxidation of the alcohol, an aldehyde of the composition $C_{10}H_{16}O$ was obtained, which is identical with Citral (Semmler's geranium aldehyde). The sp. gr. of this aldehyde is 0·898 at 15° C., boiling-point at 10 mm. pressure 105°–107° C., refraction equivalent 1·490. By the action of dehydrating agents Cymol. ($C_{10}H_{14}$) was formed.

The alcohol of lavender oil combines with four atoms of bromine. With hydrochloric acid, a liquid compound of the composition $C_{10}H_{18}Cl_2$ results. By the action of acid anhydrides the corresponding esters may be prepared. The acetic, propionic, butyric, valerianic, and benzoic acid esters were actually prepared, all of which are pleasant smelling liquids. The acetic ester arouses considerable interest, first because it occurs ready formed in lavender oil, and next because it possesses such a characteristic bergamot odour that it immediately suggested its presence in bergamot oil—an hypothesis which proved to be quite correct. Bergamot oil in fact contains about 40 per cent. of this ester.

On comparison of lavender alcohol with numerous other known

compounds of similar composition, it was eventually found to be identical with Linalool, which was discovered by Semmler in Mexican Lignaloe oil *. Linalool and its acetic ester are also contained in Petit grain oil.

ROSEMARY.

Although the genus *Rosmarinus* consists of but one species, the aspect of the plant in different countries and soils presents differences easily perceptible to the eye of the cultivator, and great differences are recognized in the fragrance of the perfume developed in its leaves. The English oil has a greater value per ounce than some of the foreign oils have per pound, and this difference is only in a slight degree due to more careful distillation. The difference in the physical characters of the English and foreign oils is also very marked.

Rosmarinus officinalis is, in its general appearance, a dense branching shrub, with linear sessile leaves, stalkless, and the edges recurved, green, smooth and shining above, and whitish beneath. The pale blue blossoms, variegated with purple and white, make their appearance early in the summer. The plant is evergreen, and rises to a height of 4 or 5 feet in some situations; but in very calcareous soil it is much smaller, although more fragrant and more hardy. When the roots are well drained the plant will stand severe frosts, although it is a native of the South of Europe; but in rich moist land it is far more susceptible to frost. The generic name *Rosmarinus* is evidently derived from the Latin—*ros*, dew, and *marinus*, in allusion to its inhabiting the sea-coast. There is a white-striped leaf variety, called the "silver rosemary," and a yellow-striped variety, called the "golden rosemary," but they are not usually cultivated.

In establishing a plantation it is not well to risk planting foreign seed, as they may produce rubbish; but preferably obtain cuttings from English-grown plants and patiently multiply them. If planted in August cuttings strike rapidly, especially if pulled off with a "heel" by bending back a young shoot until it separates from the branch. These can be struck in light loam in boxes holding about fifty each; they can then be put in the shade and

* Ber. Deutsch. chem. Ges. xxiv. p. 207.

watered, and kept under cover during the winter. The following April they can be bedded out, 18 inches apart, in rows 4 feet apart, on a dry, calcareous, sloping ground—preferably with a chalk sub-soil. The chalk, into which the roots will penetrate, holds sufficient moisture in summer, and yet provides good drainage in winter.

At the beginning of September the young shoots are carefully and evenly cut, with a strong pair of sheep-shears, right down to the wood ; and the plants soon form a compact stunted hedge, about 18 inches to 2 feet high. The old leaves remain on the plant a considerable time, not shrivelling off on the lower and inner boughs as do those of lavender. They want but very little manure, but on rich ground as much cinder-ash as possible.

The young shoots and trimmings taken off at the end of August or beginning of September are distilled precisely in the same way as lavender-flowers ; but as much as possible of the woody parts should be separated, as they uselessly fill up the still and impart a turpentiny rankness to the distillate. If rosemary is not distilled very quickly after it is cut it is liable to heat, changing its cha-racter ; and if spread out till the next day it loses some of its fragrance. When put into the still the water should just cover it. The same still can be used as for lavender ; but of course another con-denser. Also, as the particles of oil condensing are heavier and much smaller than those of lavender, it is necessary to employ a deeper Florentine receiver, so that they shall not be carried downward by the rush of water and escape by the waste-pipe. A piece of smooth wood floating in the receiver can also be used to break the down-ward current of the water. It takes rather longer to distil out the oil than does the same bulk of lavender. The oil which comes over during the first 30 minutes after starting boiling is the finest ; the receiver can then be changed, and the operation continued until no more oil is seen to come over. The oil should not be taken out of the receivers the same day, but stood in a cool place (or in water) until the next day, by which time the oil will be bright and can be removed.

The new oil requires to be carefully dried with fresh calcic chloride, as the water held mechanically in suspension, or in solution, is very unpleasant, somewhat reminding of fish-brine, or trimethylamine.

The oil is almost colourless, or with the slightest tinge of straw-colour ; it is very mobile and refractive. The sp. gr. of Surrey oil has been found to be 0·901, of Sussex oil (1889) 0·911, ditto (1890) 0·924*. It rotates polarized light to the left; several foreign oils of rosemary rotate it to the right. Between 16° and 17° C. it dissolves in all proportions in 90-volume per cent. alcohol, forming a clear solution. The admixture of turpentine, which in commercial oils is often found to the extent of 50 per cent., may be sought for by Langbeck's test †. He states that oil of rosemary should dissolve one-tenth part of salicylic acid, but if it contain 10 per cent. of turpentine the solubility is reduced to 1 in 24; oil of turpentine, when freshly distilled, dissolving only $\frac{1}{625}$ part.

Oil of rosemary is very frequently adulterated with petroleum (necessarily the heavier portions of petroleum). It may be detected and approximately estimated by exposing the oil to the heat of a water-bath in an open dish until the odour of rosemary is practically lost, the residue consisting of petroleum and a slight resinous portion which is always left by the oil. Oil so adulterated is of course less soluble in alcohol proportionately to the amount of adulteration. Pure English oil is soluble in 5 volumes of rectified spirit of 0·838; some adulterated samples tested in this way required from 20 to 30 volumes of such spirit to effect solution. Samples of rosemary oil adulterated with alcohol naturally require a less volume of the solvent than pure oil; some samples tested required 3½ and 4½ volumes, according to the amount of adulteration. Alcohol may also be detected by agitating the oil with a particle of " magenta: " if pure no colour is imparted to the oil, but if adulterated the dye dissolves, the colour being deeper the greater the quantity of alcohol present ‡. These tests may be applied to other oils; but the magenta-test is not applicable to all—for instance, oil of cloves.

The sp. gr. of pure Italian (Dalmatian) oil of rosemary has been found to be 0·901 to 0·907 at 15° C., and it requires 12 volumes of alcohol of 0·864 to effect solution §. The sp. gr. of Spanish oil has been observed as 0·892 at 15° C., and the French at various figures between 0·881 and 0·907, but there seems some diversity of opinion as to the true sp. gr. of pure French oil. Messrs. Schimmel

* Cripps, Pharm. Journ. [3] xxi. p. 937.
† Year-Book of Pharmacy, 1885, p. 126.
‡ Cripps, Pharm. Journ. [3] xx. p. 415.
§ Bericht, Oct. 1888.

aver that it should not lie below 0·900, or in any case not lower than 0·890 *. French oil of rosemary examined by Bruylants † was found to contain 80 per cent. of a lævogyrate hydrocarbon $H_{10}C_{16}$, 4 to 5 per cent. of Borneol $C_{10}H_{18}O$, and 6 to 8 per cent. of a camphor $C_{10}H_{16}O$.

All oils of rosemary are apt to turn yellow and oxidize by age, especially if kept in bottles which are badly stoppered or not full. Such resinified oils can simply be rectified by Dragendorff's process with a current of steam. Small quantities can so be operated upon in a tubulated glass retort, into which a glass tube is passed almost to the bottom; this tube, containing a little light packing of cotton, is connected with a flask in which water is heated to strong ebullition, by which means steam is passed through the oil. About 68 per cent. of oil can thus be recovered, quite colourless and mobile, and almost as good as freshly prepared oil. The residue in the retort consists principally of rosemary camphor.

The *Cedrella Rosmarinus* (Loureiro, ' Cochinchinensis,' p. 160), or " Rosemary bastard-cedar," is a shrub of about 4 feet in height, native of Cochin China and about Macao in China, and is said to yield an oil very similar to rosemary oil.

* Bericht, April 1891.
† Journ. de Pharm. [4] xxix. p. 508.

ADDENDA.

Adulteration of Santal-wood Oil.—Mesnard states * that by the application of ordinary pure sulphuric acid it is possible to recognize the presence of oil of cedar, cubebs, copaiba, or turpentine in oil of santal. If the oil be pure a viscid liquid is produced, which becomes pasty and is rapidly transformed into a solid mass adhering to the glass. This mass is characterized by its light grey-blue or greyish colour and the dusty appearance it gradually assumes. In adulterated samples the resinous mass does not solidify entirely, and remains of a deep tint and distinctly brilliant lustre. The test is said to be so delicate that it is possible to determine by the following method the approximate proportion of adulteration :—Two or three centigrams of the oil mixed with a drop of pure sulphuric acid are placed on to the surface of a piece of ground glass, and the flattened end of a small glass rod applied to the mixture. The upper end of the glass rod being attached to the lower surface of the scale-pan of a balance, the degree of adhesiveness of the oil to the glass (indicating the degree of purity) can be read off.

Source of Ambergris.—Jourdain † considers that the micro-scopical and chemical examination of ambergris show it to be of a similar nature to intestinal calculi. He was struck by the presence of numerous remains of cephalopods in the mass, and thinks this fact may have some bearing upon the existence of this odorous substance. He assumes that it may even have been produced by

* Comptes Rendus, cxiv. p. 1546, 27th June, 1892.
† Ibid. p. 1557 ; and Pharm. Journ. [3] xxiii. p. 82.

the cephalopods swallowed in large quantities by the Cetacea. As evidence, he points to the fact that several molluses exhale a pronounced odour, which lasts after their death and even desiccation. The ancients are known to have utilized various species for the sake of their odour; and M. Jourdain suggests that the action of the biliary products upon the cephalopodic perfume may modify it and result in the production of ambergris.

Lemon-grass Oil.—A sample of Ceylon oil of lemon-grass examined by Dodge * was found to have a sp. gr. of 0·886 at 22° and 0·8955 at 15°·5. On distillation, the boiling-point was found to range between 200° and 240°, a small residue not being volatile at the latter temperature. The principal fraction came over between 220° and 225° C.

The identification of a new species of *Vanilla, V. ensifolia,* Rolfe, from New Granada, a specimen of which is in the herbarium of the Pharmaceutical Society, has now been confirmed by the authorities at Kew, and its botanical analysis fully described in the 'Kew Bulletin,' No. 66, p. 141. The species is a very distinct one, easily recognized by its narrow sword-shaped leaves †.

* Am. Chem. Journ. xii. p. 558.
† Pharm. Journ. [3] xxiii. p. 85, July 30th, 1892.

ALPHABETICAL INDEX.

A.

Abir, 25.
Acacia homalophylla, 113.
„ species, 114.
Achillea moschata, 20.
Acroleïn, 140.
Adoxa moschatellina, 20.
Eranthus fragrans, 134.
African santal, 324.
Agallochum, 283.
Aglaia, 86.
Agrimonia odorata, 24.
Alau-gilan, 117.
Algum, 327.
Almonds, 175.
Aloe-wood, 283.
Aloexylon, 286. '
Aloysia citriodora, 90.
Altingia excelsa, 249.
Alyxia buxifolia, 134.
Amberboa moschata, 20.
Ambergris, 13.
„ source of, 374.
Ambrette, 19.
Ammonium salicylite, 49.
Amygdalin, 177.
Amygdalus communis, 175.
Amylbenzene, 240.
Amyris altissima, 291.
„ ambrosiaca, 291.
„ comiphora, 267.
„ Gileadensis, 270.
„ Kataf, 266.
Andropogon citratus, 89.
„ fragrans, 47.
„ Iwarancusa, 311.
„ Lanigerum, 311.
„ muricatus, 309.
„ Nardus, 50, 88.
„ Schœnanthus, 44.
Anethum fragrans, 134.
Anisic aldehyde, 140, 189.
Anona, 130.
Anthoxanthum odoratum, 134.

Antilope Dorcas, 12.
Aplotaxis lappa, 109.
Aquilaria Agallocha, 4. 283.
Argyreia Bona-nox, 230.
Artabotrys odoratissimus, 130.
„ suaveolens, 131.
Artificial benzoic acid, 237.
„ camphor, 341.
„ cinnamic acid, 199.
„ civet, 10.
„ coumarin, 137.
„ essence of hyacinth, 97–101.
„ essence of jasmine, 95.
„ heliotrope, 188.
„ musk, 6.
„ odour of geranium, 50.
„ oil of bitter almonds, 180.
„ oil of cloves, 227.
„ rose odours, 49.
„ styrolene, 245.
„ vanillin, 162.
Asarum Canadense, 48.
Atar, 24.
"Aubepine," 189.
Aveneine, 170.
Azulene, 305.

B.

Balm of Gilead, 270.
Balsam of Gilead, 270.
„ of Mecca, 270.
„ „ estimation of, 273.
„ Peru, 250.
„ „ estimation of, 254.
„ „ white, 260.
„ of Son Sonate, 261.
„ Tolu, 261.
„ „ estimation of, 264.
Balsamo Blanco, 260.
Balsamodendron Ehrenbergianum, 270.
„ Gileadense, 270.
„ Káfal, 265.

Balsamodendron Kataf, 266.
    ,,   *Mukul,* 267.
    ,,   *Myrrha,* 268.
    ,,   *Opobalsamum,* 270.
    ,,   *Playfairii,* 267.
Bdellium, African, 266.
    ,, Indian, 266.
    ,, opaque, 267.
    ,, perfumed, 266.
Bean-flower, odour resembling, 261.
Benzaldehyde, 180.
Benzoic acid, 235.
Benzoin, 232.
Benzoylacetone, 232.
Benzoyl-eugenol, 228.
Benzyl acetate, 101.
    ,, alcohol, 100.
    ,, chloride, 181.
    ,, cinnamate, 201, 253.
Bergamot orange, 75.
    ,, odour resembling, 346.
Bissa-ból, 266.
Bitter almonds, 175.
    ,, orange, 64.
Black Frankincense leaf, 243.
    ,,   ,, oil, 243.
    ,, storax, 248.
Blumea balsamifera, 343.
Borneene, 348.
Bornyl acetate, 342.
Bos moschatus, 12.
Boswellia Carterii, 275.
    ,,   *Frereana,* 278.
    ,,   *glabra,* 267.
    ,,   *serrata,* 267, 278.
Bourbon Tea, 134.
Brazilian Lign-aloe, 291.
Bulgarian method of distilling, 28.
Bursera Delpechiana, 288.
Butylbenzene, 240.

C.

Cajeput, 349.
Cajuputene, 351.
    ,, chloride, 351.
    ,, di-hydrate, 351.
Cake storax, 248.
Calcium salicylite, 49.
Camel-grass, 311.
Camphene, 341.
Camphire, 313.
Camphor, 333.
    ,, artificial, 341, 342.
    ,, Borneo, 333.
    ,, Cedar, 329.
    ,, Chinese, 336.
    ,, of Feverfew, 346.
    ,, Formosan, 336.
    ,, Japan, 336.
    ,, Laurel, 335.

Camphor, Ngai, 343.
    ,, oil, 347.
Camphoric acid, 341.
Camphorophone, 342.
Cananga odorata, 117.
Carbon disulphide, purification of, 59.
Carduus nutans, 19.
Carlines, 113.
Carpobalsamum, 274.
Caryophylline, 230.
Caryophyllus aromaticus, 219.
Cassia, 203.
    ,, buds, 208.
    ,, oil, 209.
    ,, vera, 208.
Cassie, 114.
Cedar, 328.
    ,, camphor, 329.
    ,, Florida, 332.
    ,, Honduras, 330.
    ,, Indian, 331.
    ,, Jamaica, 330.
    ,, of Lebanon, 331.
Cedrella alternifolia, 331.
    ,,   *angustifolia,* 331.
    ,,   *Bermudiana,* 331.
    ,,   *Brazilicnsis,* 331.
    ,,   *odorata,* 330.
    ,,   *Rosmarinus,* 373.
    ,,   *Toona,* 331.
Cedrene, 329.
Cedrus Deodara, 331.
    ,, *Libani,* 331.
Cerambyx moschata, 13.
Cerasus Laurocerasus, 185.
Ceratopetalum apetalum, 134.
Cestrum nocturnum, 20.
Chamædorea fragrans, 49.
Chamæleons, 113.
Champac, 124.
Champeya, 129.
Chelonia cacuana, 13.
Cherry-Laurel, 185.
Chloranthus, 87.
Chloride of sylvestrine, 77.
Cinnamaldehyde, 198, 212, 247.
Cinnameïn, 253.
Cinnamic acid, 199, 216, 246.
    ,, alcohol, 99.
Cinnamol, 245.
Cinnamomum Camphora, 335.
    ,,   *Cassia,* 206.
    ,,   *Culilawan,* 231.
    ,,   *Zeylanicum,* 191.
Cinnamon, Brazilian, 197.
    ,, Cayenne, 197.
    ,, Ceylon, 191.
    ,, China, 203.
    ,, Indian, 196.
    ,, Java, 197.
    ,, oil of, 197.

Cinnamon leaf oil, 201.
Cinnyl alcohol, 99, 247.
„ cinnamate, 244, 253.
Citral, 36, 369.
Citrine odours, 64.
Citron, 65.
Citronella, 88.
Citronellic aldehyde, 50.
Citronellol, 89.
Citrus aurantium, 64.
„ *bergamia*, 75.
„ *Bigaradia*, 64.
„ *decumanum*, 74.
„ *limetta*, 72.
„ *limonum*, 65.
„ *Medica*, 65.
„ *nobilis*, 74.
„ cultivation in America, 78.
Civet, 10.
Clove bark, 231.
Clove-Pink, 225, 230, 231.
„ odour resembling, 264.
Cloves, 219.
„ odours analogous, 230, 231.
Clove-oil, artificial, 227.
„ estimation of, 228.
„ price of, 225.
„ from the flower-stalks, 226.
Cochin-China santal, 327.
Coleus aromaticus, 307.
Colliguaja odorifera, 48.
Concrete oils, 63. See also Processes,
    various.
Coniferin, 163.
Convolvulus scoparius, 48.
Cortex Thuris, 243.
„ *Thymiamatis*, 243, 247.
Costus, 109.
Coumaric acid, 138.
Coumarin, 132.
„ artificial, 137.
„ odours resembling, 140.
„ various sources of, 135.
Cratægus oxyacantha, 189.
Culilabanus Papuanus, 231.
Cupressus, 332.
Curaçao, 85.
Cyanus orientalis moschatus, 20.
Cymol, 47.
Cytisus laburnum, 85.

D.

Daniellia thurifera, 280.
Daphnetin, 141.
Deer's tongue, 133.
Delphinium Brunonianum, 16.
„ *glaciale*, 17.
„ *heterocarpum*, 113.
„ *moschatum*, 17.

Deodar, 331.
Dhelum, 296.
Dimethylcoumarin, 139.
Dimethylprotocatechuic acid, 167.
Dipentene, 36.
Dipentylene-glycol, 96.
Dryobanalops Camphora, 333.
Dysoxylon Fraserianum, 48.

E.

Eagle-wood, 283.
Ecuelle, 66.
Elaphrium Aloexylon, 290.
Emulsin, 176.
Enfleurage, 55.
Epicharis, 327.
Eponge, 66.
Erodium moschatum, 19.
Ethyl benzene, 94.
„ benzoate, 239.
„ cinnamate, 201.
Ethylcoumarin, 139.
Ethyl hydrocinnamate, 201.
Eugenia caryophyllata, 219.
Eugenol, 169, 227.
Euryangium Sumbul, 17.
Eurybia argophylla, 19.
Evodia, 87.
Exocarpus latifolia, 326.

F.

Fanave, 134.
Fat, purified, 52.
Fiji santal, 325.
Flores cassiæ immaturæ, 208.
„ *tristes*, 21.
Florida cedar, 332.
Frankincense, 275.
„ black, 243.
„ West African, 280.

G.

Gardenia citriodora, 85.
Gendarussa vulgaris, 307.
Geonoma, species, 114.
Geraniol, 47.
Geranium, 42.
„ odour resembling, 51.
Geranyl chloride, 47.
„ ether, 48.
„ valerate, 48.
Geum urbanum, 226.
Ginger-grass, 44.
Giroflé, 226.
Gnaphalium odoratissimum, 20.

Googul, 267.
Grass oils, 308.
Guaiacol, 171.
Guarea Swartzii, 19.
Guiana Lign-aloe, 291.
Gyrinus natator, 16.

H.

Hawthorn, 189.
Heliotrope, 187.
Heliotropine, 188.
Hennah, 313.
Hibiscus Abelmoschus, 19.
Honduras cedar, 330.
Hotai, 268.
Hound's tongue, 133.
Hyacinth, 96.
,, odour resembling, 247, 253.
Hyacinthus muscari, 20.
,, species, 103.
Hyawa, 291.
Hydrocinnamic acid, 201.

I.

Icica altissima, 291.
,, *decandra*, 292.
., *Icicariba*, 292.
Ihlang, 117.
Incense of Java, 232.
,, oil, 243.
Indian cedar, 331.
., oil of geranium, 44.
Ipomœa grandiflora, 230.
Iris, species, 106.
Isobutylbenzene, 240.
Isonitrosobenzyl ether, 239.

J.

Jacinthe, 96.
Jamaica cedar, 330.
Jasmine, 91.
,, syrup of, 94.
Jonquil, 95.
Joss-sticks, 327.
Juncus odoratus, 311.
Juniperus Virginiana, 328.
Justicia Gendarussa, 307.

K.

Kafal, 266.
Kalambak, 287.
Kaya-pootie, 349.
Keora, 127.

Keora-ka-utter, 129.
Koot, 112.

L.

Laurus Camphora, 335,
,, *Culilaban*, 231.
Lavender, 354.
,, oil, estimation of, 367.
Lavandula, species, 355.
,, *vera*, 356.
Lawsonia inermis, 313.
Lemon-grass, 89, 375.
Lettsomia Bona-nox, 230.
Liatris odoratissima, 133.
Licaria Kanali, 48.
Lign-aloe, Brazilian. 291.
,, Chinese, 286.
., Guiana, 291.
,, Mexican, 288.
,, Oriental, 283.
Lime, 72.
Lippia citriodora, 90.
Liquidambar Altingiana, 243, 249.
,, *Formosana*, 249.
,, liquid, 248.
,, *Orientalis*, 242, 247.
.. *Styraciflua*, 248.
,, white, 249.
Liquid storax, 242.
Lubán Jáwi, 232.
Lubán Mayeti, 276.

M.

Macassar Santal wood, 323.
Maceration, 52.
Mal-oil, 19.
Maltese orange, 74.
Malus Medica, 65.
Malva moschata alba, 19.
Mandarine, 74.
Marlea Vitiensis, 19.
Melaleuca acuminata, 354.
,, *Cajuputi*, 349.
., *decussata*, 352.
,, *ericifolia*, 353.
,. *genistifolia*, 353.
,, *leucadendron*, 350.
,, ,, var. *lancifolia*, 354.
., *linarifolia*, 353.
,, *minor*, 349.
,, *squarrosa*, 353.
,, *uncinata*, 353.
,, *viridifolia*, 352.
,, *Wilsonii*, 353.
Melilotol, 136.
Melilotus officinalis, 135.

Mesua ferrea, 129.
Meta-cinnamein, 253.
Methyl benzoate, 238.
Methylbenzylenic ether, 49.
Methyl chloride, 60.
„ cinnamate, 201.
β Methylcoumarin, 139.
Methylumbelliferon, 142.
Metrosideros, 352.
Mexican santal-wood bark, 326.
Michelia Champaca, 124.
Mignonette, 101.
„ Jamaica, 314.
Mimulus moschatus, 19.
Mirbane, 182.
Moschus Altaicus, 2.
„ moschiferus, 1.
Mukul, 267.
Musk, 1.
„ alligator, 13.
„ antelope, 12.
„ artificial, 6.
„ Baur, 6.
„ Canadian, 11.
„ cranesbill, 20.
„ larkspur, 17.
„ mallow, 19.
„ milfoil, 20.
„ ox, 12.
„ plant, 19.
„ rat, 11, 12.
„ substitutes, 6.
„ thistle, 19.
„ tree, 19.
„ turtle, 13.
„ various sources of, 16.
„ volatility of, 5.
Myall wood, 113.
Myristic acid, 108.
Myroxocarpine, 261.
Myroxylon Pereirœ, 250.
„ *Peruiferum*, 261.
„ *Toluifera*, 261.
Myrrh, 268
Myrtus caryophyllus, 219.

N.

Nag-champa, 130.
Nag-Kesur, 129.
Narcissus Jonquilla, 95.
Nardosmia fragrans, 187.
Neroli, 80.
New-Caledonia santal, 326.
New-Zealand santal, 326.
Ngai camphor, 343.
Night-flowering plants, 21.
Niobe essence, 238.

Nitrobenzene, 182.
Nitrotoluene, 184.

O.

Odours, multiplication of, 115.
Oil of apples, 19.
„ Balsam Peru pods, 261.
„ camphor, 347.
Oils, essential, are described under heading of plants etc. from which they are respectively derived.
Olibanum, 275.
Olibene, 278.
Olivil, 172.
Oncidium inosmum, 113.
Ondrata zibethica, 11.
Opopanax, 265.
Orange-flower-water, 83.
Orange-peel, 84.
Origanum Indicum, 308.
Orris root, 106.
Otto of rose, 35.

P.

Pæonia albiflora fragrans, 49.
Palma rosa, 46.
Pandanus, 127.
Paracoumarhydrin, 139.
Patchouli, 293.
„ camphor, 305.
„ estimation of, 304.
„ oil. 306.
Pelargonium, 42.
Pentylbenzene, 240.
Peru balsam, white, 260.
Peruvin, 253.
Petasites fragrans, 187.
Phenyl benzoate, 50.
Phenylbromethylene, 99.
Phenylchlorethylene, 98.
Phenyl ethylene, 245.
Phenylnitro-ethylene chloride, 23.
Phenyl paratoluate, 49.
Philadelphus coronarius, 85.
Physocalymna floribundum, 48.
Picrasma quassioides, 113.
Pinene hydrochloride, 341.
Piperine, 188.
Piperonal, 188.
Piver's Patents and other processes, 53.
Plectranthus, spec., 307.
Pogostemon Heyneanum, 295.
„ *Patchouli*, 293.
Polianthes tuberosa, 103.
Pomme Reinnette, 23.
Protium Gileadense, 270.
„ *Kafal*, 265.

Protocatechuic acid, 165.
Prunus Laurocerasus, 185.
Pseudo-mirbane, 184.
Pterocarpus santalinus, 328.
Putchuk, 112.

Q.

Quino-quino, 261.

R.

Red Storax, 243.
Rhodinal, 36, 39.
Rhodinol, 36.
Rhodiola rosea, 48.
Rhodium, 48.
Rhus aromatica, 48.
Rosa alba, 28.
,, *centifolia*, 25, 41.
,, *Damascena*, 25.
Rose, 22.
,, culture of, in Bulgaria, 26.
,,   ,,   in France, 40.
,, Malloes, 243, 249.
,, odours, artificial, 49.
,, odours resembling, 48.
,, otto, statistics of, 32.
Rosemary, 370.
,,   Bastard Cedar, 373.
,,   Dalmatian, 372.
.,   French, 372.
,,   Spanish, 372.
Rose-root, 48.
Rose-wood oil, 48.
Rusa-grass, 45.

S.

Salicylic acid, 166.
Santal, 315.
,, African, 324.
,, Cochin-China, 327.
,, Fiji, 325.
,, Macassar, 323.
,, New-Caledonia, 326.
., New-Zealand, 326.
,, oil, estimation of, 328, 374.
,, Sandwich-Islands, 326.
,, West-Australian, 324.
,, West-Indian, 325.
,, wood bark, Mexican, 326.
Santalal, 322.
Santalene, 322.
Santalin, 328, 368.
Santalol, 322.
Santalum album, 316.
,, *myrtifolium*, 322.
Santalyl acetate, 322.

Schiff's Reagent, 38.
Sepia moschata, 14.
Shaddock, 74.
Sighala oil, 243.
Silver Wattle, 114.
Sondeli, 12.
Son Sonate Balsam, 261.
Stacte, 269.
Stereoptene, 37.
Stiptes Caryophylli, 226.
Storax, 241.
,, alcohol, 99.
,, black, 248.
,, cake, 248.
,, liquid, 242.
,, red, 243.
Styraccne, 99.
Styrax Benzoin, 232.
,, *calamita*, 242, 247.
,, *officinale*, 241.
Styracin, 244, 247, 253.
Styril alcohol, 247.
Styrol, 245.
Styrolene, 245.
,,   artificial, 245.
,,   bromide, 246.
,,   chloride, 246.
Styrolyl acetate, 94.
,,   bromide, 95.
Styrone, 99, 253.
Sumbul, 17.
Sweet orange, 64.
Sweetbriar, 23.
Sylvestrene, 77.
Syringa, 85.

T.

Tea, method of scenting, 86.
Terpine-hydrate, 96.
Terpineol, 97.
Thallin, 139.
Thermo-pneumatic extractor, 69.
Thooth, 112.
Thus Judæorum, 243.
Tolu Balsam, 261.
Toluene benzoic acid, 237.
Toluifera Balsamum, 261.
Toon, 331.
Tritelia uniflora, 113, 115.
Tubéreuse. 103.
Turkish oil of geranium, 45.
Tussilago fragrans, 187.

U.

Umbelliferon, 141.
Unona, 123.
Uvaria, 117, 130.

V.

Vanilla, 143.
  ., fertilization of, 150.
  „ new species, 375.
  „ odours analogous, 156.
Vanillin, 158.
  ,, artificial, 162.
Vanillon, 161.
Verbena odours, 90.
Vetiver, 309.
Violet, 104.
Viverra Civetta, 13.
  „ Zibetha, 10.

W.

West-Australian santal, 324.
West-Indian santal, 325.

White Balsam Peru, 260.
  „ „ of Son Sonate, 261.
  „ Cedar, 332.
  ,, Liquidambar, 249.
Winter Heliotrope, 187.

X.

Xymenia Americana, 327.

Y.

Ylang-ylang, 117.

Z.

Zambak, 91.